Enzyme Chemistry

Enzyme Chemistry

Impact and applications

SECOND EDITION

Edited by

Colin J. Suckling

Professor of Organic Chemistry
University of Strathclyde, UK

CHAPMAN AND HALL

LONDON • NEW YORK • TOKYO • MELBOURNE • MADRAS

UK   Chapman and Hall, 11 New Fetter Lane, London EC4P 4EE

USA   Chapman and Hall, 29 West 35th Street, New York NY10001

JAPAN   Chapman and Hall Japan, Thomson Publishing Japan, Hirakawacho
            Nemoto Building, 7F, 1-7-11 Hirakawa-cho, Chiyoda-ku, Tokyo 102

AUSTRALIA   Chapman and Hall Australia, Thomas Nelson Australia, 480 La Trobe
            Street, PO Box 4725, Melbourne 3000

INDIA   Chapman and Hall India, R. Sheshadri, 32 Second Main Road, CIT East,
            Madras 600 035

Second edition

© 1990 Chapman and Hall

Typeset in 10/12 Times by
Scarborough Typesetting Services
Printed in Great Britain by
T. J. Press Ltd, Padstow, Cornwall

ISBN 0 412 34970 1 (HB)

British Library Cataloguing in Publication Data

Enzyme chemistry. – 2nd. ed.
  1. Enzymes
  I. Suckling, Colin J. *1947–*
  547.7′58

    ISBN 0–412–34970–1

Library of Congress Cataloging-in-Publication Data

Enzyme chemistry: impact and applications/edited by Colin J.
  Suckling. – 2nd ed.
      p.   cm.
  Includes bibliographical references.
  ISBN 0–412–34970–1
  1. Enzymes.   I. Suckling, C. J. (Colin J.)
  [DNLM: 1. Enzymes.   QU 135 E588 E61]
  QP601.E5157   1990
  574.19′25 – dc20
  DNLM/DLC
  for Library of Congress   89–23899
                                                                    CIP

Contents

Contributors   ix
Preface   xi

1 Infant enzyme chemistry   1
   Colin J. Suckling

2 The mechanistic basis of enzyme catalysis   8
   Ronald Kluger
   2.1 The mechanistic approach   8
   2.2 Concepts of catalysis   9
   2.3 Describing a mechanism   9
   2.4 Kinetics: the measure of catalysis   12
   2.5 Stereochemistry and specificity   15
   2.6 Stereochemistry and mechanism   20
   2.7 Entropy and enzymic catalysis   22
   2.8 Acid–base catalysis   26
   2.9 Linear free-energy relationships and enzymic reactions   29
   2.10 Enzymic efficiency   31
   2.11 Examples of intramolecular catalysis   34
   2.12 Transition state analogues   36
   2.13 Multiple binding sites   40
   2.14 Biomimetic chemistry   42

2.15 Conclusion   45
References   46

3 Chemical models of selected coenzyme catalyses   50
   Seiji Shinkai
   3.1 Introduction   50
   3.2 Model investigations of nicotinamide coenzymes   51
   3.3 Flavin catalyses   64
   3.4 Catalyses relating to vitamin B_1 and analogues   75
   3.5 Pyroxidal catalyses   80
   3.6 Catalyses of thiol coenzymes   84
   3.7 Conclusion   87
   References   88

4 Selectivity in synthesis – chemicals or enzymes?   95
   Colin J. Suckling
   4.1 Introduction   95
   4.2 Problems overcome   100
   4.3 Logic and analogy in the synthetic uses of enzymes and
        micro-organisms   106
   4.4 Enzymes and chemical reagents in 'competition'   113
   4.5 Late-stage functional-group modification   128
   4.6 Biomimetic chemistry in synthesis   134
   4.7 Enzymes in organic solvents   153
   4.8 Advances in protein chemistry and molecular biology   156
   4.9 Conclusions   164
   References   165

5 Enzymes as targets for drug design   171
   Philip D. Edwards, Barrie Hesp, D. Amy Trainor and Alvin K.
   Willard
   5.1 Introduction   171
   5.2 Case studies in drug discovery   173
   5.3 Recent developments and their application to inhibitor
        design   219
   References   222

6 The impact of metal ion chemistry on our understanding of
   enzymes   227
   Donald H. Brown and W. Ewan Smith
   6.1 Introduction and general chemical principles   227
   6.2 The transition elements iron and copper   232
   6.3 Transition metal ions   243

Contents   vii

6.4   Main group elements   250
6.5   Some toxic metals   251
6.6   Metal ions as drugs   253
6.7   Modern physical methods   257
6.8   Conclusions   262
References   263

7   The enzymology of the biosynthesis of natural products   265
    David E. Cane
    7.1   Introduction   265
    7.2   Recent advances in the study of biosynthetic enzymes   267
    7.3   Terpenoid cyclases   284
    7.4   Problems and prospects   298
    References   300

8   Enzymes in the food industry   306
    David R. Berry and Alistair Paterson
    8.1   Introduction   306
    8.2   Amylases and starch hydrolysis   309
    8.3   Enzymes acting on glucose and oligosaccharides   318
    8.4   The plant cell wall and its breakdown   321
    8.5   Industrial applications of plant cell wall degrading enzymes 332
    8.6   Exogenous enzymes in cheesemaking   338
    8.7   Conclusions   348
    References   349

9   Enzymology and protein chemistry in the wider area of biology   352
    Keith E. Suckling
    9.1   Introduction   352
    9.2   Studies of enzymes by techniques of molecular biology   354
    9.3   Theoretical treatments of enzyme catalysis   356
    9.4   Protein structure, homology and genetic relationships   358
    9.5   Interactions between proteins and DNA   360
    9.6   Novel catalysts   362
    9.7   Apolipoprotein B   363
    9.8   Receptors   367
    9.9   Conclusions   372
    References   372

Index   375

Contributors

David R. Berry	Department of Bioscience and Biotechnology, University of Strathclyde, Glasgow, UK
Donald H. Brown	Department of Pure and Applied Chemistry, University of Strathclyde, Glasgow, UK
David E. Cane	Department of Chemistry, Brown University, Rhode Island, USA
Philip D. Edwards	ICI Pharmaceuticals Group, ICI American Inc., Wilmington, Delaware, USA
Barrie Hesp	ICI Pharmaceuticals Group, ICI Americas Inc., Wilmington, Delaware, USA
Ronald Kluger	Department of Chemistry, University of Toronto, Ontario, Canada
Alistair Paterson	Department of Bioscience and Biotechnology, University of Strathclyde, Glasgow, UK
Seiji Shinkai	Department of Industrial Chemistry, Nagasaki University, Nagasaki, Japan
W. Ewan Smith	Department of Pure and Applied Chemistry, University of Strathclyde, Glasgow, UK

Colin J. Suckling   Department of Pure and Applied Chemistry, University of Strathclyde, Glasgow, UK

Keith E. Suckling   Smith, Kline and French Research Ltd., Welwyn, UK

D. Amy Trainor   ICI Pharmaceuticals Group, ICI Americas Inc., Wilmington, Delaware, USA

Alvin K. Willard   ICI Pharmaceuticals Group, ICI Americas Inc., Wilmington, Delaware, USA

Preface

As the first edition of this book was going through the publication process, a revolution was taking place in the technologies available for the study of enzymes. The techniques of molecular biology, especially in genetic engineering of organisms and in site specific mutagenesis of genes, were established and were being brought into use to solve many problems in fundamental and applied science, not least in enzymology. Added to these advances the possibility of generating catalysts from antibodies has become a topic of major interest. These major innovations have changed the emphasis of much bioorganic research; whereas in the past, the protein was often the 'sleeping partner' in a study, its detailed function is now the major focus of scientific interest. Similarly in industry, the potential of genetically manipulated organisms to satisfy the needs for the production of chemicals and foodstuffs has been widely recognised. The second edition of 'Enzyme Chemistry, Impact and Applications' takes on board these new developments whilst maintaining the overall aims and views of the first edition. Many of the chapters have been completely rewritten to take account of advances in the last five years especially with regard to the impact of biologically based technologies. Although the book continues to approach its subject matter from the point of view of the chemist, the increased interdisciplinary content of much modern science will be obvious from the discussion. The scope of the book has also been extended to include further discussion of areas of industrial significance in relation to chemical synthesis and, through a new chapter, the food industry. This chapter has a significantly greater applied scientific bias than many of the others and, because of the macromolecular nature of the substrates involved, is less specific with regard to detailed structural organic chemistry. Such an extension is nevertheless necessary to give an adequate view of the breadth of impact of enzyme chemistry today. For those chemists unfamiliar with the new biological technologies and their associated jargon, an overview is provided through the discussion in the final chapter. I hope that readers will continue to find the articles in this book helpful to their appreciation of an important field of chemical science whether they are seeking an introduction to a topic or are reflecting from a position of experience.

C. J. Suckling
University of Strathclyde

1 | Infant enzyme chemistry

Colin J. Suckling

When this book was first planned, the idea in mind was to review, through a series of personal but related essays, the major impact that the study of enzymes has had upon some important fields of chemistry in the last thirty years. It was therefore something of a surprise to discover in the nineteenth century literature that enzymes had already prompted a great deal of chemical research, some of it with a remarkably modern ring, as I shall try to show in the next few pages. As early as 1833 observations had been made of the phenomenon of the natural hydrolysis of potato starch but with vitalistic concepts still much in people's minds, it was difficult to accept the existence of biological catalysts. The idea that enzymes are chemicals provoked prolonged scepticism and controversy. During the first half of the nineteenth century further naturally occurring reactions were recognized, in particular fermentations involving yeasts. On the one hand, it was held that the enzymic activity responsible for these fermentations was a property inseparable from living cells. Pasteur, amongst others, took this view. On the other hand, Liebig and, not surprisingly, Wöhler, regarded enzymes as chemical catalysts, albeit of unknown constitution, that could be separated from cells. Indeed these two may well have conspired to lampoon vitalism in an anonymous paper in Liebig's *Annalen der Pharmacie* (Anon., 1839). In this amusing article we read of chemical reactions brought about by

'small animals which hatch from eggs (yeast) in sugary solution and which on microscopic examination are seen to take the form of a Beindorf distillation apparatus, without the condenser . . . these animals, which have neither teeth nor eyes, but possess a stomach, a bladder which, when full, looks like a champagne bottle . . . devour the sugar with the production of excrement as alcohol and carbon dioxide.'

Eventually the argument was settled by experiment. In 1897, Buchner demonstrated that a yeast extract was capable of sustaining the fermentation of sugar but a few years earlier, a remarkable series of contributions began to appear from the laboratory of Emil Fischer (1894). The papers make enthralling reading, not only for their scientific content, but also because they convey great enthusiasm, sometimes naïve, but always evident. The main subject to which Fischer addressed his powerful experimental skills and penetrating intellect was stereoselectivity in enzymic catalysis, a field still of current significance; the ability of enzymes to select between stereoisomers has proved one of their most alluring properties.

Fischer's paper (1894) is remarkable for its discoveries themselves and also for the insight of a man of genius into future developments. He was, of course, uniquely well placed to tackle the problem of stereoselectivity because he had available an extensive series of stereoisomeric sugars which he had synthesized to determine their configurations. Derivatives of these compounds served as substrates for glycosidases which even in those days were available in crude cell-free form. His paper begins

'The different properties of the stereoisomeric hexoses with respect to yeast led Thierfelder and I to the hypothesis that the active chemical agent of yeast cells can only attack those sugars to which it possesses a related configuration.'

The hypothesis was supported by demonstrating, amongst other things, that the enzyme that hydrolyses sucrose, called 'invertin' by Fischer, acts only upon α-D-glucosides: β-D-glucosides and L-glucosides were completely untouched. There was no doubt that this was not just a chance phenomenon because a second enzyme, emulsin, was found to hydrolyse β-D-glucosides of both synthetic and natural origin. The complementary nature of these results is conclusive and, of course, still important in modern stereochemical studies. Fischer's assessment of his results is fascinating reading. It also makes an admirable preface to this book because it foreshadows much of what follows. When you have read further, you may be interested to reflect upon these lines:

'But the results suffice in principle to show that enzymes are choosy with respect to the configuration of their substrate, like yeast and other micro-organisms. The analogy between both phenomena appears so complete in this respect that one may assume the same origin for them,

and accordingly, I return to the abovementioned hypothesis of Thierfelder and myself. Invertin and emulsin have many perceptible similarities
and consist doubtless of an asymmetrically built molecule . . . To use an
image, I would say that the enzyme and glucoside must fit each other like a
lock and key to be able to exert a chemical influence upon each other . . .
The facts proven for the complex enzymes will soon also be found with
simpler asymmetric agents. I scarcely doubt that enzymes will be of use for
the determination of configuration of asymmetric substances . . . The
earlier much accepted distinction between the chemical ability of living
cells and the action of chemical agents with regard to molecular
asymmetry does not in fact exist.'

Although the last sentence quoted was directed to his contemporaries,
much of the preceding extract reads remarkably freshly to modern chemists
nearly a century later. We have the advantage over Fischer in techniques,
but some of the concepts that he advanced have still to be realized in
perfection as we shall see. However, Fischer was by no means the only
scientific prophet in the field of stereochemistry and his work depended
much upon the understanding developed by Pasteur. There is little in
modern stereochemical research that does not derive something from the
experimental and conceptual contribution of these two great scientists (see
Robinson, 1974).

It is remarkable how much was achieved in Fischer's time with impure
enzyme preparations. A parallel in today's research might be the study of
preparations containing unpurified neurotransmitter or hormone receptors,
although these too are now amenable to purification by modern chromatographic techniques. As Fischer predicted, enzymes have become widely
used for the determination of configuration but it is only in recent years that
'simpler asymmetric agents' have been able to reproduce enzymic stereoselectivity (see Chapters 3 and 4). Not surprisingly, the ever enthusiastic
Fischer even had a go at asymmetric synthesis himself (Fischer and Slimmer,
1903, and see Fig. 1.1). Knowing that glucose is chiral, Fischer hoped that
the naturally occurring glycoside, helicin, would undergo asymmetric
addition at the carbonyl group guided in some way by the asymmetric
environment created by the glucose ring. This strategy has since proved
successful (Chapter 4) and had Fischer used a more bulky nucleophile, he
too might have been successful. His first attempt was to add hydrogen
cyanide to helicin and to hydrolyse the product carefully. An optically
inactive product resulted. So Fischer tried again using diethyl zinc and this
time the product obtained from vacuum distillation was optically active. In
the exhilaration of discovery he wrote '. . . with this we thus believed that we
had solved the problem of asymmetric synthesis.'! Then came the snag,
Gilbertian 'modified rapture'. Rigorous control experiments clearly showed

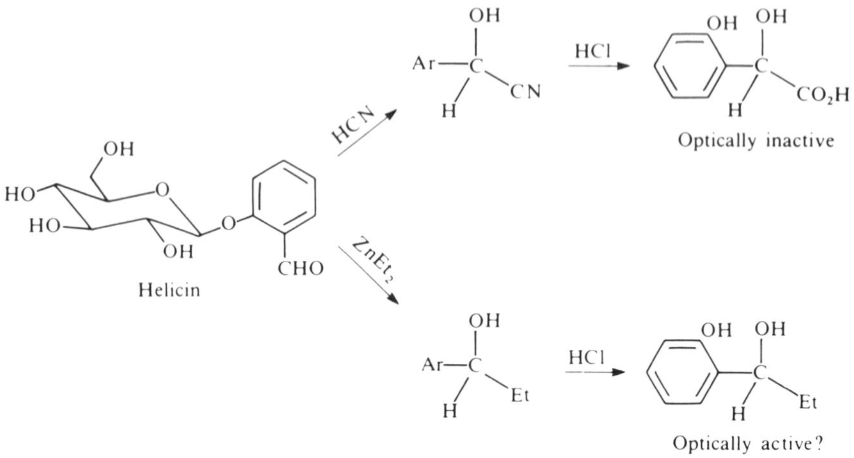

Fig. 1.1

that the apparent asymmetric induction was due to an impurity derived from glucose during distillation and no further attempts were reported. Many people have had similar, but unpublished, experiences.

The turn of the century also marked the first steps in the synthetic use of enzymes. Croft-Hill (1898) demonstrated that yeast enzymes could be used synthetically and Emmerling (1900) reported a synthesis of the glycoside amygdalin using enzymes. These, and other pioneering contributions, are cited by Hoesch (1921) in a special edition of *Berichte* devoted entirely to a biography of Fischer. Many of Hoesch's comments are equally apt today more than sixty years later. For instance, in summarizing Fischer's contribution to enzyme chemistry, Hoesch remarks 'Pure chemists may certainly not feel at home with the enzymatic studies of Emil Fischer'. Another notable comment was that Fischer's lock and key metaphor describing enzymic specificity was much appreciated in his day. From Hoesch's review and Fischer's own writings, it seems possible that Fischer never intended this image to be a scientific hypothesis but used it to illuminate the concept of stereochemical biospecificity to an audience totally unfamiliar with the new idea. Modern work, of course, makes it clear that the physical rigidity of a lock and key do not make an appropriate description of a conformationally mobile enzyme–substrate interaction. Once he had demonstrated bio-specificity with enzymes, similar complementary interactions were enthusiastically discussed for the behaviour of other biosystems such as toxins. However naïve the metaphor, it was certainly seminal.

Yet another part of our story began in the 1890s. Scientists were not only studying microbial enzymes but mammalian systems were also beginning to be investigated. In 1898, the kidney was shown to contain proteolytic activity (Tigerstedt and Bergmann, 1898). It was further demonstrated that an enzyme named renin hydrolyses a large plasma peptide, which today we know as angiotensinogen, to angiotensin I. We now know that angiotensin I has very little activity in the central or peripheral nervous system; it is further hydrolysed to a smaller peptide, angiotensin II, by an enzyme known as angiotensin converting enzyme. Angiotensin II has powerful effects on the circulatory system and studies of inhibitors of this enzyme have recently developed into one of the classics of modern drug invention (see Chapter 5).

Although much current work was foreshadowed or even initiated at the turn of the century, yet from that time, chemists' contact with enzymes became more remote as for the next five decades, chemists, with some notable exceptions, pursued the systematic study of the reactivity of organic compounds. Sir Robert Robinson was one such exception. Whilst contributing greatly to natural product chemistry and of course to ideas concerning reactivity, he realized that enzymes catalyse reactions under very mild conditions and sought laboratoy analogues in syntheses of alkaloids (Robinson, 1917). Meanwhile, biochemists wanted to find out in detail what enzymes are and set about their purification. The first systematic attempts were begun by Willstaetter in the 1920s but the first substantive success came from Sumner who in 1926 reported the crystallization of urease. Perhaps because he couldn't believe that someone else had done it first, Willstaetter disputed that Sumner actually had an enzyme. Nevertheless, proteolytic enzymes were soon purified to crystallinity and it became clear that enzymes are, as Fischer had surmised, proteins.

Although purified enzymes were available from that time on, chemists were by no means ready to accept the idea of macromolecules, let alone macromolecular catalysts. Staudinger, one of the fathers of polymer chemistry, had great difficulty in persuading the Swiss Chemical Society, at a meeting which ended in uncharacteristic Swiss uproar, that macromolecules can exist. A similar scepticism greeted the ideas of a young physical chemist, McBain, concerning the nature of micelles at a meeting of The Royal Society in London. He was told that his notions of molecular aggregation were 'nonsense'. In Germany too Hans Fischer, who established the structures of porphyrins by degradation and synthesis, as late as 1937 appeared to be unaware of the wide physiological importance of porphyrins although the isolation of the porphyrin-containing proteins, cytochromes, had been described in the mid 1920s by Keilin. Despite their temporary but acute myopia with regard to enzymes, chemists at this time were making great strides in understanding the basis of mechanistic organic chemistry. In time, the synthesis of artificial polymers was demonstrated and natural macromolecules too became respectable. The conceptual basis for a symbiotic growth of organic chemistry and enzyme chemistry was founded. This book recounts some of the branches of this growth.

What in particular amongst the properties of enzymes has been most significant for chemistry? In the first place, enzymes are such excellent catalysts. Indeed it has been argued that enzymes have evolved to perfect their catalytic function (see Chapter 2). If this is so, then it is a formidable challenge for chemists to understand the chemical basis for enzymic catalysis and a still greater one to mimic it effectively. However, in addition to these purely scientific aims, there are also extremely important practical consequences of the properties of enzymes. Selectivity in catalysis, as Fischer surmised, is one of the most important and it can be applied in a direct sense to perform both regioselective and stereoselective transformations in organic synthesis (Chapter 4). In addition, selective enzyme inhibitors are immensely important as drugs for the treatment of bacterial and viral diseases (Chapter 5).

In the last twenty years, great strides have been made in our understanding of the chemical basis of enzymic catalysis and it is the application of this and related enzyme chemistry that is developing apace. In the chapters that follow, a team of authors from many different countries and backgrounds discuss enzyme chemistry in relation to two broad themes, firstly synthetic organic chemistry. You will read how the study of the mechanism of action of coenzymes has led to a number of novel synthetic reactions (Chapter 3, Seiji Shinkai). Coenzymes are a good starting point for organic chemists because they are relatively small molecules with some innate catalytic activity even in the absence of enzymes. The wide range of reactivity observable is in itself fascinating and Professor Shinkai reviews much recent work for the first time. How conventional synthetic reactions compare in selectivity with enzyme-catalysed reactions and biomimetic systems is the topic of my own contribution (Chapter 4). The second theme, by way of contrast, concerns more biologically significant topics. The relevance of enzyme chemistry to chemotherapy in organic and inorganic aspects is discussed respectively in Chapters 5 (Philip Edwards, Barrie Hesp, Amy Trainor, and Alvin Willard) and 6 (Donald Brown and Ewen Smith). Biosynthetic studies have always built a bridge between organic chemistry and biochemistry and recent developments in this field are reviewed by David Cane in Chapter 7. A major development since the first edition of this book is the impact of genetic engineering and molecular biological techniques in both chemistry and biology. All fields of study described in this book have felt the benefit of the new technology. The basic concepts and methods are outlined in Chapter 9. Enzyme chemistry has borrowed much from and given much to biochemistry and the book ends with a consideration of future interactions between the disciplines (Chapter 9, Keith Suckling). To provide a basis for these discussions and to demonstrate the depth of thought into catalysis itself that enzyme properties have provoked Ron Kluger begins the essays with some thoughts on the mechanistic basis of enzyme catalysis (Chapter 2). As the scientific understanding of enzymes has improved, so has the extent of the appreciation of their industrial potential. This is recognized in Chapter 4 (synthesis) and in Chapter 8 which concerns the food industry (David Berry and Alistair Paterson). Interestingly and, to some extent, coincidentally several subject areas are discussed from different points of view by several contributors. These topics include enzyme stereochemistry (Chapters 2–5), prostaglandin chemisty (Chapters 4 and 5), β-lactam antibiotics (Chapters 4, 5 and 7), cyclodextrins (Chapters 2–4), and genetic engineering (Chapters 4, 7 and 8). Further areas of chemistry could also have been selected but these seven essays will give the reader insight into the impact of enzyme chemistry upon laboratory and industrial chemistry and the contacts of chemistry with the life sciences.

   Now is the time to let each author speak for himself. In editing this book, I have learned much from the thoughts of my fellow contributors. I am sure that they will convince you too that enzyme chemistry has contributed much to chemistry and is still vibrant and vital. Whilst the subject continues to develop, new challenges for biological chemistry are emerging, challenges that can be met all the better because of what the chemist has learned from enzymes in methods, concepts and techniques. As was alluded to earlier, other proteins can now be purified, in particular antibodies and neuro-transmitter and hormone receptors. In ten years' time, perhaps someone will be writing the closing lines of an introduction to the impact of receptor and antibody chemistry.

REFERENCES

Anon. (1839) *Annal. Pharm.*, **29**, 100.
Fischer, E. (1894) *Chem. Ber.*, **27**, 2985.
Fischer, E. and Slimmer, M. (1903) *Chem. Ber.*, **36**, 2575.
Hoesch, K. (1921) *Chem. Ber.*, **54**, 375.
Robinson, R. (1917) *J. Chem. Soc.*, 762.
Robinson, R. (1974) *Tetrahedron*, **30**, 1477.
Sumner, J. B. (1926) *J. Biol. Chem.*, **69**, 435.
Tigerstedt, R. and Bergmann, P. B. (1898) *Skand. Arch. Physiol.*, **8**, 223.

2 | The mechanistic basis of enzyme catalysis

Ronald Kluger

2.1 THE MECHANISTIC APPROACH

Enzymes and the reactions they catalyse can be studied effectively using the mechanistic principles which form the basis of physical organic chemistry. This chapter will identify and illustrate some of the important connections between the mechanistic concepts of physical organic chemistry and the solution of problems in enzymology. The principles of these connections guide us in the design of experiments, the interpretation of results, and the generation of new ideas.

2.1.1 Origins of the mechanistic approach

During the development of physical organic chemistry as a discipline in the 1930s and 1940s, a few chemists and enzymologists saw the possibilities of a connection between enzymology and physical organic chemistry. That era saw the rapid development of the kinetic approach to the elucidation of reaction mechanisms and the beginning of systematic approaches (For examples of these approaches see Hammett, 1970; Ingold, 1953). Evidence that enzymes are composed of polypeptide chains with specific structures was widely accepted and analytical determinations confirmed the general idea in many cases. The determination of the three-dimensional structure of enzymes by X-ray crystallography later dramatically confirmed and expanded the information concerning structure. In an informative historical review, Westheimer (1985, 1986) illustrates the developments with significant case histories, particularly with respect to the function of coenzymes. This historical perspective permits the reader to understand the concepts and their origins in depth.

At present, the sound application of principles of mechanistic chemistry to problems in enzymic catalysis continues to be the major intellectual force in advancing our understanding of enzymes (Kirsch, 1987). Future developments will require an understanding of the principles that have been developed. As with any powerful tool, care in its application and interpretation are required.

2.2 CONCEPTS OF CATALYSIS

The reactions catalysed by enzymes are remarkably diverse, fitting into all the major categories of organic and inorganic reaction types, including substitution, elimination, addition, oxidation and reduction. Such reaction patterns classify the relationship between the starting materials and products; mechanisms provide the means by which a connection occurs. Although the enzyme-catalysed reaction types are obvious, the mechanisms are often conjectural. Where sufficient information is available, it is useful if the reactions can be systematically divided into mechanistic types. Walsh (1978) provides a useful survey of enzyme-catalysed reactions in terms of reaction type and mechanisms. Other critical and extensive evaluations of mechanisms relevant to enzymic catalysis are also available. (Some examples are Bruice and Benkovic, 1966; Jencks, 1969; Jencks, 1975; Bender, 1971; Page, 1984; Sinnott, 1988) as well as books that place the mechanistic approach into a larger biochemical framework such as Scrimgeour, 1977; Metzler, 1977; Cunningham, 1978; Dugas and Penney, 1981; Fersht, 1985; Zubay, 1988.)

2.3 DESCRIBING A MECHANISM

A full description of the events that occur during a chemical reaction, especially one that occurs in solution, whether catalysed or uncatalysed, is unattainable because of the large number variables which necessarily describe it. In addition to changes that the substrates undergo during the transformation, catalysts, solvent and the medium also undergo reorganization (Ritchie, 1969). Thus, we first seek a very approximately defined path and fill in necessary details with information from our experiments.

We would like to know the pathway of a reaction in terms of the involvement of specific intermediates and modes of catalysis. How does the substrate change during the reaction and what is the molecular function of the enzyme in assuring efficiency and specificity? Can the transformation be quantitatively compared to a reaction which does not involve the enzyme? By having such information we can begin to understand the function of the enzyme in promoting and controlling the reaction. We will also be able to begin to design catalysts which can be based on what we have learned from analysing the enzymic system (see Chapters 3 and 4).

2.3.1   The three-dimensional energy diagram

A remarkably effective approach to understanding mechanistic alternatives and arriving at conclusions about the nature of the key transition state has been advocated by Jencks (1985) based on the specific application to analysis of elimination reactions reported by More O'Ferrall (1970). In this treatment a reaction path is considered as an optimized minimum which results from potentially competing routes between the sets of reactants and products. The competing paths involve passage through different intermediates or through concerted routes that avoid the intermediates.

For comparison, a conventional reaction co-ordinate diagram is a monotonic plot of the free energy of a reaction as a function of progress toward completion ('reaction coordinate'). Empirical insights as to the nature of the transition state, such as the Hammond postulate, developed from the monotonic view of a reaction. In that scheme, the progress of the reaction depends on a single reaction coordinate. The nature of the transition state is directly related to structures along that co-ordinate. In the More O'Ferrall–Jencks diagram, the transition state is also affected by the nature of the path itself and this is an effect that is orthogonal to the optimized reaction path. (Surprisingly, the approach is not a recent development. The utility of a second co-ordinate for analysing the nature of a transition state was proposed by Ogg and Polanyi (1935). Since this work preceded the more widespread interest in mechanisms, the generality may not have been appreciated.) Thus, for a simple chemical reaction, a single reaction co-ordinate diagram does not convey an adequate sense of the competition and the resulting larger picture. In the case of an enzyme-catalysed reaction, with even more potential for catalytic alternatives, the use of multidimensional plots will be especially appropriate.

The literal application of this method would require a detailed knowledge of the points on the free-energy surface represented by the diagram. Such a detailed knowledge is of course not attainable for a complex reaction in solution. Instead, all the information that is obtained is centred on determining the properties of the rate-determining transition state. Trends that are revealed by studies of free-energy relationships based on substituent effects, solvent effects, and isotope effects give complementary information which is utilized synergistically. For the practical problem of deciding between mechanisms, logical guidelines help to define the source of information needed to determine the relative merits of two routes. Jencks (1985) provides an excellent guide through the use of this procedure and the interested reader will be well rewarded in studying that review.

Nucleophilic substitution mechanisms provide an illustration of the use of the multidimensional diagrammatic approach. The S_N1 mechanism is a two step reaction while the S_N2 reaction is a single step (or concerted) mechanism. The More O'Ferrall–Jencks approach illustrates how these mechanisms are related and that reactions can exist that fall in an area that is borderline for either mechanism. The S_N1 mechanism can be simply modelled as two sequential conventional monotonic diagrams, the first producing the carbocation from the reactant and the second yielding the product from the carbocation and nucleophile. Placing these two processes as adjoining sides of a rectangle establishes the basic approach of a More O'Ferrall–Jencks rectangular diagram. The corner is the common point of the two steps: the carbocation is the intermediate. The rectangle is thus defined in three corners: reactants, products, carbocation. The remaining corner is a hypothetical intermediate which would result if the nucleophile adds prior to the departure of the leaving group.

The S_N2 reaction occurs by a path which avoids the corners of the diagram. These corners represent intermediates. If the S_N1 reaction can occur, then the reaction proceeds through the carbocation intermediate. If the inter-mediate is highly destabilized or has too short a lifetime to exist (less than the time for a single vibration), then the path diverges from the corner and a reaction occurs which involves no intermediate. Such a reaction is then the result of concerted processes: two or more bonds change in a single step and no intermediate forms. The concerted process (S_N2) is no longer considered as simply a competing alternative to a stepwise reaction series but rather is the consequence of the energetics of the entire reaction surface, anchored by

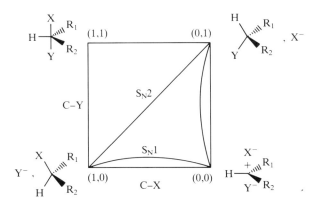

Fig. 2.1   A hypothetical More O'Ferrall–Jencks diagram for competing S_N1 and S_N2 reactions.

the nature of potential intermediates. The effects of substitutents on rates and value of slopes in linear free energy relationships locates the position of the transition state of the rate-determining step on this diagram. Recent work by Williams on the mechanisms of substitution in phosphates provides an especially clear illustration (Bourne *et al.*, 1988).

2.4   KINETICS: THE MEASURE OF CATALYSIS

Kinetic methods are essential for analysing the effectiveness of an enzyme as a catalyst. We can usually determine an experimental rate law for an enzyme-catalysed reaction and rate constants associated with the rate law can be evaluated by the use of steady-state methods (Segel, 1975). Thus, a logical sequence of events in terms of conversions between stable species and intermediates can be established.

This general description of an enzymic reaction can be further refined by the use of such techniques as isotope exchange (Segel, 1975), pre-steady-state analysis (Fersht, 1985) and other rapid techniques. With this quantitative set of information, we should be able to calibrate the effectiveness of the enzyme as a catalyst against a standard in order to be able to interpret the results. One useful comparison is with a similar reaction in the absence of enzyme. We know that the enzyme probably catalyses the reaction through a series of steps. It is most helpful if the reaction we are comparing can be set up to coincide step by step with the enzyme reaction. Then, by comparing similar enzymic and non-enzymic kinetics, we can describe the function of the enzyme in terms of specific energetics (see Hall and Knowles, 1975; Kluger *et al.*, 1981, Lim *et al.*, 1988). In order to do this, we need to be able to analyse the enzymic and non-enzymic systems on a common kinetic basis. The comparison must involve considerable information on both processes, including knowledge of the identity of the rate-determining step.

For these reasons, kinetic methods are especially important and we shall begin by reviewing some of the concepts used by physical organic chemists to simplify the analysis of multistep systems. First, the rate at which reactants are converted to products in any multistep reaction is described by a complex rate equation which can often be simplified using the steady-state assumption (see Hammett, 1970, pp. 77–79). If a reactive intermediate or a reactive complex with a catalyst is on the reaction path, then long before the system reaches its final equilibrium destination, it will reach what is called a steady state. Intermediates present in low concentrations in the steady state undergo changes in concentration that are very small compared to those of the observed reactants or products. This method of simplification is particularly appropriate for catalytic systems, since complexes of the substrate and catalyst will usually be present in very low concentrations (Klotz, 1976). In the case of an enzymic reaction, the concentration of enzyme is necessarily low (since its molecular weight is so high). In other reactions, reactive intermediates will also be high in energy and as a result will also be present at low concentrations. A steady state develops as changes in concentrations become undetectable. By this definition, an equilibrium is a steady state that is permanent. However, the steady state we deal with here will be that which is a non-equilibrium condition and will dissipate as components providing the 'fuel' for the steady state are used up. (Life is a steady state; death approaches equilibrium).

2.4.1   Rate-determining step

Murdoch (1981) has noted that in any multistep reaction, steps occurring after the rate-determining step have no effect on the rate of conversion of reactants to products, provided that no intermediate is more stable than the reactant or product. If an intermediate is more stable, then the system should be treated as two separate processes and not as a steady-state system. The rate-determining step is the slowest in rate, not the step with the smallest rate constant. Since rate depends on concentration of reactants and the rate constant, the rate-determining step is a function of concentration of species undergoing that step as well as the barrier to reaction from that step. The rate of the reaction is dependent only on the energy required to promote the reactants to the transition state energy of the rate-determining step. Steps preceding the rate-determining step provide a flux of material to the state prior to the rate determining transformation. This means that one can identify a rate-determining step by finding the step with the highest energy-transition state. After a mechanism has been established, we can compare the relative rates of any two processes that an intermediate can undergo. The faster process cannot be rate determining. Comparisons can be made for each intermediate until a self-consistent answer is found.

2.4.2   Mechanism is not a function of direction

The principle of microscopic reversibility requires that the lowest energy path in the forward direction is the same as that of the reverse reaction under the same conditions. If there are two competing paths in the forward direction, they will also compete in the reverse direction and to the same extent. Neglecting to take this into account may lead to erroneous conclusions which will give contradictory results. The principle of microscopic reversibility does not require that all molecules react by one pathway – it is a statistical statement that things will distribute themselves according to energy (Hammett, 1970). If the principle can determine with certainty which step is the slowest in one direction, we will know it to be the same for the other direction as well. Competing pathways must compete to the same extent in both directions under the same conditions. We can take advantage of this fact by choosing to examine steps from the direction which is most convenient experimentally. We must confirm what we have learned by studying the reverse process and observing quantitative agreement.

An example of the utility of this principle is the work of Breslow and Wernick (1976) on the mechanism by which carboxypeptidase A catalyses the hydrolysis of peptides. It had been reported by others that the enzyme catalyses the incorporation of oxygen from isotopically labelled water into the terminal carboxyl group of N-benzoylglycine. In any peptide hydrolysis, water must be incorporated into the product carboxylic acid. If the reaction occurs via an acyl-enzyme intermediate, then the enzyme should be able to catalyse exchange of the oxygen atoms of the terminal carboxyl group with those of solvent water in the absence of any free amine (Fig. 2.2). If an amine is required, then exchange does not prove the existence of an acyl-enzyme, since synthesis of peptide bond also removes oxygen from the carboxylic acid. Subsequent hydrolysis introduces the isotope of the solvent into the carboxyl group (Fig. 2.2). Breslow and Wernick showed that ^{18}O is incorporated from water into the carboxyl group of N-benzoylglycine only when a contaminating amino acid is present, and, in the absence of amino acids, exchange into the carboxyl group does not occur at a kinetically competent rate. Therefore, exchange occurs only by complete hydrolysis and its complete reversal.

The extension and implications of this principle are particularly interesting in the case of racemase enzymes. In these systems the product and reactant must be of equal energies but since an enzyme is chiral, the paths need not mirror each other and non-equivalent enzyme forms may be invoked for the specific binding of each enantiomer (Cardinale and Abeles,

$$(1)\ RC(O)NHR' + EXH \longrightarrow RC(O)XE + H_2NR'$$

$$(2)\ RC(O)XE + H_2O^* \dashrightarrow RC(O)^*H + EXH$$
- -
$$RC(O)OH + H_2O^* \longrightarrow RC(O)O^*H + H_2O$$
or

$$RC(O)OH + NH_2R' \longrightarrow RC(O)NHR' + H_2O$$
then

$$RC(O)NHR' + H_2O^* \longrightarrow RC(O)O^*H + NH_2R'$$

Fig. 2.2 Distinction between two mechanisms of catalysis of carboxyl oxygen exchange catalysed by carboxypeptidase, as reported by Breslow and Wernick (1976).

1968; Rudnick and Abeles, 1975; Belasco *et al.*, 1986). Thus, although it would appear that the enzyme is acting in a single direction, in fact the reaction is kinetically controlled by structural changes of the enzyme and the reverse route is still minimized by following the path of the forward reaction.

2.4.3   Conformational changes are usually fast

The Curtin–Hammett principle (Curtin, 1954; Hammett, 1970) is a rigorous application of transition-state theory in the testing of mechanistic assumptions. It tells us that the ratio of products of two reactions depends only on their relative rates of formation (Eliel, 1963, pp. 149–156). If the two reactants are conformational isomers then the barrier to interconversion of the two is usually small relative to the barrier to reaction. The yield of products from each conformer depends only on the energy of the transition states and not on the energy difference between the conformations. As a result, knowledge of the product ratio tells us the relative transition-state energy levels but not the relative amounts of starting conformations. On the other hand, if the barrier to interconversion of conformers is larger than the barrier to reaction, then the reaction product ratio does tell you about the conformational population. Product ratios in cases where interconversion of conformers is slow relative to the rate of product formation can be useful if detailed rate information is available but the mathematics of the kinetics of such a system are complex.

In the case of an enzymic reaction, in which bond formation and breakage occur rapidly, the movement of the enzyme may be a slow process (Fisher *et al.*, 1986) but the single bond rotations of the substrate remain rapid. Consequently, the advantages of rigid binding of a particular conformation of an intermediate by an enzyme will be small if the conformation is otherwise freely accessible by thermal motions. However, conformational effects on enzyme-catalysed reactions can be significant as a basis for reaction specificity. If several products can form from a single intermediate, an enzyme must assure that the product will be the one that is needed and not an uncontrolled distribution. Stereoelectronic control of product selection is a consequence of enzymic control of conformation.

2.5   STEREOCHEMISTRY AND SPECIFICITY

The term 'stereoelectronic effect' refers to a local structural or reactivity property in a molecule that varies in magnitude with changes in conformation of the molecule and is attributed to the properties of occupied orbitals. The original application of such ideas led to the notion of the restricted rotation within molecules about double and triple bonds. The application to epiphenomena due to interactions between electrons is non-bonded orbitals and sigma electrons (Kirby, 1980) has not been demonstrated convincingly (Sinnott, 1988).

Stereoelectronic analysis can deal with understanding the ability of an enzyme to catalyse a specific reaction for a substrate which could undergo competing reactions when subject to catalysis of the desired reaction. How can enzyme catalysis exclude reaction paths that would be favourable under non-enzymic reaction conditions from activation of the same intermediate? For example, the enzyme-catalysed decarboxylation of acetoacetate has been shown by Westheimer and his co-workers (Westheimer, 1969) to proceed *via* an imino-enzyme derived from the amino group of a lysine residue and the carbonyl group of the substrate ((2.1), Fig. 2.3). The product formed after carbon dioxide has been lost is an enamine (2.2) derived from the enzyme and acetone. In order for acetone to leave the active site, the enamine must tautomerize to the imine, which then hydrolyses. Both steps were shown to be enzyme-catalysed since the enzyme catalysed the incorporation of deuterium into acetone from deuterium oxide solvent (Tagaki and Westheimer, 1968). Therefore, a base must be available on the enzyme to remove the proton from the acetone imine to form the enamine and for the reverse process. These observations appear to suggest that the enzyme might be capable of having dual specificities with a single substrate. If formation of the imine from acetoacetate and the enzyme produces an adduct which will lose carbon dioxide readily, imine formation would normally also increase the acidity of the alpha hydrogens derived from the substrate promoting tautomerization of the imine to an enamine.

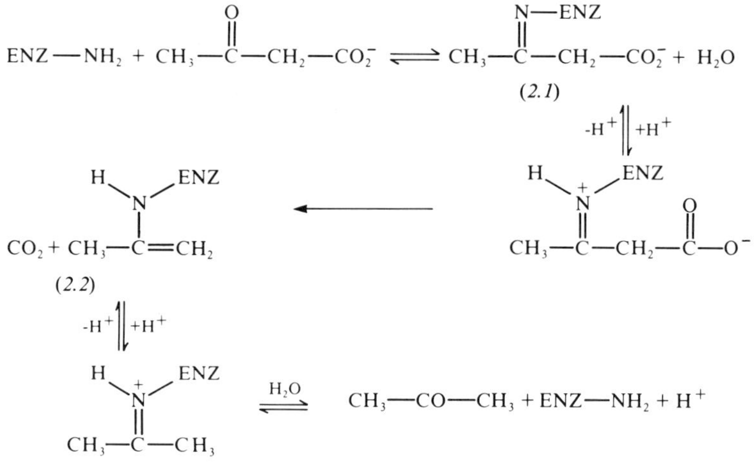

Fig. 2.3 Mechanism of decarboxylation of acetoacetate catalysed by acetoacetate decarboxylase.

(The detailed process is described in the next paragraph for a related system.) The enamine would not decarboxylate since conjugative unsaturation is no longer present (2.3). How then does the enzyme prevent this non-productive pathway from being followed?

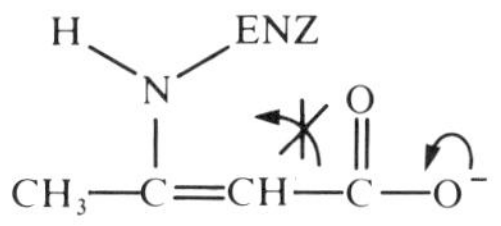

$$
\begin{array}{c}
H\diagdown \quad \diagup ENZ \\
N \\
| \\
CH_3\!-\!C\!=\!CH\!-\!C\!-\!O^-
\end{array}
$$

(2.3)

Dunathan (1966) proposed that orientation of orbitals controls the specificity of pyridoxal phosphate-dependent enzymes. The enzymes function by forming iminium derivatives between the amino acid substrate and the aldehyde group of pyridoxal phospate group of (PLP). Once the iminium group has formed, the pi system of that group can potentially activate several reactions (Fig. 2.4). For example, protons attached to carbon are normally very strongly bonded to carbon and will not exchange in aqueous base. If, however, the carbon atom to which the proton is attached is adjacent to a carbonyl group or iminium group, exchange occurs fairly readily. The product of removal of the proton is the corresponding enolate or enamine in which there is considerable double bond character between the two carbon atoms with delocalization to the heteroatom. This pi delocalization of the negative charge which would have been localized on carbon in the transition state for proton removal stabilizes the conjugate base of the carbon acid. The transition state leading to this delocalized system must reflect the overlapping character of the orbitals which constitute the pi system of the product. Therefore, the conformation of the transition state must be able to provide for the overlap. Looking at it the other way round, if such a conformation is inaccessible, then labilization of the C–H bond would be prevented.

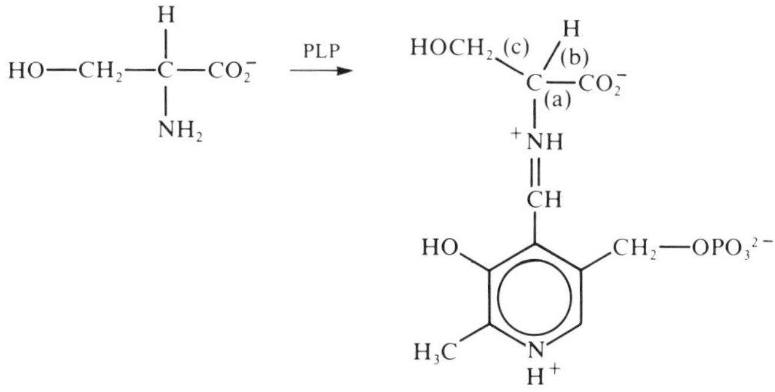

Fig. 2.4   Condensation of pyridoxal 5′-phosphate with an amino acid.

An enzyme can take advantage of this phenomenon very readily. If the substrate is bound in a relatively rigid conformation by electrostatic interactions, then this conformation will control which bonds will be activated when an iminium derivative forms. The enzyme can control the conformation by electrostatic interaction between a positive charge on the protein and the negative charge of the carboxylate of the substrate. In the case of PLP derivatives of amino acids, there are three bonds which could be activated by overlap. The first ((a) in Fig. 2.4) is the C–C bond to the carboxylate; overlap with this bond will cause the enzyme to be a decarboxylase. Overlap of the C–H bond ((b) in Fig. 2.4) will activate the alpha proton; an adjacent base can remove the proton prior to deamination or racemization of the amino acid. Overlap of the pi system and the sigma bond to the chain ((c) in Fig. 2.4) will permit the enzyme to function as an aldolase. Convincing evidence has been obtained that PLP-dependent enzymes indeed utilize this feature to control specificity (Dunathan (1967) provides an early review). A number of very specific enzyme inactivators has been designed to take advantage of the mechanism of the PLP-dependent reactions (Abeles and Maycock, 1976; see Chapter 5).

The specificity of acetoacetate decarboxylase can be explained in terms of this stereoelectronic control of specificity. If the enzyme binds the iminium derivative of acetoacetate in a conformation that prevents the C–H bond from overlapping the C–N of the iminium functional group, then the proton is not activated for transfer to a base (2.4). Even if there is a base adjacent to the proton, transfer of the proton will not occur because the transition state cannot be stabilized. However, the bond to the carboxylate group does overlap the pi system, and it will be activated so that decarboxylation can occur without a competitive reaction to slow the overall turnover or destroy the enzyme's specificity.

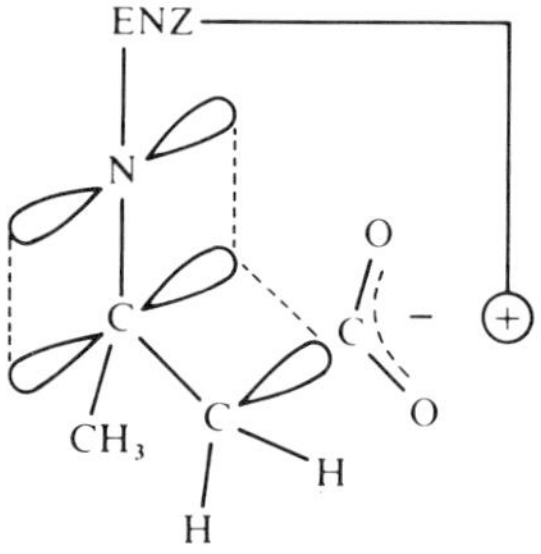

(2.4)

Evidence for this interpretation comes from studies of diverse reactions catalysed by acetoacetate decarboxylase. It was shown that butanone binds to the enzyme's normal substrate-binding site. For example, when butanone was incubated with the enzyme in deuterium oxide, the pro-*R* proton at C–3 was exchanged much faster than the pro-*S* proton (Hammons *et al.*, 1975; Benner *et al.*, 1981). The C–1 methyl protons are in homotopic environments and exchange at equal rates. However, since the exchange of the protons of the methylene group involves an enantiotopic selection, that reaction must be occurring in the chiral environment of the active site (*2.5*). The exchange is much faster than in the absence of enzyme, so the enzyme must be actively catalysing the exchange process, presumably via formation of the iminium derivative. Kluger and Nakaoka (1974) reported that the direction of binding of a substrate analogue can account for the nature of the exchange reactions catalysed by this enzyme. Ketones with an anionic substituent necessarily bind in a charge-defined array analogous to acetoacetate. Uncharged analogues have no defined direction with respect to the site that normally binds the carboxylate of acetoacetate and thus do not show a directional distinction in their exchange reactions.

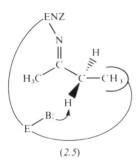

(2.5)

Rozzell and Benner (1984) studied the stereochemistry of the process in the reaction catalysed by acetoacetate decarboxylase in which the proton from solvent replaces the released carbon dioxide at the methylene group of acetoacetate. With substrates other than acetoacetate, the reaction is stereospecific but with acetoacetate, the reaction gives racemic products. This is not surprising in terms of orbitals, since the pi orbital which is generated upon loss of carbon dioxide has a local plane of symmetry. However, since the active site of the enzyme provides a chirotopic environment, it would be expected to favour the production of a single epimer of the adduct of enzyme and acetone imine. The source of this phenomenon of lack of stereospecificity in this case has not been identified. Rozzell and Benner (1984) suggest that protonation to form acetone may occur outside the chirotopic environment. Alternatively, we would suggest that the pair of adjacent lysine residues at the active site share in the functions of providing a nucleophile and a cation with a dynamic proton exchange rapidly interconverting these functions.

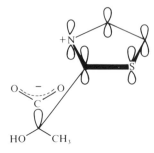

Fig. 2.5   Adduct of thiamine and pyruvate. In this conformation, either CO_2 or pyruvate can be eliminated (from Kluger, 1982). Reprinted with permission of copyright holder (New York Academy of Sciences).

All enzymic reactions that involve breaking or formation of C–H or C–C bonds should be subject to similar analysis. The most likely structures of adducts of thiamine diphosphate and pyruvate derive from extrapolations from the X-ray crystal structure of the phosphonate analogue of the adduct (Turano *et al.*, 1982). The structure that results from the maximal overlap direction in formation of the C–C bond between pyruvate and thiamine also aligns the C–C bond of the carboxylate group so that it will overlap the pi system of the thiazolium ring of thiamine ideally placed for decarboxylation (Kluger and Gish, 1988). Thus, the two main catalytic steps are structurally correlated (Fig. 2.5) in the catalytic cycle of pyruvate decarboxylase with the most efficient overlap of orbitals at every stage of the reaction.

In summary, although conformational factors in the substrate have little effect on accelerating any step in enzymic catalysis, they are often crucial in controlling selectivity.

2.6   STEREOCHEMISTRY AND MECHANISM

As the previous section has shown, the application of stereochemical concepts to the study of enzyme-catalysed reactions gives useful mechanistic information. The proper use of stereochemical concepts is important. The monographs by Bentley (1969) and Alworth (1972) provide extensive background for the application of stereochemical principles to enzyme-catalysed reactions. More recently, the topic of internal and external molecular stereochemistry has been further refined through a very useful combination of concept and formalism (Mislow and Siegel, 1984). Although the terminology is presented in reference to structural properties of organic molecules, it is especially useful for dealing with the stereochemical properties of substrates and intermediates in an enzymic environment.

The concept of *chirality* or handedness properly applies to objects and not to parts of objects. In the Mislow-Siegel analysis, molecules are treated as objects with combined dynamic and static properties. If a molecule is not chiral, it is designated as prochiral if a single substitution will make it chiral. The molecular structure of ethanol is *prochiral*: replacement of one of the methylene protons at C–1 by an isotope of hydrogen produces a chiral species. Methanol is neither chiral nor prochiral since *two* substitutions are required to convert it to a chiral species (it is pro(prochiral)).

Chiral objects differ from their mirror image objects (enantiomers) in a way that leads each to have a different energetic interaction with other chiral objects, such as enzymes. Prochiral molecules react selectively with chiral molecules and catalysts if the reaction converts the prochiral molecule into stereo isomers. *Meso* compounds can be considered as general members of the class of prochiral substances.

Internal distinctions within molecules are treated separately from the concepts of chirality. The important concepts in this matter are related to distinguishability. Can similar groups or atoms within a molecule be distinguished? If two groups exist in a pair, then they are indistinguishable if a symmetry operation defined by a rotation axis interconverts these groups.

In a chiral molecule there are no twofold axes although there may be threefold axes, such as those through a methyl group. If the rotation of the methyl group on the threefold axis is rapid, then the hydrogens are indistinguishable. Only in the extremely rare case where rotation is slow relative to the reaction of interest can the hydrogens be distinguished (Mislow and Siegel, 1984). The environment within a chiral molecule is said to be *chirotopic*. If the local environment does not possess a plane or axis of symmetry, the region is chirotopic. If it does possess such a plane or axis, the environment is *achirotopic*. Presumably an achirotopic region in a prochiral molecule could be considered to be prochirotopic. The two hydrogens at the 4 position of NADH are in chirotopic environments due to the fact that the molecule, including its sugar component, is chiral (You *et al.*, 1978). If the dihydronicotinamide is cleaved from the ribose and reattached to a methyl group, then an achiral species is produced and the environment of the two hydrogens is achirotopic. In the achirotopic environment, distinction between the two hydrogens can be made by a chiral species. In the chirotopic environment of the intact coenzyme, distinction between the two hydrogens does not require other sources of stereoheterotopicity.

The generalization of the concept of the *stereogenic element* (McCasland, 1953) is another important part of the Mislow–Siegel terminology. Such an element, or stereocentre, is a point within a molecular structure (usually but not necessarily coincident with an atom) where the permutation of ligands (whether bonded to that occupant of the centre or not) generates a new stereoisomer. Carbon atoms with four chemically or stereochemically different groups attached may reside at a stereocentre (If ligands differ only as to their being *R* or *S*, the distinction is sufficient if the other conditions are met). Mislow and Siegel note that neither a carbon atom nor a centre are chiral objects and so the expressions 'chiral carbon' and 'chiral centre' are inappropriate as are 'prochiral carbon' and 'prochiral centre'. Interestingly, stereocentres need not coincide with an atom: chiral molecules, such as *trans*-cyclooctene or substituted biphenyls, possess only a stereocentre that does not coincide with any atom. Thus, one should not be surprised to find a chiral species without a 'chiral carbon' nor should it be surprising that achiral compounds may possess mirror image 'chiral carbons' (e.g. *meso* compounds are achiral).

The earlier notation of 'stereotopicity' (Mislow and Raban, 1967; Eliel, 1975, 1980) has proven to be very useful for analysing enzyme-catalysed reactions. Pairs of chemically identical ligands are compared by their location within a molecule. If the pair is interchanged by the symmetry operation of a rotation, the ligands are in *homotopic* environments and are equivalent or indistinguishable, even by an enzyme. If the ligands are interchanged by a process requiring a reflection operation, they reside in mirror image environments which are called *enantiotopic*. These will interact with different energies with chiral materials, such as enzymes but equivalently with achiral materials. If the paired groups are in different environments, with no symmetry operation interconverting them, then they are in *diastereotopic* environments and will interact differently with both chiral and achiral species.

2.7   ENTROPY AND ENZYMIC CATALYSIS

Another application of physical organic chemistry to enzyme catalysis is the analysis of the means by which enzymes accelerate reactions of bound substrates. Most current discussions on the subject include a consideration of the relationship of the loss of entropy on binding of the substrate to the enzyme to gains in catalytic efficiency (see Jencks (1975) for a thorough discussion). Detailed treatments are presented which relate entropy effects to catalytic rate constants and later in this chapter we will present such a discussion. In order to make this more useful, however, it is probably a good idea at this point to review the general concepts and equations related to the relevant aspects of entropy.

Thermodynamics and statistical mechanics arrive at expressions for free energy in terms of enthalpy and entropy. Since free energy is related directly to chemical equilibria, it is the thermodynamic state function of most use in studying chemical phenomena. The change in free energy with temperature at constant pressure is defined as entropy. The definition of Gibbs free energy (G) is given by

$$G = H - TS$$

where H is enthalpy, T the Kelvin temperature, and S is entropy. Thus the derivative of free energy with respect to T is S. The relationship of this type of equation to molecular phenomena and mechanisms comes from applications of statistical mechanics (Denbigh, 1964, part III)

$$G = E - RT\ln Q + RT$$

The quantity Q is a partition function which is the sum of probabilities of molecular energy states, E is thermal energy, and R is the gas constant. Solving for G in the two expressions for free energy given above yields a single equation which can be used for defining entropy

$$S = (E/T) + R\ln Q$$

Since E/T is held constant, all that is needed in order to find a numeric value for the entropy change is the difference in the $R\ln Q$ term. To go from A to B, the entropy change will be $R\ln(Q_A/Q_B)$. If the partition functions for the two species can be derived, so can the contribution of entropy to the change in free energy.

The partition function, Q, can be factored into the components that sum to the total energy of a chemical species i.e. its translational, vibrational, and rotational energy contributions at constant electronic energy. The translational contribution can be calculated by using a statistical analysis of the number of ways particles can be distributed among energy levels. A particle of mass m in a container of volume V has a partition function

$$Q = (2\pi mkT)^{3/2}V/h^3$$

where h is Planck's constant.
Since the arrangement of species and energy levels is independent of the identity of the particles, the expression for translational entropy of N molecules is statistically corrected.

$$S = R[5/2 + \ln(2\pi mkT)^{3/2}V/h^3]$$

This expression is known as the Sackur–Tetrode equation and its validity in providing accurate values for the translational entropy of atoms and molecules has been confirmed experimentally in a large number of cases. Thus we can use this expression to estimate the translational entropy of a molecule before it comes in contact with an enzyme.

The Sackur–Tetrode equation predicts that a species such as a substrate of molecular weight 100 in solution at a concentration of 10^{-6} M will possess about 60 cal mol^{-1} °C^{-1} of translational entropy. At 300 K, therefore, the immobilization of the molecule by its becoming bound to an enzyme leads to the loss of translational entropy corresponding to deficit in free energy of 18 kcal mol^{-1} (75 kJ mol^{-1}). This must be compensated to a significant extent by the favourable interactions which lead the substrate to associate with the enzyme.

Rotational and vibrational entropy can also be calculated from knowledge of the appropriate partition functions. Rotational contributions to the overall entropy of a molecule are considerably smaller than translational contributions. The binding of a substrate to an enzyme therefore does not usually result in a significant change in this quantity. Vibrational contributions are usually small enough to be ignored.

In summary, the translational entropy of a molecule can be approximated from the Sackur–Tetrode equation on the basis of the derivation presented above in which the statistical and classical definitions of free energy are equated.

2.7.1   The relationship between association and catalysis

It is obvious that the chance of a bimolecular reaction occurring between species in dilute solution is less than it is if those two species are adjacent to one another. In order for them to be adjacent for a significant interval of time they must be held there by a third party which assumes the role of catalyst. One very important role for an enzyme is to function in this way. It must be able to encourage two species to react by ensuring that they are in proximity to one another and that they are aligned in such a manner that their contact can lead to a reaction.

The substrate-binding sites of enzymes possess functional groups at which it is energetically favourable for the substrates to associate. A considerable loss of entropy occurs when a molecule which has been free to move about in solution becomes associated in a specific manner with an enzyme. This is compensated for by favourable binding interactions (Westheimer, 1956). This form of enthalpic 'trapping out' of entropy by an enzyme applies to any number of different types of substrate molecules which interact with the enzyme and which eventually must interact with one another. If the entropy loss on binding has been compensated for by the favourable enthalpic interactions which also result, then the molecules associated with the enzyme have undergone a productive transformation from randomly moving individuals in a solution to productively aligned reacting partners on a catalytic surface. They can be considered to be appendages of the enzyme surface and as such two reacting species have become parts of the same molecule. If they are held in a manner that encourages reaction, then a considerable part of the entropic barrier to reaction has been removed

(Jencks, 1975; Herschlag, 1988). How to establish whether, and to what extent, a situation favourable to reaction will be created in any particular case was the subject of considerable controversy in the 1960s and 1970s. Several approaches were popular and then went out of favour as inaccuracies were discovered. Although the question may not be fully settled yet, it appears that a general consensus has been reached that a logical and justifiable analysis can be made by considering the entropic consequences of binding by comparing the situation to that of an intramolecular reaction as explained in the following paragraphs.

2.7.2 The intramolecular reaction: analogy to enzymic catalysis

In a bimolecular reaction, two molecules possessing translational entropy lose a considerable amount of that entropy when they interact. Since the two species must not separate from one another if they are to react, for the purposes of entropy analysis the two molecules can be treated as a single entity. As such, the equivalent of the translational entropy of one of the two is lost. In considering the reduction in translational entropy caused by binding of a substrate to an enzyme, we saw that we can use the Sackur–Tetrode equation to arrive at a numerical value for the entropic change. A similar change should take place when a bimolecular reaction occurs.

In an enzymic reaction the entropic loss caused by binding is compensated for by the favourable enthalpic interactions which occur when the substrate associates with the enzyme. For the reactions between two free species in solution, in the general case, no significant compensating enthalpic interactions occur. In an intramolecular reaction, the immobilization of a reactant that is present at a concentration of 1 M corresponds to a loss in translational and rotational entropy of about 35 entropy units (10.5 kcal-mol^{-1} (44 kJ mol^{-1} at 300 K)). If the species is previously immobilized by being covalently associated with its reaction partner in a rigid manner, then this entropy loss can be avoided. This corresponds to an increase in an equilibrium constant of about 4×10^7.

A convenient way of calibrating an intramolecular reaction against its bimolecular counterpart is through the quantity called effective molarity (see Kirby (1980) for a thorough discussion on this topic). A bimolecular reaction is characterized by a rate constant for a reaction that obeys second-order kinetics. The rate constant k has units M^{-1}s^{-1} but the corresponding intramolecular reaction has units of s^{-1}. The ratio of the rate constant for the intramolecular process to that for the bimolecular process will not be a dimensionless number but will have units of molarity. The particular intramolecular analogue of the bimolecular reaction being studied will be characterized by this ratio, the effective molarity, with units of molarity (Kirby, 1980). Since the bimolecular reaction can be brought to comparable units by multiplication by molarity, the intramolecular reaction can be said to be accelerated relative to the bimolecular reaction by an apparent concentration factor. The meaning of this factor in terms of physical reality has been controversial. Values for 10^8 M have been found for the effective molarities of intramolecular reactions, yet this cannot correspond to a physically attainable concentration. Does this then mean that the intramolecular phenomenon must be considered from a basis other than this apparent concentration?

The question of the basis of the intramolecular effect has been addressed in detail by Page and Jencks (1971) and this discussion is based primarily on their conclusions. Since effective molarity is really only the ratio of rate constants, it need not correspond to a phenomenon that is achievable only by increasing concentration; other factors are obviously involved. Two reacting species which are held together have the translational entropy of a single species and roughly the rotational entropy of the two separate species. Two separate reactants each possess the full complement of translational entropy. Combining two reacting species into a single molecule eliminates the translational entropy of one molecule. Since this translational entropy will have to be lost anyway in order for the bimolecular process to occur, what actually happens in the intramolecular process is a preactivation towards the transition state for the bimolecular reaction. This measures the entropic advantage of the intramolecular process in units of molarity.

2.8   ACID–BASE CATALYSIS

Reactions carried out in the laboratory are often subject to catalysis by acids or bases but reactions involving intramolecular general acid catalysis or general base catalysis never show effective molarities over about 60 M. Bernasconi and coworkers (Bernasconi *et al.*, 1982) have provided an incisive analysis of the basis of effective internal proton transfers. Factors which control the effectiveness of the transfer process include ring size for the internal proton transfer (six-membered ring is most effective), acidity and basicity of the donor and acceptor sites, acidity of the reaction medium, and the sensitivity of the proton-transfer rate to the basicity of the acceptor. An important conclusion with respect to enzymic catalysis is that large rate accelerations do not occur by facilitation of single proton-transfer steps. However, in combination with other steps, the factor is multiplicative. An ineffective proton transfer will have a major retarding effect that is also multiplicative.

The transfer of a proton between a Brønsted acid and a Brønsted base occurs as part of most enzyme-catalysed reactions. The details of proton-transfer reactions have been worked out and volumes devoted to the subject

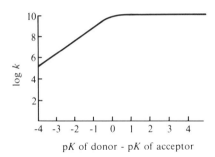

Fig. 2.6 Second-order rate constant ($M^{-1}s^{-1}$) for transfer of a proton from an oxygen or nitrogen acid to a series of oxygen or nitrogen bases. The phrase 'pK acceptor' refers to the conjugate acid of the acceptor. The rate constant reaches a maximum value which is that for diffusion of a proton in water.

have appeared (for example, Bell, 1973). In the case of enzymic reactions, proton transfer is usually coupled to some other bond-forming or bond-breaking process. Eigen and his coworkers established the fundamental properties of proton-transfer reactions (Eigen, 1964). An acid–base re-action is an equilibrium which is classified as being thermodynamically favourable, unfavourable, or neutral. The equilibrium constant for the reaction determines the classification. If the value is greater than unity, it is favourable in the forward direction. Eigen showed that for the transfer of proton between species capable of forming hydrogen bonds with water, in the thermodynamically favoured direction, the reaction occurs at a rate which has diffusion of the proton in water as its rate-limiting step. Proton transfers that occur at the diffusional limit therefore will have comparable rate constants. If the equilibrium constant for the proton-transfer reaction is known (from the ratio of acid dissociation constants of the two partners in the reaction), then the rate constant for proton transfer in the 'slow' direction can be calculated. If the reaction is thermodynamically neutral, then neither direction will be at the diffusional limit and both the forward and reverse reaction rates will be comparable. A plot of the rate constant for transfer of a proton from an acid to a series of conjugate bases is called an Eigen plot (Fig. 2.6). In the case where one or both of the reaction partners is incapable of hydrogen bonding, the situation is subject to different limits since the properties of diffusion in water are no longer as significant as the solvation of the reactants.

2.8.1 Proton transfer from carbon acids

The transfer of a proton from a carbon acid is limited by the inability of most carbon acids to form hydrogen bonds with water. Furthermore, the maximal rate is orders of magnitude slower than the rate of diffusion if the conjugate base is a delocalized carbanion (Kresge, 1975). If the anion is localized, as in the case of formation of cyanide ion from HCN, then the rate of formation can achieve larger limiting values (Bednar and Jencks, 1985). Thus, in an enzyme-catalysed reaction where transfer produces a delocalized carbanion, the inherent limits of proton transfer rates may become significant and alternatives to direct transfer may increase the efficiency of the process.

A model for an enhanced mechanism for proton transfer has been developed (Kluger *et al.*, 1984). The rate constants for transfer of a proton from a 3-ketophosphonate to a series of substituted pyridines follows a normal Brønsted relationship if the pyridine contains no substituents on the carbon adjacent to nitrogen, the site to which the proton is transferred. If the pyridine contains alkyl substitutents at the 2 or 6 positions of the ring, the rate constants for proton transfer fall below the line of the Brønsted plot. This is consistent with a mechanism for proton transfer from the carbon acid in which the amine accepts the proton directly, without the intervention of solvent, and thus the reaction is sensitive to steric effects. Addition of magnesium ion or manganese ion acclerates the reaction and the rate constants for the metal catalysed reactions follow Brønsted plots which parallel the plot for data in the absence of metals. However, the rate-decreasing effect of substituents at the 2 and 6 positions is not observed for the metal-assisted processes. This requires that the accelerated reaction of the sterically hindered bases does not proceed by the direct transfer mechanism but by another mechanism with decreased steric effects. We propose that the transfer involves reaction of the base with co-ordinated water of the metal complex with the concomitant abstraction of the proton from the carbon acid by the metal-bound hydroxide ion (Fig. 2.7).

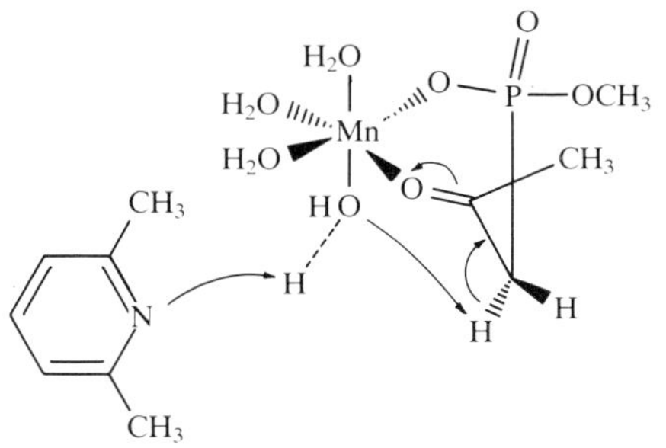

Fig. 2.7   Mechanism of facilitated proton transfer from carbon.

2.8.2   Proton transfer in concert with other processes

For reactions in which proton transfer occurs in conjunction with some other bond making or breaking process (a concerted system), the analysis becomes more complex. How can we be sure that the two steps are really occurring together and not in rapid succession? (For a review see Jencks (1980).) The problem is a significant one because it sets the limits to the efficiency of acid–base catalysis.

A reaction that is accelerated by the presence of Brønsted acids other than protons is said to be general-acid-catalysed. This is easily diagnosed since addition of a Brønsted acid (i.e. a buffer) at constant acidity will increase the rate of the reaction. Since a buffer consists of an acid and base component, the effect of changing the buffer ratio must be assessed to determine whether the acid or the base component is catalytic. Occasionally both components are active. For general acid catalysis, the stronger the acid catalyst, the larger should be the observed rate constant. A plot of the logarithm of the observed rate constant versus the pK_a of the catalytic acid will normally be a straight line provided that the mechanism and rate-determing step remain constant. This plot, called a Brønsted plot, is an extremely useful tool for analysing mechanisms related to enzymic catalysis.

The catalysis of a reaction by buffers suggests that transfer of a proton is involved in the rate-determining step, but it does not identify that step. The rate law that is obtained describes kinetically equivalent mechanisms (see Bell, 1973). Furthermore, is proton transfer occurring alone or in conjunction with some other process? Jencks (1980) has argued that proton transfer may occur in a single step along with another bond breaking or forming process if that is a lower-energy pathway than one in which the processes occur in separate steps. These rules must apply to enzyme-catalysed reactions as well since proton transfers often occur in conjunction with other processes. A detailed analysis of enzyme reactions at a similar level of detail is becoming possible through the interpretation of phenomena such as solvent isotope effects (see Schowen, 1977).

2.9   LINEAR FREE-ENERGY RELATIONSHIPS AND ENZYMIC REACTIONS

As we have seen, the determination of linear free-energy relationships can provide a valuable probe of structure of the transition states for a series of closely related chemical reactions. The rate constants for the reactions (and therefore differences in free energies of activation) are calibrated by their relationship to an equilibrium in a well-defined reaction series. One of the most reliable means of probing a catalytic system involves the use of the Brønsted catalysis law (Hammett, 1970). This requires one to vary the catalysts rather than the substrate. The slope of the linear relationship between the pK_a of the catalyst and the logarithm of the rate constant for the catalysed reaction involving a proton transfer has a value which is directly related to the extent of proton transfer in the transition state. The method can be extended to other group transfer reactions through correlations of rate or equilibrium free energies against a standard set of equilibrium free energies. Thus, since enzymes are catalysts, a Brønsted-type study would give information about the energetics of the reaction. Since an enzyme is a single catalyst rather than a set of catalysts, what series could be used? For enzymes with low specificity, substrates can be varied in some way. The problem with such an approach is that the catalyst (enzyme) is the object of interest and the variation in the substrate does not give specific information about the catalyst but rather about how the substrate responds to the catalyst. Furthermore, the specificity of most enzymes removes the possibility of such an approach. Clearly, it would be desirable to be able to modify an enzyme in a systematic way so that calibrated rates or equilibria of reactions on the set of enzymes with a common set of substrates could be studied.

Direct chemical methods for alteration of an enzyme to change its catalytic properties in a systematic way have not yet been devised. The possibility of altering the nature of an enzyme through the use of site-specific mutagenesis (see section 9.11) provided the chance to relate the kinetic changes in an enzymic process to a related thermodynamically controlled property (Kirsch, 1987). Fersht and coworkers (1986, 1987) utilized a self-calibrating reaction in which the free-energy difference of intermediates in a series of altered enzymes is the calibration for the free energy of activation for the same step. For a series of modified tyrosyl t-RNA synthetases, the relationship

$$\log k_1 = a \times (\log K_1), \text{ or}$$

$$d(\log k_1)/d(\log K_1) = a$$

The value of a is about 0.8. What can be concluded from this? First, as Fersht notes, one must be cautious since the defining reaction and the reaction to be calibrated are the same.

$$K_1 = k_1/k_{-1}$$

$$d(\log K_1)/d(\log K_1) = d(\log k_1)/d(\log K_1) - d(\log k_{-1})/d(\log K_1).$$

$$1 = 0.8 - d(\log k_{-1})d(\log K_1) = 0.8 - a'$$

Since the 'a' term is dominant, any value near unity will give a reasonable correlation if the variation in 'a' is small. Estell (1987a and b) reports that a plot of random numbers for k_1 and K_1 with these characteristics will also give a straight line with what appears to be good statistical properties. Estell suggests that a plot of $\log k_{-1}$ versus $\log K_1$ would be more useful. In fact, before Fersht's work was reported no information was available on the range of either value. Estell's random numbers are not random but fall within the range of Fersht's data. As a result of this work, the tools of the mechanistic chemist are most directly brought to bear in the context of an enzymic reaction. Fersht concludes that the data reveal that the transition state for the transfer reaction on the enzyme resembles the product of the reaction and that enzymic catalysis is based on this property.

2.10 ENZYMIC EFFICIENCY

Let us consider an enzyme that catalyses a multistep process which converts a single substrate to a single product. Just how 'good' is this particular enzyme as a catalyst compared to other proteins which might catalyse the same reaction? It has been argued convincingly by Hanson and Rose (1975) that enzymic efficiency is an evolutionary process subject to mutation and improvement. We can see how to set a standard if we consider the basic facts of the system.

The thermodynamic relation of reactants and products is independent of the catalytic process. Thermodynamic analysis of the reaction tells us whether the conversion will result in the net gain or loss of chemical energy to the system, which will be converted to heat and/or work. An endergonic process will require an energy input and will always have a barrier at least equal to the thermodynamic difference between the starting material and product. An exergonic process will not have a thermodynamic barrier to overcome but it will be limited by the rate of diffusion of the substrates to and of the products away from the enzyme. The endergonic process has this diffusion barrier superimposed upon the thermodynamic barrier. To simplify our analysis of efficiency, we shall eliminate consideration of the thermodynamic barrier but will remember that it must be superimposed upon any conclusion we reach.

We know that in any multistep reaction, the most significant barrier is that of the rate-determining step. If that barrier is lowered, the reaction will proceed faster. (The rate expression is always reducible to the rate constant for the rate-determining step times the concentration of the species reacting in the forward direction in that step adjusted for the extent of reverse reaction.) An enzyme can become more efficient if it can lower the barrier of the rate-determining step (by such devices as more reactive functional groups or better relative placements of binding sites, for example). However, there must be a limit to this since, as the barrier for that step becomes sufficiently low, another step will then have a higher absolute barrier and thus will become rate determining. This step will then be subject to evolutionary pressure, and it will improve until another step takes over. The limit of this improvement is reached when all steps have transition states which are comparable in energy (corrected for thermodynamic barriers). That is, each is partially rate-determining (for an example see O'Leary and Baughn, 1972). The lowest absolute barrier that can exist for any reaction in solution is that for diffusion; therefore, the ultimately most efficient, isolated enzyme will be that which catalyses all steps with barriers equal to that for diffusion.

Albery and Knowles (1977) developed a mathematical function that describes the efficiency of an enzyme as a catalyst based on the assumption that the ultimate rate-controlling factor for any single enzyme-catalysed reaction must be the rate of diffusion. The inherent thermodynamics of the reaction being catalysed also place a limit on its ultimate efficiency: equilibria of bound substrates on an enzyme are influenced by the energetics of the reaction (Chin, 1983). The efficiency of an enzyme can then be compared to that expected for a perfect or completely evolved enzyme, using parameters derived from kinetic measurements. This efficiency function is a quantitative measure of the 'goodness' of catalysis in analogy to parameters which measure the goodness of a statistical fit of data to an equation. Although the method is very stringent and not amenable to data thus far available, it promotes conceptual organization when examining the properties of an enzymic reaction (Kluger, 1982; Nambiar et al., 1983).

Knowles and his coworkers (Lim et al., 1988) applied the idea of efficiency in enzymes in a further refinement by the combination of site specific mutagenesis and variation of the viscosity of the medium through which the substrate diffuses. The enzyme triose phosphate isomerase has evolved to perfection by the Albery–Knowles criteria (Albery and Knowles, 1977). That is, aside from the constraints placed by the laws of thermodynamics, the rates are limited only by the diffusional properties of the substrates (or products). If the enzyme is made less perfect by mutational processes, then the diffusion process becomes less significant in terms of kinetics than an internal step. Diffusion cannot be more or less perfect but an internal process can be. Thus if the rate of diffusion is decreased by changing the solvent in which the substrate is dissolved, then in the case of a perfect enzyme, the diffusional step becomes more rate determining.

For a two step ready state process

$$k = k_1 k_2 / k_{-1} + k_2$$

If k_2 becomes smaller and k_{-1} and k_1 do not change, then k becomes larger. The free-energy barrier associated with the second step becomes larger and thus significant. Albery and Knowles had proposed that triose phosphate isomerase has evolved to the state of perfection defined in their theoretical papers. In the case of this evolved enzyme, changes in the diffusional properties of the solution surrounding the enzyme then will have a significant effect on the enzyme's functions as a catalyst and this is the case (Lim *et al.*, 1988). These workers also produced a less efficient variant of triose phosphate isomerase by site-directed mutation of the bacterial gene encoding this enzyme. The rate for diffusion is independent of the enzyme but rather is a physical property of the substrate, product and medium. Internal rates for the enzymic processes, on the other hand, are affected by the mutation. As a result, in a 'less perfect' enzyme, diffusion is less kinetically significant and therefore changes in the rate of diffusion by changes in the medium have less of an effect on observed catalytic properties than do such changes in a 'perfect' enzyme. This is precisely what was observed by Knowles and coworkers. In the mutant enzyme, the catalytic rate constant is about 1/1000 that of the wild type. The addition of glycerol to the reaction medium reduced k_{cat}/K_M for the reaction of glyceraldehyde phosphate with the wild-type enzyme by an amount proportional to the change in rate expected for the change in viscosity. For the mutated, less efficient enzyme, k_{cat}/K_M does not vary with viscosity.

2.10.1   Other effects of diffusion

Substrates undergoing enzyme-catalysed reactions must be able to arrive and depart from the active site of the enzyme without difficulty. We have considered that the rate of diffusion of substrates and products ultimately determines the productivity of catalysis. If a conversion is to occur, the reactants must be present at the active site simultaneously. The products must also be able to escape readily after a reaction has occurred. An enzyme with a very high affinity for the reaction products would cause the reaction to 'back up' toward the transition state of the step prior to the diffusion step. If this barrier is lower than that for diffusion, the efficiency of the catalytic process is compromised.

This may be a serious problem in catalysis for very efficient enzymes or those that deal with large substrates which can interact strongly at many sites, for example, peptidases and nucleases. One way an enzyme can turn the affinity problem into an asset is to use the binding energy in a productive way to achieve catalysis. A tightly bound product will be released at a rate less than the diffusion-controlled maximum. However, if the energy that is available through the interaction of the product and the enzyme could be used to cause the enzyme to isomerize to a form that has a lower affinity for the product, then catalysis would be facilitated. However, once the product was adhered to the form of the enzyme for which it has an affinity, it locks out the chance for a change. The isomerization should take place before the last step has occurred to avoid the product blockage. The formation of covalent intermediates in enzymic reactions may help to overcome this problem.

Consider the case of pepsin, an enzyme that catalyses the hydrolysis of peptide bonds in large substrates at an early stage of digestion in the stomach. Evidence has been accumulated that suggests that the reaction may occur by the formation of sequential covalent intermediates between the enzyme and the two portions of the peptide bond (see the discussions in Kluger and Chin (1982); Hofmann and Fink (1984), Somayaji *et al.*, (1988)). Since the substrates are large, they have a high affinity for the enzyme and diffusion of the products away must be a problem. If sequential intermediates (covalent or non-covalent) form, then the production of the intermediate provides the energy that enables an isomerization of the enzyme to occur. The isomerized enzyme can then release the substrate readily.

2.11   EXAMPLES OF INTRAMOLECULAR CATALYSIS

One of the most effective procedures for gaining information about the nature of interactions between the functional group of a substrate and a catalytic group on an enzyme is the use of intramolecular model reactions. Our general analysis of the basic characteristics of enzymic catalysis suggests that entropic effects play a major role in an enzyme's processing of its substrates. The prior immobilization of interacting functional groups relative to one another enables them to react without first overcoming the barrier that is due to losses of translational entropy in bimolecular reactions. Many interactions which do not have very high enthalpic barriers are masked by the built-in entropic barrier. Thus one may falsely conclude, on the basis of the low reactivity of bimolecular models, that a particular substrate is unreactive towards the functional groups at the active site of an enzyme.

For example, one mechanism proposed for the ATP-dependent carboxylation of biotin involves nucleophilic attack of the ureido group of biotin upon the terminal phosphate of ATP to produce *O*-phosphobiotin (Calvin and Pon, 1959). The reactivity of a urea towards the phosphate functionality was unknown because any bimolecular reactions between those functional groups had been too slow to observe. However, this can be overcome by immobilizing the groups on a single molecule so that no loss of translational entropy occurs when the two functional groups interact. We found that the urea reacted readily with the adjacent phosphate derivative. The entropic barrier to bimolecular reactions had obscured the likely interaction of these two functional groups (Kluger *et al.*, 1979). The mechanism has thus been established as a serious possibility for the enzymic carboxylation of biotin (Wood and Barden, 1977; Hansen and Knowles, 1985) (Fig. 2.8).

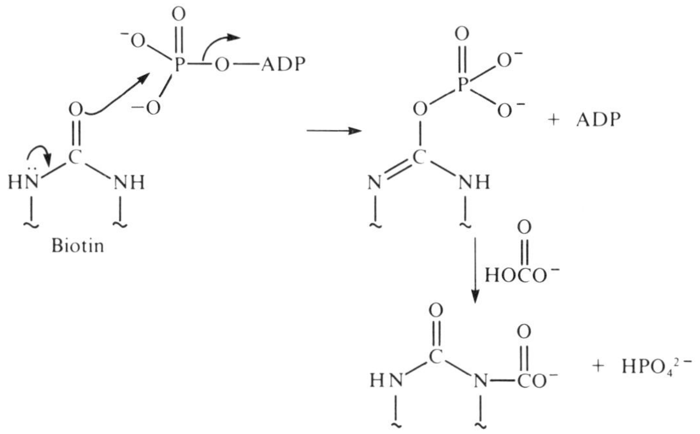

Fig. 2.8 Carboxylation of biotin *via* formation of phosphorylated biotin (see Kluger *et al.* 1979).

Carboxyl derivatives also exhibit strong neighbouring group effects. For example, an amide that can react with a neighbouring carboxyl is hydrolysed much more rapidly than an isolated amide (Bender, 1957). The amide reacts first with the carboxyl group to form an anhydride with release of the amine portion of the amide (Fig. 2.9). The anhydride then reacts with water to produce the carboxyl group derived from the amide and also to regenerate the carboxyl group that was originally present.

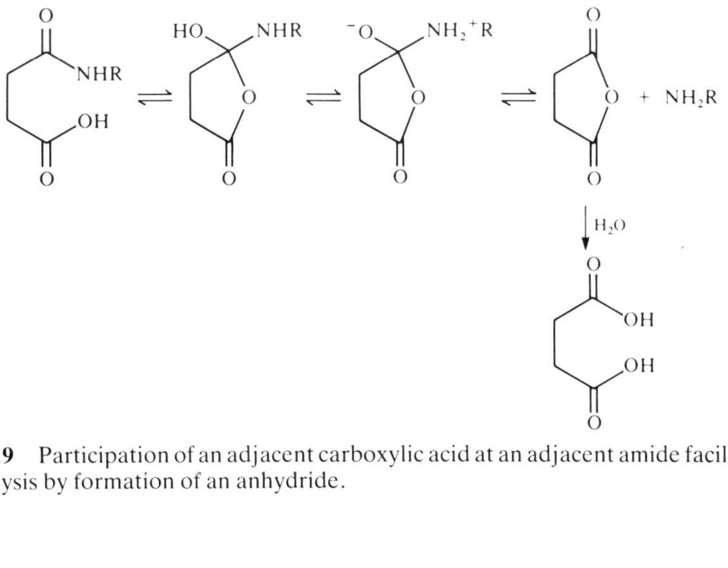

Fig. 2.9 Participation of an adjacent carboxylic acid at an adjacent amide facilitates hydrolysis by formation of an anhydride.

The first compounds of this class that were subjected to mechanistic study were derivatives of phthalamic acid. Bender and his coworkers used isotope labels and kinetic analysis. They found that catalysis of the hydrolysis of the amide occurs by intramolecular formation of an acylated orthoamide by addition of the carboxyl group to the neighbouring amide (Bender, 1957; Bender *et al.*, 1958). The form of the molecule in which the carboxylic group is undissociated, or some tautomeric equivalent, is reactive whereas the conjugate base is stable. It was proposed that the carboxyl group adds to the amide group in a four-centre reaction to produce the cyclic orthoamide intermediate (Bender *et al.*, 1958). This intermediate decomposes to anhydride and amine. The key to the efficiency of this route is the ease of formation of the orthoamide intermediate.

Kirby and his coworkers studied the reaction in greater detail (Kirby and Lancaster, 1972; Aldersley *et al.*, 1972). Through a series of rigid substituted derivatives of maleamic acid, they demonstrated that ground-state strain dramatically increases the rate by promoting cyclization, giving effective molarities of up to 10^{10}. They also showed that formation of the intermediate is a faster step than is its decomposition to products. In other words, the conversion of the amic acid to anhydride and amine involves rate-determining breakdown of the intermediate.

2.12   TRANSITION-STATE ANALOGUES

At the beginning of this chapter, we argued that the organic chemical concept of reaction mechanism should be logically applied to enzyme reactions, but we have also seen that the problem is even more complex than in the usual organic reaction because the important and large enzyme molecule must be accounted for. Since the enzyme is much larger than any usual substrate it provides the environment of the reaction. In a complete analysis, accounting for the enzyme's course during the reaction is as important as accounting for the transformations of the substrate. However, it is much easier to follow structural changes of the substrate than it is to follow the conformational changes of the enzyme. Since the enzyme is a catalyst, its structure before and after a reaction has occurred should be constant. The substrate, on the other hand, has undergone a series of changes which have converted it into the product. Examination of the enzyme before and after the reaction obviously gives no information. Observing the enzyme directly during catalysis is important and possible but usually difficult. One way of simplifying this task is to monitor the enzyme by using a substrate that reports the nature of its interactions with the enzyme. Spectroscopic observation of a substrate or of a modified substrate often gives extremely useful information in this connection. (For an example of a modern spectroscopic approach see Huber *et al.*, 1982.) An alternative approach is to prepare inhibitors which are substrate variants designed to interact with a transient but catalytically significant form of the enzyme. The effects of the inhibitor can then be examined kinetically or spectroscopically. Such strategies are most valuable in medicinal chemical studies (see Chapter 5). Let us consider the various steps involved in the catalytic process to see where the approach can be used.

Enzyme-catalysed reactions involving a single substrate which is converted into a single product present a basis for analysis of reaction mechanisms. Multiple substrate reactions add problems that must be dealt with from the basis established for unimolecular processes. For a unimolecular conversion, we know that the substrate must bind to the enzyme, be converted to the product during the catalytic process and then separate from the enzyme.

With the substrate bound at the active site of the enzyme we can assume, for simplicity, that the loss of entropy due to binding is more than compensated for by favourable interactions between the substrate and the enzyme. This initial state must be such that the substrate is relatively unaltered from its structure before binding. However, to undergo reaction the substrate must proceed to some higher energy form, towards the structure of the transition state. The enzyme must encourage this process. Pauling originally proposed that an enzyme could catalyse a reaction by assuming a conformation that would stabilize the transition state particularly well (see Wolfenden (1972) for a discussion of this point). If the enzyme can alter its conformation to a form with a high affinity for the transition state, then the reaction will be catalysed if the price of the isomerization in terms of energy is less than the amount of transition-state stabilization that will be achieved.

Since the transition state occurs at the highest energy point on the reaction surface, it is necessarily an unstable arrangement and no method will be available to isolate it. Yet, if the goal of the enzyme is to stabilize this species, the enzyme should be complementary in form to the transition state whose formation it is promoting. If we could isolate the form of the enzyme that stabilizes the transition state, then we would have a 'negative image' of the transition state. This means of analysis was proposed by Wolfenden (1972) and by Lienhard (1973). They also recognized that the enzyme will exist as a collection of forms that bind substrate and transition state. The formation of the enzyme–substrate complex can provide the energy necessary to isomerize the enzyme to the form that will stabilize the transition state. (Jencks (1975) has called this the Circe effect and has written an extensive analytical review on the subject.)

Experimental evidence has also suggested that species that bind much more tightly to enzymes than the natural substrate might be doing so because they resemble the transition state for the reaction (Cardinale and

Abeles, 1968). For example, the transition state for the interconversion of D- and L-proline, catalysed by the enzyme proline racemase, should resemble the planar species, pyrrolecarboxylic acid (Fig. 2.10). The observation that pyrrolecarboxylic acid binds more tightly to the enzyme than does either substrate is consistent with its being a structural analogue of the transition state for the interconversion catalysed by the enzyme (Jencks, 1969, pp. 300–301). Wolfenden (1972) and Lienhard (1973) realized that in addition to this being a useful *ad hoc* explanation of a single phenomenon, it is also a basis for a general approach to analysing transition-state structure and to inhibitor and drug design (see Chapter 5 and Wolfenden, 1979).

This approach has been very useful for designing inhibitors of very specific enzymes and numerous examples are cited in review articles. However, we must be careful to realize that we are undertaking both the experiment and interpretation with considerable prejudice. We assume that tight binding is occurring because a molecule resembles the expected transition state for the reaction. Even if we are right and can exclude other transition states from contention, we must still make sure that we are really dealing with a proper cause and effect relationship; we must exclude all other possible reasons why the analogue and the enzyme would have a high affinity for one another. However, since we cannot know what all other possible reasons are, we are left with experimental evidence that at its best can only be suggestive and never conclusive.

For a mechanistic example, let us consider one of the classic cases of a successfully designed transition state analogue. Pyruvate dehydrogenase catalyses the oxidative decarboxylation of pyruvate. The single substrate for decarboxylation is pyruvate which forms a covalent adduct with thiamine

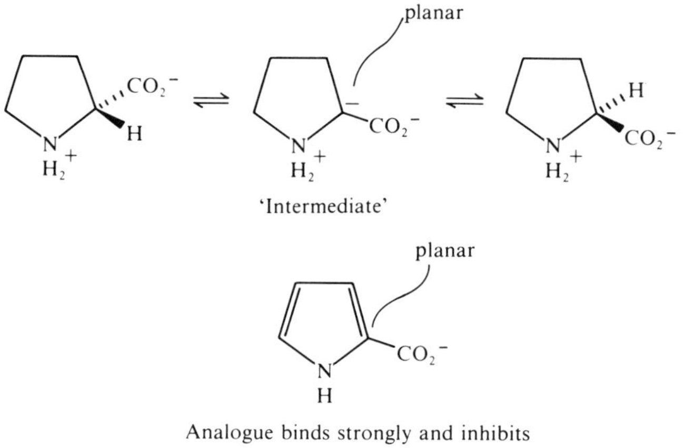

'Intermediate'

Analogue binds strongly and inhibits

Fig. 2.10   Enzyme-catalysed interconversion of D- and L-proline proceeds *via* a planar transition state which can be mimicked by pyrrole-2-carboxylate.

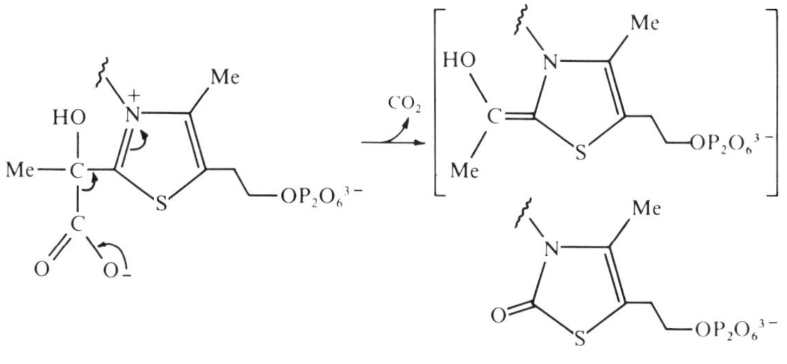

Fig. 2.11 Conversion of adduct of pyruvate and TDP involves a transition state that probably has character associated with neutral enamine formed as CO_2 is lost (Kluger *et al.*, 1987). TTDP is an analogue of the enamine.

diphosphate (TDP) (Kluger and Smyth, 1981) at the active site of the enzyme. The transition state for the decarboxylation reaction involves conversion of a zwitterion to a species that has much less charge separation. The transition state therefore will be best reflected in a molecule whose structure does not contain the zwitterionic features of the starting adduct. Gutowski and Lienhard (1976) reasoned that thiamine thiazolone diphosphate (TTDP) resembles the transition state for the decarboxylation process in terms of its charge distribution (Fig. 2.11).

In accordance with the prediction, TTDP was found to bind very tightly to pyruvate dehydrogenase, binding much more tightly than do thiamine diphosphate or pyruvate, perhaps because it is a transition-state analogue. However, as we argued above, there may be other unanticipated factors that cause it to bind tightly. For example, the TDP-binding site of this enzyme and other TDP-dependent enzymes appears to be low in polarity, so that analogues of TDP that are not zwitterions in general bind well to the TDP-binding sites. For example, thiochrome diphosphate (*2.6*) is a tricyclic planar molecule that is not a close structural analogue of TDP or of a transition state, yet it binds well to the TDP-binding site of pyruvate decarboxylase (Wittorf and Gubler, 1970). Analysis of the kinetics of

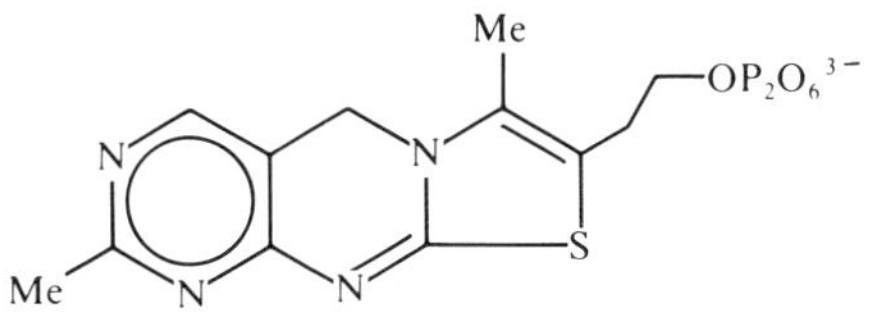

(*2.6*)

binding of TTDP to pyruvate decarboxylase revealed that the observed high affinity is a combination of fast reversible processes that generate adducts in which TTDP has a relatively low affinity for the apoenzyme and a slower, irreversible process (Kluger *et al.*, 1984).

It is also difficult to tell a transition-state from a reactive intermediate, since they may be close in energy and structure (the Hammond postulate). Thus, the phrase 'transition-state analogue' is often replaced with the more conservative 'reaction intermediate analogue' (Byers, 1978). However, since the two types of species should resemble one another, the difference is not serious, for the purpose of analysis. Wolfenden (1974) has suggested that a transition-state analogue may not bind to an enzyme if the enzyme isomerizes as a result of forming the transition state. The substrate must bind and then both enzyme and substrate change together to the transition state. Enzyme alone does not isomerize so that the form that would recognize the transition-state analogue is not present in the absence of substrate. As long as these limitations to the method are considered, transition-state analogues can be very useful tools in elucidating enzymic mechanisms and in the design of inhibitors. By observing an enzyme–analogue complex spectroscopically, it is potentially possible to observe the structure of the enzyme form that is specifically responsible for catalysis.

2.13 MULTIPLE BINDING SITES

Reaction sites on substrates in enzyme-catalysed processes are often small parts of larger molecules. The affinity of an enzyme for a small molecule is limited, since the attractive interactions between enzyme and substrate depend on the sum of contributions of each structural component of the small molecule with the enzyme (Jencks, 1975). The larger and more complex the substrate molecule, the more possible sites for favourable energetic interactions with the enzyme are likely to exist. The interactions of portions of the substrate with the enzyme are primarily enthalpic and include hydrogen-bonding, electrostatic attractions, dipolar alignments and apolar solvent exclusion.

The organic coenzymes provide good examples of molecules with many sites that can interact favourably with enzymes (Bruice and Benkovic (1966), volume 2; Dugas and Penney, 1981, ch. 7). It is likely that their evolutionary success has been partially due to their multiple sites for interacting with the protein with which they associate. For example, the nicotinamide adenine dinucleotides contain phosphate groups that can make strongly favourable electrostatic interactions with cationic sites on the enzyme to which they bind. In addition, the ribosyl moieties of the coenzyme contain hydroxyl groups that can hydrogen bond to acceptor sites. Similar nucleotide-like units in other coenzymes, such as FAD and thiamine diphosphate, can also assist in binding (Metzler (1977), ch. 8).

The binding interactions of substrates and enzyme-bound thiamine diphosphate are illustrative. Thiamine diphosphate binds non-covalently to the apoenzyme with which it is associated with a relatively high affinity (Krampitz, 1970). In the case of pyruvate decarboxylase, the substrate pyruvate and the product acetaldehyde are small molecules and have little to offer for favourable enthalpic interaction with the catalytic site of the enzyme. During the catalytic cycle of the enzyme, thiamine diphosphate and pyruvate combine to form an adduct, lactylthiamine diphosphate which exchanges carbon dioxide for a proton, yielding hydroxyethylthiamine diphosphate (Kluger, 1982) (Fig. 2.11). This product decomposes to acetaldehyde and thiamine diphosphate (Krampitz *et al.*, 1961). The thiamine diphosphate molecule remains associated with the apoenzyme while acetaldehyde is released into solution. During the course of catalysis, the small, weakly bound pyruvate molecule is converted into a fragment of the larger lactylthiamine diphosphate molecule. The formation of this adduct between thiamine diphosphate and pyruvate is thus the key step of the catalytic cycle.

The importance of binding in catalysis has been emphasized in this chapter and we might expect that formation of lactylthiamine diphosphate would be a process that increases the affinity of the enzyme and pyruvate. However, evidence has now accumulated that the adduct does not improve the affinity of pyruvate for the enzyme (Kluger and Smyth, 1981). In fact, it has a markedly lower affinity than does thiamine diphosphate itself. Where, then, is the catalytic advantage of adduct formation? The adduct provides a route to the transition state for the decarboxylation process, which, as we already have shown is more like a neutral thiazole (Breslow, 1962; Gutowski and Lienhard, 1976). Clearly, if the enzyme has a high affinity for a neutral thiazole it cannot also have a high affinity for zwitterionic lactylthiamine. Experimental evidence for this remarkable situation was obtained as follows.

The affinity of the apoenzyme of wheatgerm pyruvate decarboxylase for the lactylthiamine diphosphate adduct must be very low (Kluger and Smyth, 1981) because we could detect no activation of the enzyme by the adduct above the level of a control. However, the t-butyl ester of lactylthiamine diphosphate binds and activates the enzyme with K_m about twice that of thiamine diphosphate. Thus, these affinity measurements do not reflect a steric sensitivity to the bulk of the pyruvate group. If thiamine diphosphate serves as a high-affinity 'anchor' (Jencks, 1975), then a decrease in affinity of lactylthiamine diphosphate relative to the unsubstituted species indicates that the lactyl group has a negative affinity for the apoenzyme. Since the ester binds, this indicates that the negative effect is due to the presence of an unesterified carboxyl group on lactylthiamine diphosphate.

Herschlag (1988) has provided a detailed examination of the energetics of enzymes which undergo structural changes which promote catalysis. The isolation of a transition state for the enzyme-catalysed reaction from the surrounding solution, as originally proposed by Wolfenden (1972), is a major force in producing effective catalysis.

2.14 BIOMIMETIC CHEMISTRY

A logical extension of the acquisition of principles that we have discussed so far is the application of these principles to new situations. If we can explain how an enzyme works in terms of chemistry, then we should be able to design other chemical catalysts whose properties are predictable from what we know about enzymes. Breslow coined the expression 'biomimetic chemistry' to deal with those aspects of the field whose goal is to design and build molecules whose properties rival enzymes as catalysts. It has become a very popular activity to design molecules which have predictable enzyme-like characteristics i.e., they are efficient and specific catalysts which have a high affinity for their substrates. We will illustrate some of the progress in the design of catalysts aimed at promoting catalytic efficiency (see also Chapter 4).

2.14.1 Synthetic catalysts

A major effort of physical organic chemists who pursue research in biomimetic chemistry is the production of 'model enzymes'. Their goal is to produce an efficient and specific catalyst by rational design. Initially, the approach that has been taken has involved the conversion of a naturally occurring species that has some ability to bind the desired substrate. For example, a series of oligosaccharides that are cyclic (1–4) hexamers of glucose, known as Schardinger dextrins, can be obtained by microbial degradation of starch. Variants containing six glucosyl units (cyclohexa-myloses), seven units (cycloheptamyloses) and eight units (cyclo-octamyloses) can be obtained. They are generally called cyclodextrins when no ring size is specified. The structures of these species have been carefully studied and it appears that in general they exist as toroids (doughnut-shaped solids). The hydrophilic hydroxyl groups of the glucose units are directed toward the solution while C–H and C–O–C groups are directed toward the innter cavity (2.7). Apolar molecules have an affinity for the inner cavity of the cyclodextrin and will form stable complexes. Considerable effort has gone into making catalysts based on the cyclodextrin molecules (Tabuishi, 1982).

Early examples of catalysis by cyclodextrins usually involved reaction of a hydroxyl group of the cyclodextrin with an unsaturated electrophilic centre on the substrate. Thus Bender found that the acylation of a cyclodextrin by 3-t-butylphenyl acetate occurs with a rate constant that is 250 times larger

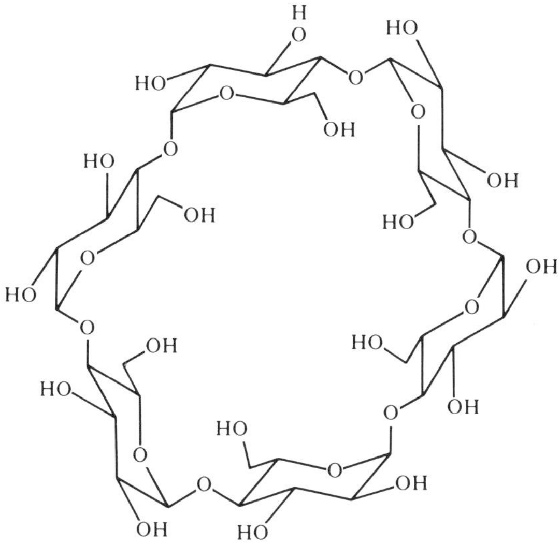

(2.7)

than the rate constant for hydrolysis (van Etten *et al.*, 1967) (Fig. 2.12(a)). Although this is a significant acceleration, it is not typical of accelerations of k_{cat} expected for an enzyme-catalysed process. Breslow and his co-workers (Breslow *et al.*, 1980) carefully analysed the structural features of the cyclodextrin and the substrate to try to design improvements into the catalytic system. They found that the reaction Bender had studied is slowed down because, although the substrate is bound effectively within the cyclodextrin cavity, the hydrolysis intermediate tends to move out of the cavity. To prevent the intermediate from leaving the cavity, the cyclodextrin was modified by having a synthetic capping moiety placed across one side of the cavity. Further, substrates were found that fit the cavity better and thus would have a smaller tendency to leave. After examination of the molecular models, it was concluded that ferrocene should bind especially well to cyclodextrins so that esters derived from ferrocene should make excellent substrates. This led to increases in the catalytic rate constant of 1000-fold over those that had been observed by Bender and his coworkers (Fig. 2.12(b)). The net result is that the hydrolysis of *p*-nitrophenyl 3-*trans*-ferrocenyl-propenoate is accelerated a millionfold by the cyclodextrin, a rate enhancement that is comparable with that produced by enzymes. Specifically designed receptors have increased the range and nature of specificity and catalysis by synthetic species (Stoddart, 1983). Rebek *et al.*, (1985) have developed systems with dynamic properties in which structural changes due to binding produce catalytic effects.

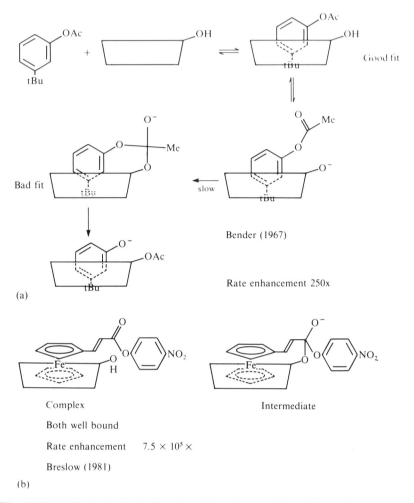

Fig. 2.12(a)   The hydrolysis of m,t-butylphenyl acetate is facilitated by a cyclodextrin but steric problems prevent maximal efficiency. **(b)** A molecule designed to fit the cyclodextrin cavity in the catalytic transition state is subject to more rapid reaction.

Lehn has developed the idea of *supramolecular catalysis* in which large molecules which are specifically designed as receptors to provide catalysis (Lehn, 1978, 1985, 1986). For example, the hydrolysis of acetyl phosphate is catalysed by a macrocyclic polyamine–ether and produces pyrophosphate (Hosseini and Lehn, 1987). In this case, two molecules of catalyst and a phosphorylated catalyst intermediate are involved. In related work, Cram designed host molecules called spheranels, hemispherands and cryptos-pherands, all of which provide analogous catalytic possibilities (Cram and Lam, 1986; Cram, 1988).

2.14.2 Catalytic antibodies

As noted earlier, the concept of transition state analogues as enzyme inhibitors led to the design of molecules which indeed functioned as inhibitors. Recently, this concept has been utilized in the development of new *catalysts*, based on mechanistic principles. If an enzyme is regarded (in an oversimplified way) as a protein capable of specifically binding a transition state, then a protein that is designed to bind a molecule resembling a transition state should have catalytic activity. The task of designing a protein from first principles has not yet been accomplished but the idea of utilizing the variability of antibodies has been exploited (Schultz, 1988; Napper *et al.*, 1987 see also Chapters 4 and 9). The procedures are based on using the techniques of immunology combined with the principles of physical organic chemistry. The strategy involves the synthesis of an analogue of the desired transition state. This analogue is covalently attached to a larger molecule which makes the species immunogenic.

The synthetically-induced antigen then effects the production of mono-clonal antibodies which can be produced in quantity. Some of these antibodies have catalytic properties which parallel those of enzymes as to specificity and binding, although the efficiency is far from optimal (these would be good cases for study by the Albery–Knowles criteria). Since enzymes must be able to bind substrates and products as well as promoting the formation of transition states through isomerization (Herschlag, 1988), the formation of an antibody to the transition state is not sufficient to assure that the antibody will be catalytic. However, it is reasonable to expect that if antibodies are selected for their ability to catalyse a specific reaction, these individuals will be those that can bind substrates and products. An example of such a system is the antibody produced to an antigenic tetrahedral phosphate derivative. This catalyses the hydrolysis of trigonal carboxylic esters which react by way of tetrahedral transition states (Schultz, 1988).

2.15 CONCLUSION

We have seen that some of the central problems involved in understanding enzymic catalysis can successfully be approached using the principles of physical organic chemistry. Although enzymes are not understood in mechanistic detail at the level of some simple organic reactions, we do know more about many enzymic reactions than we do about most organic reactions. Enzymes direct the substrate in a much more specific manner than is possible in an uncatalysed system. Therefore, the techniques of physical organic chemistry can be applied to enzyme reactions in a very productive way.

The mechanism of interaction between substrates and enzymes remains one of the most fascinating mechanistic problems in modern organic chemistry and our knowledge is still very limited. We still cannot deduce the catalytic function of an enzyme from its three-dimensional structure but if we know the enzyme's function, knowledge of its structure permits important insights in the area of mechanism. While synthetic catalysts do not yet match all the properties of an enzyme, the application of principles of mechanism to the design of catalysts has led to very powerful effects. The very practical consequences of these ideas have been the development of principles to guide in the invention of materials which will function as drugs (Castelhano and Krantz, 1987) and agrochemicals (Baillie *et al.*, 1988). Based on recent history, it is apparent that understanding the combination of principles of mechanistic chemistry and enzymology will provide continuing opportunities for exciting discoveries.

REFERENCES

Abeles, R. H. and Maycock (1976) *Acc. Chem. Res.* **9**, 313.
Albery, W. J. and Knowles, J. R. (1977) *Acc. Chem. Res.*, **10**, 105.
Aldersley, M. F., Kirby, A. J. and Lancaster, P. W. (1972) *J. Chem. Soc. Chem. Commun.*, 570.
Alworth, W. L. (1972) *Stereochemistry and Its Application in Biochemistry*, Wiley, New York.
Baillie, A. C., Wright, K., Wright, B. J., and Earnshaw, C. G. (1988) *Pesticide Biochem. Physiol.*, **30**, 103.
Bednar, R. A. and Jencks, W. P. (1985) *J. Am. Chem. Soc.*, **107**, 7126.
Belasco, J. G., Bruice, T. W., Fisher, L. M., Albery, W. J., and Knowles, J. R. (1986) *Biochemistry* **25**, 2564.
Bell, R. P. (1973) *The Proton in Chemistry*, 2nd edn., Cornell University Press, Ithaca, New York.
Bender, J. L., Chow, Y. and Chloupek, F. (1958) *J. Am. Chem. Soc.*, **80**, 5380.
Bender, M. L. (1957) *J. Am. Chem. Soc.*, **79**, 1258.
Bender, M. L. (1971) *Mechanisms of Homogeneous Catalysis from Protons to Proteins*, Wiley Interscience, New York.
Benner, S. A., Rozzell, J. D. Jr., and Morton, T. H. (1981) *J. Am. Chem. Soc.*, **103**, 993.
Bentley, R. (1969) *Molecular Asymmetry in Biology*, Vol. 1, Academic Press, New York.
Bernasconi, C. F., Hibdon, S. A. and McMurry, S. E. (1982) *J. Am. Chem. Soc.*, **104**, 762.
Bourne, N., Chrystiuk, E., Davis, A. M., and Williams, A. (1988) *J. Am. Chem. Soc.* **110**, 1890.
Breslow, R. (1962) *Ann. N.Y. Acad. Sci.*, **98**, 445.
Breslow, R. and Wernick, D. (1976) *J. Am. Chem. Soc.*, **98**, 2519.

Breslow, R., Czarniecki, M. F., Emert, J. and Hamaguchi, H. (1980) *J. Am. Chem. Soc.*, **102**, 762.

Bruice, T. C. and Benkovic, S. J. (1966) *Bioorganic Mechanisms*, W. A. Benjamin, New York.

Byers, L. D. (1978) *J. Theor. Biol.*, **74**, 5501.

Calvin, M. and Pon, N. G. (1959) *J. Cell. Comp. Physiol. Suppl.* 1, **54**, 51.

Cardinale, G. J. and Abeles, R. H. (1968) *Biochemistry*, **7**, 3970.

Castelhano, A. L., and Krantz, A. (1987) *J. Am. Chem. Soc.* **109**, 3491.

Chin, J. (1983) *J. Am. Chem. Soc.*, **105**, 6502.

Cram, D. J. (1988) *Science* (Washington, D.C.), **240**, 760.

Cram, D. J. and Lam, P. Y. S. (1986) *Tetrahedron*, **42**, 1607.

Cunningham, E. B. (1978) *Biochemistry*, McGraw-Hill, New York.

Curtin, D. Y. (1954) *Rec. Chem. Prog.*, **15**, 111.

Denbigh, K. (1964) *The Principles of Chemical Equilibrium*, Cambridge University Press, Cambridge.

Dugas, H. and Penney, C. (1981) *Bioorganic Chemistry*, Springer-Verlag, New York.

Dunathan, H. C. (1966) *Proc. Natl. Acad. Sci. U.S.A.*, **55**, 712.

Dunathan, H. C. (1967) *Adv. Enzymol, Relat. Areas Mol. Biol.*, **35**, 79.

Eigen, M. (1964) *Angew. Chem. Int. Ed. Engl.*, **3**, 1.

Eliel, E. (1963) in *Stereochemistry of Carbon Compounds*, McGraw-Hill, New York, pp. 149–156.

Eliel, E. (1975) *J. Chem. Educ.*, **52**, 56.

Eliel, E. (1980) *J. Chem. Educ.*, **57**, 52.

Estell, D. A. (1987a) *Protein Eng.*, **1**, *445*.

Estell, D. A. (1987b) Protein Eng., **1**, 446.

Fersht, A. R., Leatherbarrow, R. J., and Wells, T. N. C. (1986) *Nature*, **322**, 284.

Fersht, A. R., Leatherbarrow, R. J., and Wells, T. N. C. (1987) *Biochemistry*, **26**, 6030.

Fersht, A. R. (1985) *Enzyme Structure and Mechanism*, W. H. Freeman and Sons, New York.

Fersht, A. R. (1987a) *Protein Eng.*, **1**, 442.

Fersht, A. R. (1987b) *Protein Eng.*, **1**, 466.

Fisher, L. M., Albery, W. J., and Knowles, J. R. (1986) *Biochemistry*, **25**, 2538.

Gutowski, J. A. and Lienhard, G. E. (1976) *J. Biol. Chem.*, **251**, 2863.

Hall, A. and Knowles, J. R. (1975) *Biochemistry*, **14**, 4348.

Hammett, L. P. (1970) *Physical Organic Chemistry*, McGraw-Hill, New York.

Hammons, G., Westheimer, F. H., Nakaoka, K. and Kluger, R. (1975) *J. Am. Chem. Soc.*, **97**, 1568.

Hansen, D. E. and Knowles, J. R. (1985) *J. Am. Chem. Soc.*, **107**, 8304.

Hanson, K. and Rose, I. A. (1975) *Acc. Chem. Res.*, **8**, 1.

Herschlag, D. (1988) *Bioorg-Chem.*, **16**, 62.

Hofmann and Fink (1984) *Biochemistry*, **23**, 5247.

Hosseini, M. W. and Lehn, J. M. (1987) *J. Am. Chem. Soc.*, **109**, 7047.

Huber, C. P., Ozaki, Y., Pliura, D. H., Carey, P. R. and Storer, A. C. (1982) *Biochemistry*, **21**, 3109.

Ingold, C. K. (1953) *Structure and Mechanism in Organic Chemistry*, Cornell University Press, Ithaca, N.Y.

Jencks, W. P. (1985) *Chem. Rev.*, **85**, 512.

Jencks, W. P. (1969) *Catalysis in Chemistry and Enzymology*, McGraw-Hill, New York.

Jencks, W. P. (1975) *Adv. Enzymol. Relat. Areas Mol. Biol.*, **43**, 219.
Jencks, W. P. (1980) *Acc. Chem. Res.*, **13**, 161.
Kirby, A. J. (1980) *Adv. Phys. Org. Chem.*, **17**, 183.
Kirby, A. J. and Lancaster, P. W. (1972) *J. Chem. Soc. Perkin Trans.*, 2, 1206.
Kirsch, J. F. (1987) *Protein Eng.*, **1**, 148.
Klotz, I. M. (1976) *J. Chem. Educ.*, **53**, 159.
Kluger, R. (1982) *Ann. N.Y. Acad. Sci.*, **378**, 63.
Kluger, R. and Chin, J. (1982) *J. Am. Chem. Soc.*, **104**, 2891.
Kluger, R. and Gish, G. (1987) Stereochemical aspects of thiamin catalysis, in *Thiamin Pyrophosphate Enzymes* (eds R. L. Schowen and A. Schellenberger), CRC Press, Boca Raton, Florida, pp. 3–9.
Kluger, R. and Nakaoka, K. (1974) *Biochemistry*, **13**, 910.
Kluger, R., Chin, J., Smyth, T. (1981) *J. Am. Chem. Soc.*, **103**, 884.
Kluger, R., Davis, P. P. and Adawadkar, P. D. (1979) *J. Am. Chem. Soc.*, **101**, 5995.
Kluger, R., Karimian, K. and Kitamura, K. (1987) *J. Am. Chem. Soc.*, **109**, 6368.
Kluger, R., Wong, M. K., and Dodds, A. K. (1984) *J. Am. Chem. Soc.*, **106**, 1113.
Kluger, R. and Smyth, T. (1981) *J. Am. Chem. Soc.*, **103**, 214.
Krampitz, L. O. (1970) *Thiamin Diphosphate and its Catalytic Functions*, Marcel Dekker, New York.
Krampitz, L. O., Suzuki, I. and Greull, G. (1961) *Fed. Proc. Fed. Am. Soc. Exp. Biol.*, **20**, 971.
Kresge, A. J. (1975) *Acc. Chem. Res.*, **8**, 354.
Lehn, J. M. (1978) *Acc. Chem. Res.*, **11**, 49.
Lehn, J. M. (1985) *Science*, **227**, 849.
Lehn, J. M. (1986) *Ann. N.Y. Acad. Sci.*, **471**, 41.
Lienhard, G. E. (1973) *Science*, **180**, 149.
Lim, W. A., Raines, R. T., and Knowles, J. R. (1988) *Biochemistry*, 27, 1158.
McCasland, G. E. (1953) *A New General System for the Naming of Stereoisomers*, Chemical Abstracts Service.
Metzler, D. E. (1977) *Biochemistry*, Academic Press, New York.
Mislow, K. and Raban, M. (1967) *Top. Stereochem.*, **1**, 1.
Mislow, K., and Siegel, J. (1984) *J. Am. Chem. Soc.*, **106**, 3319.
More O'Ferrall, R. A. (1970) *J. Chem. Soc. B*, **274**.
Murdock, J. R. (1981) *J. Chem. Educ.*, **58**, 32.
Nambiar, K. P., Stauffer, D. M., Kolodziej, P. A., and Benner, S. A. (1983) *J. Am. Chem. Soc.*, **105**, 5886.
Napper, A. D., Benkovic, S. J., Tramontano, A., and Erner, R. A. (1987) *Science*, **237**, 1042.
O'Leary, M. H. and Baughn, R. L. (1972) *J. Am. Chem. Soc.*, **94**, 626.
Ogg, R., and Polanyi, M. (1935) *Trans. Farad. Soc.*, **31**, 607.
Page, M. I. (1984) *The Chemistry of Enzyme Action*, Elsevier, Amsterdam.
Page, M. I. and Jencks, W. P. (1971) *Proc. Natl. Acad. Sci. U.S.A.*, **68**, 1678.
Pedersen, C. J. (1967) *J. Am. Chem. Soc.*, **89**, 2495.
Rebek, J. R. Jr, Costello, T., Marshall, L., Wattley, R., Gadwood, R. C., and Onan, K. (1985) *J. Am. Chem. Soc.*, **107**, 7481.
Ritchie, C. D. (1969) in *Solute-Solvent Interactions*, Vol. 1 (eds J. F. Coetzee and C. D. Ritchie), Marcel Dekker, New York.
Rozzell, J. D. Jr. and Benner, S. A. (1984) *J. Am. Chem. Soc.*, **106**, 4937.
Rudnick, G. and Abeles, R. H. (1975) *Biochemistry*, **14**, 4515.
Schowen, R. L. (1977) *Isotope Effects on Enzyme-Catalyzed Reaction*, University Park Press, Baltimore.

Schultz, P. G. (1988) *Science*, **240**, 426.
Scrimgeour, K. G. (1977) *Chemistry and Control of Enzyme Reactions*, Academic Press, London.
Segel, I. H. (1975) *Enzyme Kinetics*, John Wiley and Sons, New York.
Sinnot, M. L. (1988), *Adv. Phys. Org. Chem.*, **24**, 113.
Somayaji, V., Keillor, J., Brown, R. S. (1988) *J. Am. Chem. Soc.*, **110**, 2625.
Stoddart, J. F. (1983) *Annu. Rep. Prog. Chem. Sect. B.*, **80**, 353.
Tabushi, I. (1982) in *Frontiers of Chemistry* (K. J. Laidler, ed.), IUPAC, London.
Tagaki, W. and Westheimer, F. H. (1968) *Biochemistry*, **7**, 891.
Turano, A., Furey, W., Pletcher, J., Sax, M., Pike, D. and Kluger, R. (1982) *J. Am. Chem. Soc.*, **104**, 3089.
van Etten, R. L., Sebastian, J. F., Clowes, G. A. and Bender, M. L. (1967) *J. Am. Chem. Soc.*, **89**, 3242.
Walsh, C. (1978) *Enzymatic Reaction Mechanisms*, W. H. Freeman and Co., San Francisco.
Westheimer, F. H. (1956) *Adv. Enzymol. Relat. Subj. Biochem.*, **24**, 441.
Westheimer, F. H. (1969) *Methods Enzymol.*, **14**, 231.
Westheimer, F. H. (1985) *Adv. Phys. Org. Chem.*, **21**, 1.
Westheimer, F. H. (1986) *J. Chem. Educ.*, **63**, 409.
Wittorf, J. H. and Gubler, C. J. (1970) *Eur. J. Biochem.*, **14**, 53.
Wolfenden, R. V. (1972) *Acc. Chem. Res.*, **5**, 10.
Wolfenden, R. V. (1974) *Mol. Cell. Biochem.*, **3**, 207.
Wolfenden, R. V. (1979) *FEBS Symp.*, **57**, 151.
Wood, H. G. and Barden, R. E. (1977) *Annu. Rev. Biochem.*, **46**, 385.
You, K., Arnold, L. J., Allison, W. S., and Kaplan, N. O. (1978) *Trends Biochem. Sci.*, 265.
Zubay, G. (1988) *Biochemistry*, 2nd edn., Collier Macmillan, Toronto.

3 | Chemical models of selected coenzyme catalyses

Seiji Shinkai

3.1 INTRODUCTION

An enzyme is born through repeated 'trial and error' in nature over an enormous length of time. Active sites of enzymes evolved to allow the enzymes to mediate biological reactions under ambient conditions and thus serve as excellent biological 'catalysts'. Nevertheless, enzymes have two inevitable limitations when viewed by organic chemists: first, enzymic catalyses are often too specific and do not allow the reactions that organisms do not require and secondly, the catalytic activities appear only under ambient physiological conditions. Although many useful synthetic reactions can be executed using enzymic catalysis (see Chapter 4), reaction conditions as well as enzyme specificity limit the direct applicability of enzymes. Therefore, one has to mimic the essence of the enzymic catalyses in more simplified systems in order to utilize the enzyme-like catalyses in a more versatile manner. This is the primary aim of an enzyme model study. We thus consider that the enzyme model study consists of two main targets: first, exploitation of more versatile, enzyme-like catalysts and secondly, clarification of the reaction mechanisms in more simplified systems.

The specific concern of this chapter is with coenzymes which are prosthetic groups in many enzymes and bind to their apoenzymes via covalent bonds and/or secondary valence forces. Importantly, most co-enzymes are capable of catalysing reactions, although weakly, even in the absence of apoenzymes, while apoenzymes take charge of enhancing the reactivity and controlling the stereoselectivity in the catalytic processes of coenzymes in the manner described in the previous chapter. Therefore, if one could control the reactivity and the stereoselectivity of coenzymes by the environmental effect or by the direct modification of the coenzyme structure, these would become expedient and practical enzyme-modelled catalysts.

In this chapter, we review a number of efforts devoted to the application of selected coenzyme-catalysed systems to organic chemistry but will not refer to the detailed mechanisms of the coenzyme catalyses which have already been surveyed by other bio-organic chemists (Jencks, 1969; Bruice and Benkovic, 1966). The present review is the first comprehensive account of much coenzyme chemistry.

3.2   MODEL INVESTIGATIONS OF NICOTINAMIDE COENZYMES

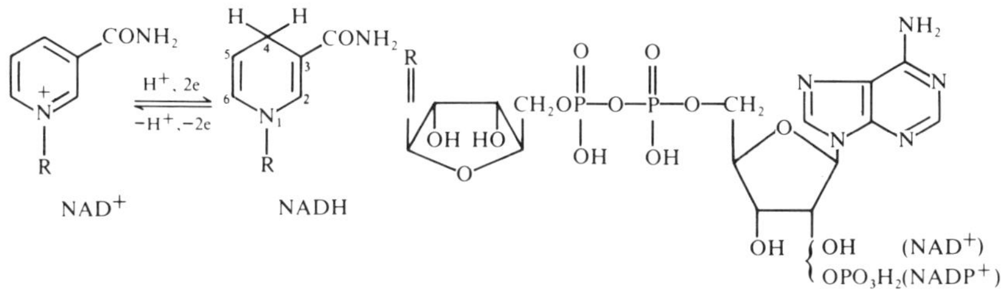

Fig. 3.1

The nicotinamide-adenine dinucleotide and its reduced form (NAD^+ and NADH) serve as coenzymes in a large number of enzymic oxidation–reduction reactions (Fig. 3.1). NADH-dependent enzymes include the dehydrogenases, transhydrogenases, diaphorases, phosphorylases and oxidases. Although the enzymic mechanisms are not simple, we would like to consider, from a viewpoint of model studies, that these reactions simply take place by hydrogen exchange between the substrate and the pyridinium cation or 1,4-dihydropyridine in the nicotinamide moiety. The essential characteristics of NADH-dependent enzymes can be summarized as follows: (i) the reactions take place with the ternary (apoenzyme–coenzyme–substrate) complexes, (ii) the transfer of hydrogen is direct and apparently involves no exchange with solvent proton, and (iii) the reactions are stereospecific with respect to both the coenzyme and substrate. These characteristic targets have been a great challenge for model investigations of NADH chemistry. Such investigations are important because they provide the chemical mechanistic background for an understanding of catalysis by dehydrogenases, a reaction that has both intrinsic interest and applications to synthesis (Chapter 4). Indeed the mechanisms of model reactions and those of enzyme-catalysed reactions have been brought into close comparison through the medium of cyclopropane-containing compounds (Suckling, 1988). Since the dihydronicotinamide moiety of NADH (or NAD$^+$) is responsible for the redox reactions, 1,4-dihydronicotinamides (3.1) with simple N-substituents (e.g., R = n-pr, benzyl, etc.) are used in model studies. Hantzsch esters (3.2) are also employed because of their preparative convenience.

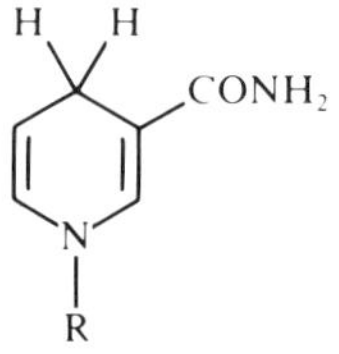

(3.1) 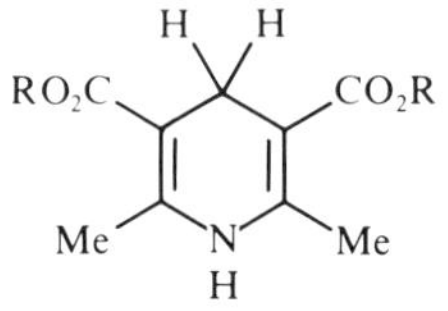   (3.2)

3.2.1   Metal ion catalysis

Alcohol dehydrogenase is one of the most representative NADH-dependent enzymes and has been a typical object of biomimetic NADH chemistry. In contrast to the ability of alcohol dehydrogenase to reduce aldehydes of widely varying structure (Jones and Beck, 1976), NADH model compounds reduce only the most activated carbonyl substrates such as hexachloroacetone, trifluoroacetophenone, chloranil etc., a fact which has made drawing mechanistic conclusions from model reactions difficult (Bruice and Benkovic, 1966; Kill and Widdowson, 1978; Sigman et al., 1978;

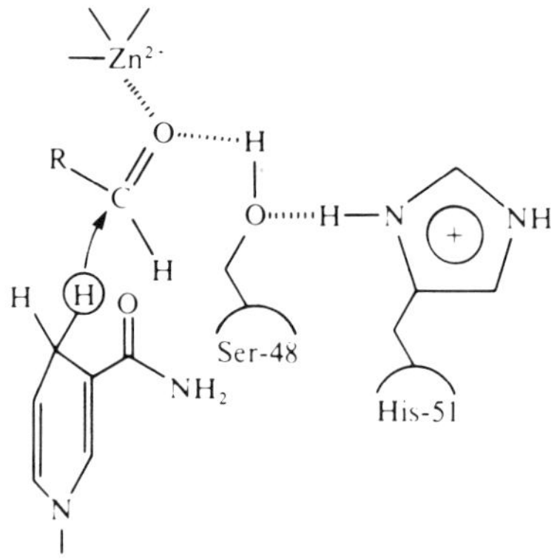

Fig. 3.2   Ternary complex of alcohol dehydrogenase.

Dittmer and Fouty, 1964; Steffens and Chipman, 1971). Alcohol dehydro-genase is known to contain zinc at the active site which plays a decisive role in the mechanism of action. Pattison and Dunn (1976) suggested that the zinc ion serves as an electrophilic catalyst in the reduction process of the ternary complex (Fig. 3.2). The concept is very suggestive to attain the biomimetic reduction of non-activated carbonyl substrates.

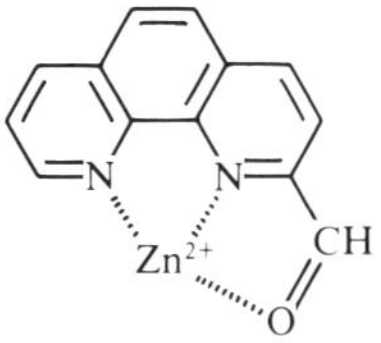

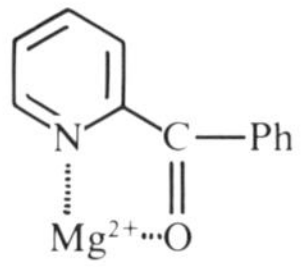

In 1971, Creighton and Sigman reported that 1,10-phenanthroline-2-carboxaldeyde (3.3) can be reduced by (3.1) in the presence of zinc ion. Probably this was the first example of metal-assisted NADH model reduction of carbonyl groups. Subsequent studies have established that metal catalysis is quite versatile in the reduction of a variety of double bonds (C=O, C=N, C=C, etc.): for example, (3.4)–(3.8) were reduced by NADH model compounds in acetonitrile in the presence of metal ions (mainly $Mg(ClO_4)_2$) (Gase et al., 1976; Gase and Pandit, 1977; Shinkai et al., 1979a; Ohnishi et al., 1975a; Ohno et al., 1980). Tintillier et al. (1986) and Dupas et al. (1982) demonstrated that (3.1) as well as polymer-immobilized (3.1) can reduce inactivated carbonyl substrates in good yield (70–100%). Their success appeared to be a consequence of the use of pure reagents in hyper-dry solvents. This means that a trace amount of water concomitant in solvents induces the acid-catalysed decomposition of (3.1) (see Fig. 3.5) and lowers the yield.

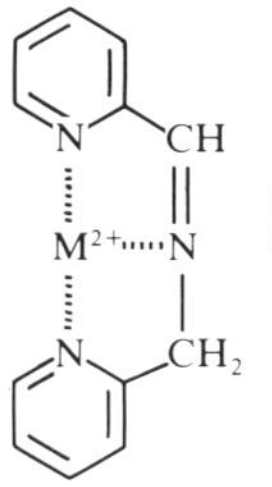

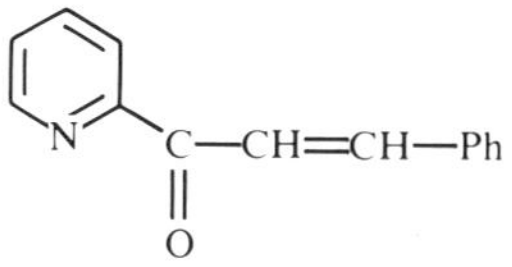

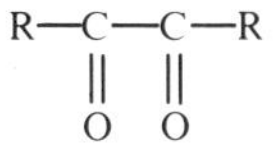

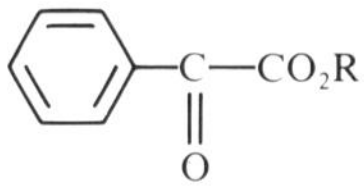

Added metal ions sometimes divert the reaction pathway. Ohnishi et al. (1976) found in the reduction of (3.9) by (3.10) that both (3.11) and (3.12) are produced in the absence of Mg^{2+} whereas (3.11) becomes the sole product in the presence of Mg^{2+} (Fig. 3.3). Since (3.12) is probably produced owing to dimerization of a radical intermediate, they considered that Mg^{2+} suppresses the radical nature of 1,4-dihydronicotinamide.

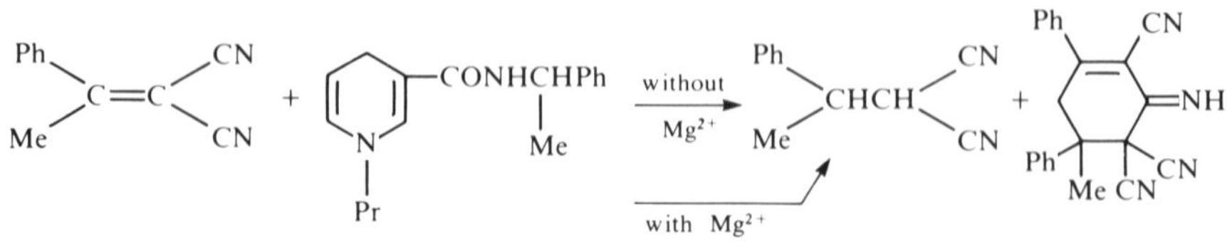

(3.9)   (3.10)   (3.11)   (3.12)

Fig. 3.3.

3.2.2   Proton acid catalysis

It is known that the active sites of D-glyceraldehyde 3-phosphate dehydro-genase and lactate dehydrogenase (NADH-dependent enzymes) have an imidazole function of the histidine residue instead of zinc. X-ray-crystallo-graphic studies (Adams et al., 1973; Moras et al., 1975) suggest that the protonated imidazole acts as a general acid during the reduction process (Fig. 3.4). The finding means that proton acids, in addition to Lewis acids such as Zn^{2+} and Mg^{2+}, would also be useful as electrophilic catalysts in the NADH model reduction. It has been reported that the ortho-hydroxyl group plays a crucial role in the dihydronicotinamide reduction of the C=O, C=S and C=N double bonds attached to the aromatic ring in (3.13)–(3.16) (Pandit and Mas Cabre, 1971; Sinkai and Bruice, 1973a; Shinkai et al., 1976a; Abeles et al., 1957; Singh et al., 1985). Similarly, Shinkai and Kunitake (1977) noticed that the aldehyde group in (3.17) is readily reduced

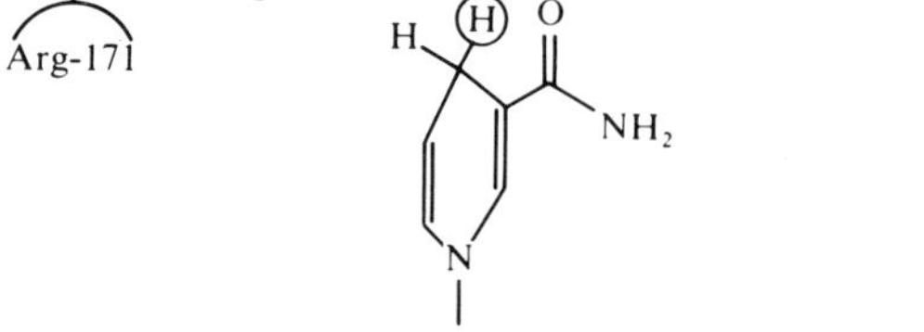

Fig. 3.4   Ternary complex of lactate dehydrogenase based on an X-ray crystallogra-phic study (Adams et al., 1973).

by (*3.1*) with the aid of the intramolecular imidazole function. These studies, which are model systems relevant to D-glyceraldehyde 3-phosphate de-hydrogenase, demonstrate the importance of the hydrogen-bonding with the reaction centre of the substrate. Van Eikeren and Grier (1976) found that the reduction rate of tri-fluoroacetophenone by (*3.1*: R = n-Pr) is markedly facilitated in protic solvents rather than in aprotic solvents. Singh *et al.* (1982, 1985) found that the reduction of imines proceeds smoothly in glacial acetic acid. It is also known that the addition of trace amounts of proton acids facilitates NADH model reductions (Shinkai and Kunitake, 1977; Wallenfels *et al.*, 1973). These results consistently support the idea that the transition state of NADH model reductions is favourably stabilized in protic or acidic media.

(*3.13*)   (*3.14*)   (*3.15*)   (*3.16*)   (*3.17*)

Here one should note the dilemma that conventional NADH model compounds rapidly decompose in acidic media to 1,4,5,6-tetrahydronico-tinamide derivatives (*3.18*). Thus, reduction in the presence of proton acids is always accompanied by the acid-catalysed decomposition of the NADH model compounds. To circumvent this undesired decomposition, we (Shinkai *et al.*, 1978a, 1979b) synthesized a new NADH model compound (*3.19*), in which the acid-sensitive 5,6-double bond of 1,4-dihydronicotina-mide is protected by the aromatic ring. We found that (*3.19*) is 3.0×10^3 times more stable in acetic acid than (*3.1*: R = benzyl). With this acid-stable NADH model, we reduced inactivated carbonyl substrates (benzaldehyde, cyclohexanone, etc.), α-keto acids and α-imino acids (products are α-amino acids), in the presence of strong acids (HCl, benzenesulphonic acid, etc.) (Shinkai *et al.*, 1979c, 1979d, 1980a). We also synthesized an acid-stable NADH model (*3.20*) which reduces carbonyl substrates with the aid of an intramolecular acidic function (Shinkai *et al.*, 1979e). The acid stability was also observed for *N*-methylacridan (*3.21*) and hindered dihydropyridines (*3.22*) and (*3.23*) (Awano and Tagaki, 1985; Singh *et al.*, 1986; Fukuzumi *et al.*, 1985). Although the reducing ability of these acid-stable NADH models should be weaker than (*3.1*), they could reduce the carbonyl and nitroso substrates with the aid of added proton acids.

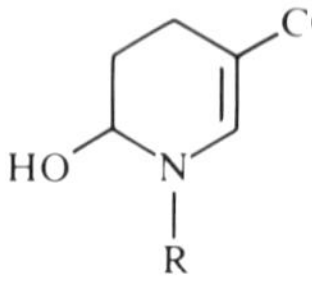

(3.18) 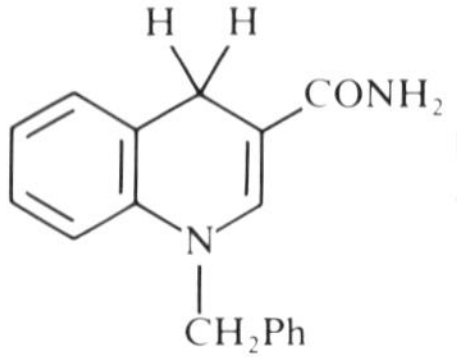   (3.19) 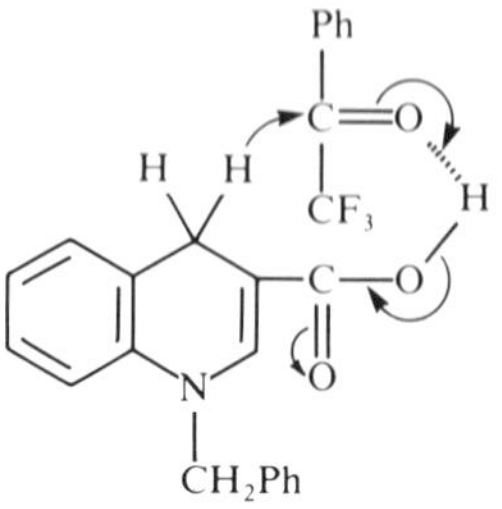   (3.20)

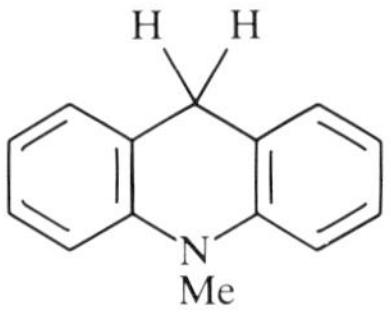

(3.21) 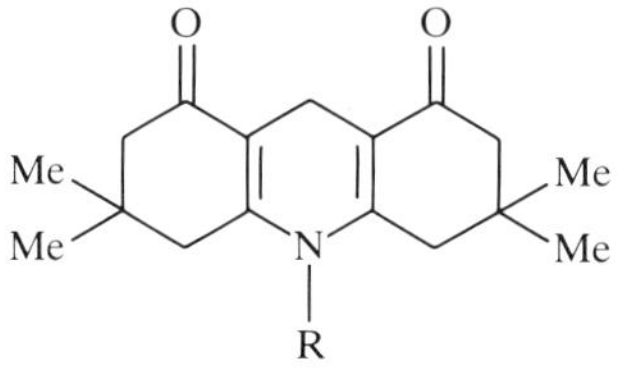   (3.22) 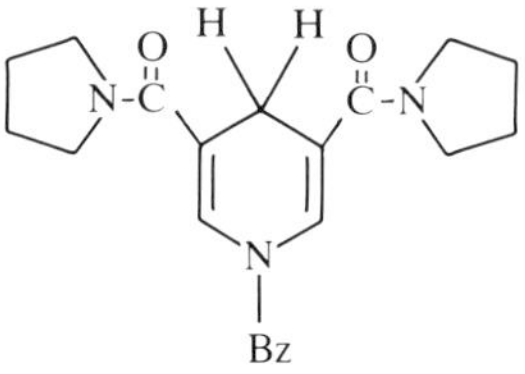   (3.23)

3.2.3 Catalysis through non-covalent interactions

Hadju and Sigman (1975) showed that dihydronicotinamides containing neighbouring carboxylate groups (3.24) can reduce N-methylacridinium ion in non-aqueous solution orders of magnitude more rapidly than can homologous derivatives lacking the free carboxylate group. For example, (3.24) exhibits a 100-fold rate acceleration relative to its methyl ester in acetonitrile; (3.25), which features the more conformationally inflexible carboxylate group, exhibits a still larger rate enhancement (10^3-fold) (Hadju and Sigman, 1977). The rate acceleration probably reflects the electrostatic stabilization of a positive charge which develops on a nicotinamide nitrogen in the reduction pathway.

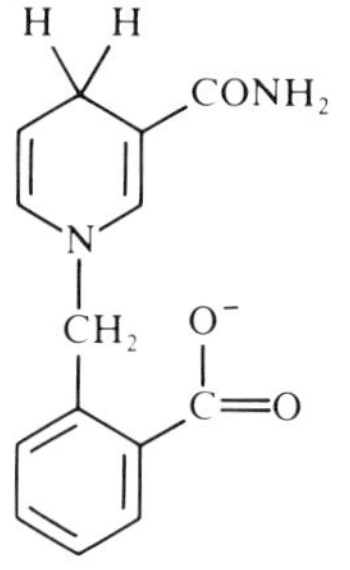

(3.24) 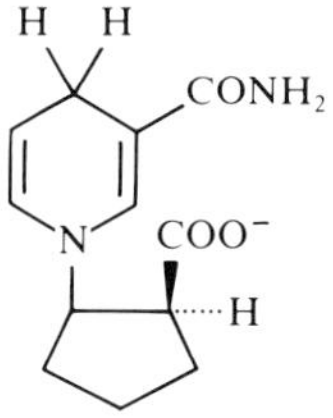   (3.25)

Similar electrostatic stabilization may be anticipated for an NADH model compound (*3.1*: R = n-dodecyl) bound to anionic micelles. However, Shinkai *et al.* (1978b) reported that the rate constants for the reduction of *N*-methylacridinium ion and isoalloxazines are enhanced to a smaller extent than in the intramolecular examples (*3.24*) and (*3.25*). On the other hand, the acid-catalysed decomposition of (*3.1*: R = n-dodecyl) is markedly accelerated by anionic micelles and suppressed by cationic micelles (Shinkai *et al.*, 1975, 1976b; Bunton *et al.*, 1978). The rate acceleration by anionic micelles is rationalized in terms of the electrostatic stabilization of a positively charged intermediate (*3.26*) by the micelle charge and the concentration of H_3O^+ on the micelle surface (Fig. 3.5).

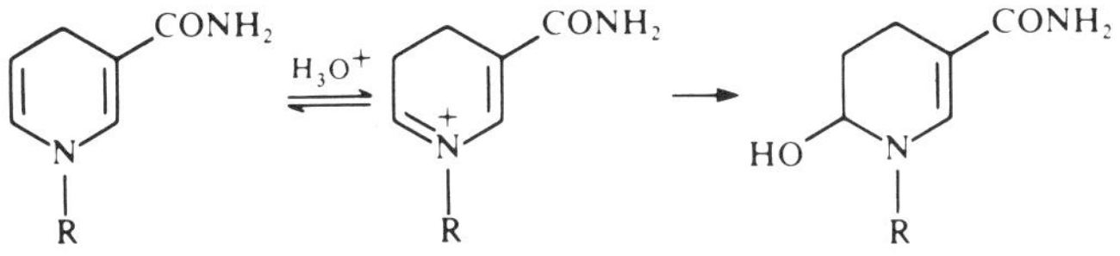

Fig. 3.5

3.2.4 NADH models as electron donors

NADH model compounds usually transfer hydride (or its equivalent) to substrates with the aid of metal ions or proton acids (MacInnes *et al.*, 1982). On the other hand, they can act as stepwise one-electron donors when the substrates exhibit the characteristics of one-electron acceptors such as $Fe(CN)_6^{3-}$ and haemin (Kill and Widdowson, 1978). It is known that good leaving groups (e.g. NO_2^-, Cl^-, Br^-, $PhSO_2^-$, etc.) are easily eliminated from the sp^3-carbon in the presence of NADH model compounds (Kill and Widdowson, 1976; Ono *et al.*, 1980, 1981; Inoue *et al.*, 1974). For example, (*3.27*) undergoes reductive debromination via a radical intermediate (*3.28*) (Fig. 3.6.). Similarly, in (*3.29*)–(*3.31*) the leaving groups (indicated by underlining) are reductively replaced by hydrogen. It is considered that the reactions take place by one-electron transfer from NADH model compounds via a nicotinamide radical cation. These reactions are closely akin to the S_{RN}-type reactions involving one-electron transfer followed by elimination of anionic groups.

An NADH model reduction that does not proceed thermally sometimes takes place under photoirradiation. For example, Singh *et al.* (1978) found that imines are photochemically reduced to amines by (*3.2*) in reasonable yields (60–92%). Dimethyl maleate and fumarate are reduced to dimethyl

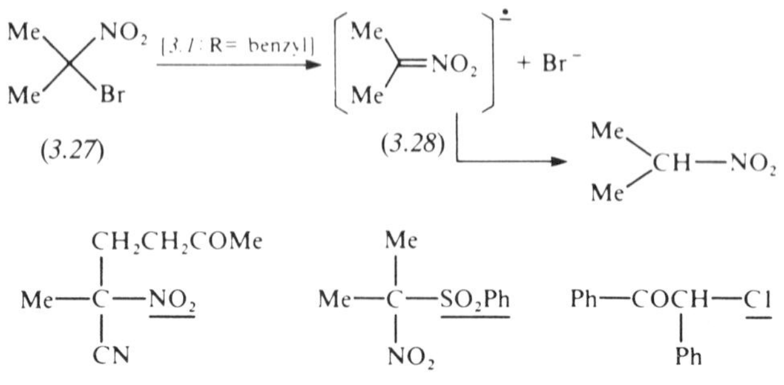

Fig. 3.6

succinate by (3.1: R = benzyl) or (3.2) under photoirradiation (> 350 nm), the yield being 60–70% (Ohnishi et al., 1975b). The photochemical reduction of C=C double bonds is further facilitated in the presence of Ru(bpy)$_3^{2+}$ (bpy = 2,2'-bipyridine) (Pac et al., 1981; Ishitani et al., 1983, 1985). In the thermal reduction hydrogen is transferred directly from 1,4-dihydropyridines to substrates, whereas in the photochemical reduction a significant amount of solvent protons is incorporated into the products (Pac et al., 1981). Also several products expected for the radical-coupling reactions were isolated (Ishitani et al., 1983, 1985). Hence, the photoactivation of NADH model compounds is related to facile one-electron transfer at the excited state.

3.2.5   NADH models attached to host molecules

Biomimetic catalyses involving host–guest complexes have been reviewed in Chapter 2; host molecules bearing functional groups frequently mimic the essential functions of enzymic catalysis (Shinkai, 1982). They bind substrates to form the equivalent of a Michaelis complex which, as has been discussed, is considered to be the origin of high activity and selectivity of enzyme catalysis. For example, β-cyclodextrin bearing a NADH model function (3.32) has been synthesized (Kojima et al., 1981). It can reduce ninhydrin with a large rate enhancement (about 40-fold) relative to a simple NADH model and the reaction proceeds according to Michaelis–Menten-type saturation kinetics. Van Bergen and Kellog (1977) synthesized a crown ether NADH mimic (3.33) which is able to reduce a sulphonium ion R^{1}COCH$_2$S$^+$R^2(CH$_3$) to R^{2}SCH$_3$, the increase in rate relative to (3.2) being 2700 times. On the other hand, the activity is efficiently quenched by NaClO$_4$. The marked rate enhancement is due to the interaction between

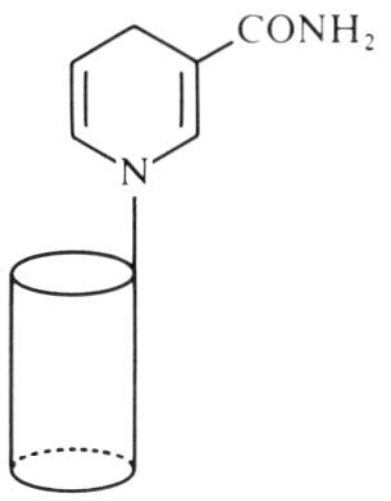

(3.32) 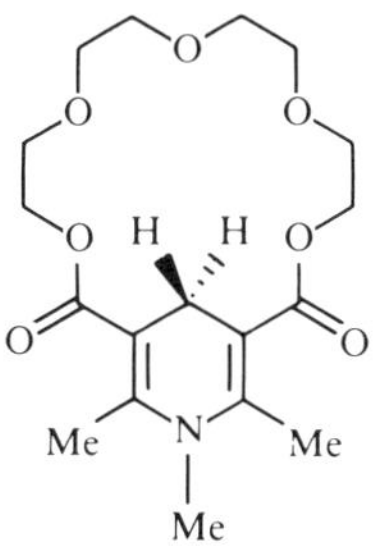   (3.33) 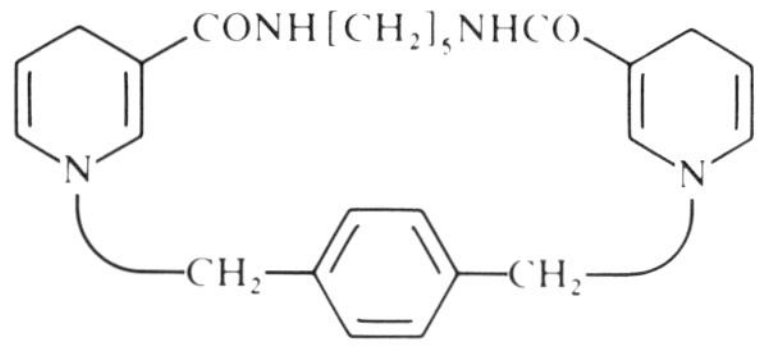   (3.34)

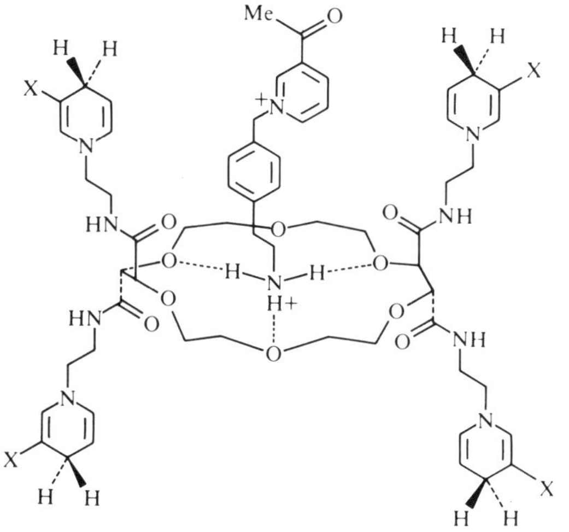

(3.35)

the sulphonium ion and the crown ring, and the inhibition is due to the competitive binding of Na^+ to the crown ring. The phenomenon reminds us of competitive inhibition in enzyme chemistry. Similarly, the reactivity of NADH models immobilized in a cyclophane structure (3.34) was investigated by Murakami *et al.* (1981a, b).

More specific design of NADH model catalysis has been reported by Behr and Lehn (1978). As illustrated in (3.35) the substrate is bound through the interaction between the crown ether of the catalyst and the ammonium group of the substrate, and hydrogen transfer occurs intramolecularly from dihydropyridine to the 3-acetylpyridinium ion. Examples of host–guest chemistry in synthesis will be found in the following chapter

3.2.6 Stereochemistry and asymmetric reductions with NADH model compounds

In NADH, the two protons at C(4) of the 1,4-dihydronicotinamide moiety occupy diastereotopic positions. Discrimination between these two protons in enzyme-bound NADH is possible from their ^{1}H-NMR chemical shifts, but the difference (if any) becomes minimal in the free coenzymes (Sarma and Kaplan, 1970; Oppenheimer *et al.*, 1971). In an NADH model system Rob *et al.* (1984) and de Kok and Buck (1985) demonstrated that these protons show quite different chemical shifts when the nicotinamide skeleton is included in a ring structure (i.e., (*3.36*) and (*3.37*)). This implies that the 1,4-dihydropyridine ring adopts a boat-shaped conformation resulting in axial H and equatorial H and the rate of the ring inversion is suppressed by the ring structure. The tracer experiments established that in the reduction of these NAD$^+$ models the incorporated hydrogen occupies almost exclusively ($> 95\%$) an axial position at C(4). Therefore, these systems constitute the model examples of NADH models able to mimic the diastereodifferentiating course of hydride exchange at pyridine nucleotides under enzymatic conditions. The axial preference is accounted for by the stereoelectronic effect (Rob *et al.*, 1984).

One of the most noteworthy characteristics of NADH-dependent enzymes is stereospecific hydrogen transfer, and, in considering asymmetric reductions, we come close for the first time in this chapter to a key interest of the synthetic chemist (see also Chapter 4). A mimic of the asymmetric

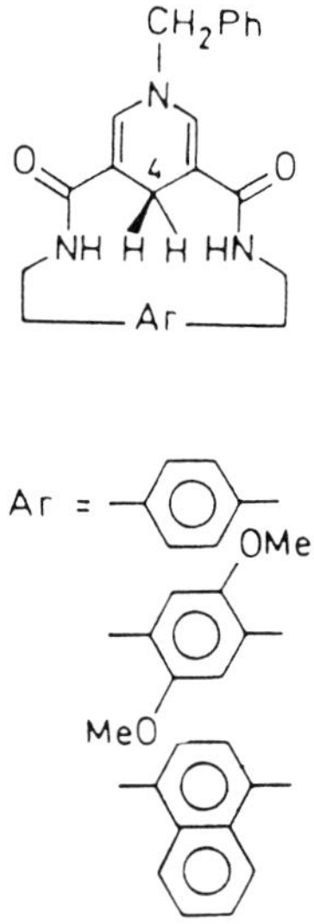

(3.36) 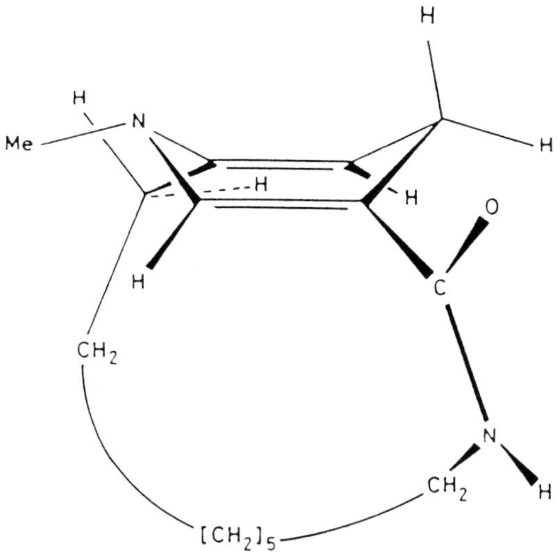   *(3.37)*

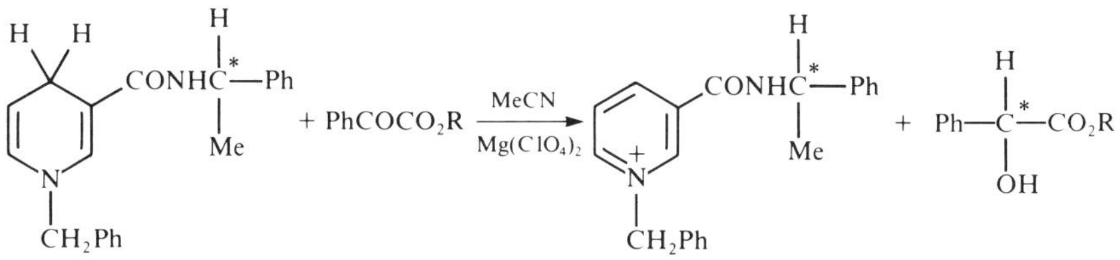

Fig. 3.7

reduction of a carbonyl substrate by a NADH model compound was first reported by Ohnishi *et al.* (1975c). They introduced a chiral α-phenethylamine to the 3-amido group of (*3.1*) and carried out the reduction of benzoylformic acid esters to mandelic acid esters (*3.38*) in the presence of Mg^{2+} ion (Fig. 3.7). The configuration of (*3.38*) was in accord with that of the 3-amido group, and the optical yields were 11–20%.

Considerable efforts to improve the optical yield have since been made. Representative examples are summarized in Table 3.1. The last four examples particularly warrant considerable attention because of the high enantioselectivity. In (*3.44*), the 4-position which releases a hydrogen is covered by an asymmetric crown ether, so that the orientation of the substrate is highly controlled in the transition state. In (*3.45*) and (*3.47*), the hydrogen used for reduction is linked directly to the asymmetric carbon. In (*3.46*) a Mg^{2+}-bridged cyclic structure is proposed as an active species. (*3.44*) and (*3.46*) are reusable by reducing the oxidized forms with sodium dithionite, suggesting possible use as chiral multiplication reagents. These examples demonstrate that the *in-vitro* enantioselectivity of the NADH model reduction is now as high as that of the enzymic systems.

3.2.7 Oxidation by NAD$^+$ model compounds

In alcohol dehydrogenases, the interconversion of aldehydes (ketones) and alcohols occurs in conjugation with that of NADH and NAD$^+$ coenzymes. In contrast to a great number of investigations on the NADH model reduction of carbonyl substrates, there are few examples of the NAD$^+$ model oxidation of alcohol substrates. Wallenfels and Hanstein (1965) showed that 9-fluorenol is oxidized to fluorenone (8% yield) by (*3.48*). Shirra and Suckling (1977) have studied the oxidation of benzyl alkoxides to benzaldehydes with a variety of pyridinium salts (e.g. (*3.49*)), but the product analysis was not fully conducted. Ohnishi and Kitami (1978) carried out the oxidation of lithium alkoxides by (*3.50*), the yields of the 1,4-dihydro-pyridine being 3.5–28%. More recently, Shinkai *et al.* (1984a) found that magnesium alkoxides RCH$_2$OMgX are smoothly oxidized by a NAD$^+$ model (*3.51*) to

Table 3.1 Optical yield for the asymmetric reduction of benzoylformic acid esters

Catalyst	Optical yield (%)	Reference
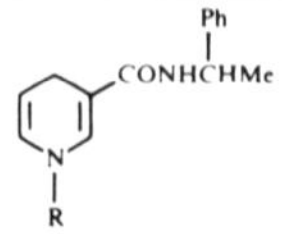 (3.39)	11–20	Ohnishi *et al.* (1975c)
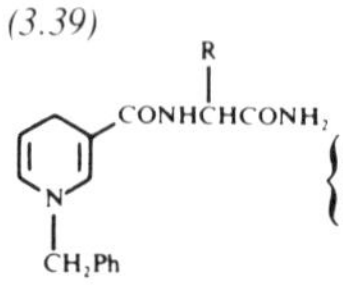 R = Me R = CH₂CHMe₂ R = CH₂Ph (3.40)	47 26 5	Endo *et al.* (1977)
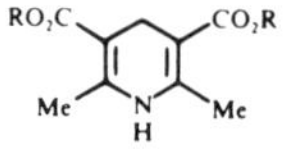 R = (−)-menthyl (3.41)	17–21	Nishiyama *et al.* (1976)
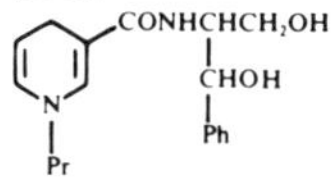 (3.42)	26.3	Makino *et al.* (1979)
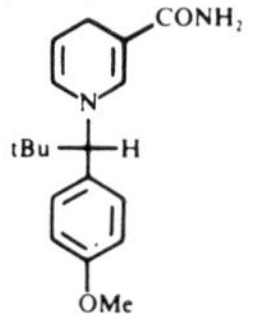 (3.43)	27	van Ramesdonk *et al.* (1978)
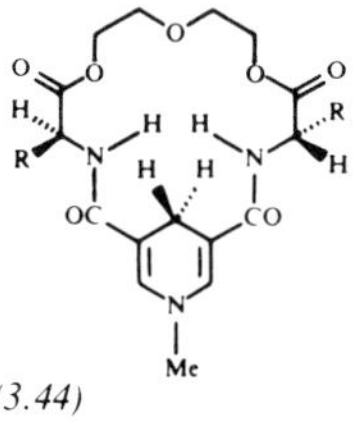 (3.44)	86	de Vries and Kellog (1979) Talma *et al.* (1985)
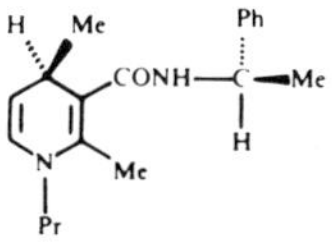 (3.45)	94.7–97.6	Ohno *et al.* (1979)
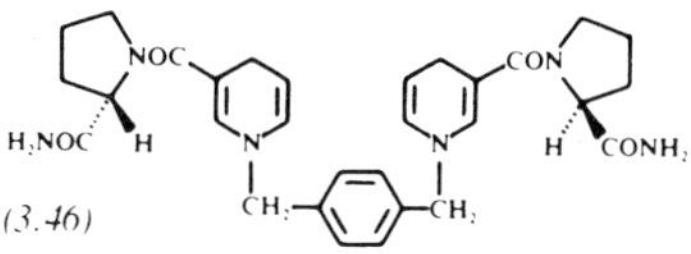 (3.46)	93–98	Seki *et al.* (1981)
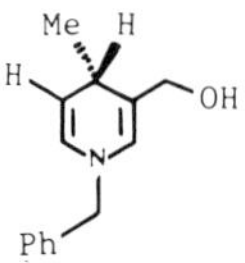 (3.47)	95	Meyers *et al.* (1986)

RCH=O, giving the dihydronicotinamide in reasonable yields (40–62%). This reaction $NAD^+\cdots H_2C(R)-O^-\cdots Mg^{2+}$ corresponds to the reverse process of the carbonyl reduction by a NADH model $NADH\cdots RCH=O\cdots Mg^{2+}$. These still unsatisfactory results suggest that the N-substituted pyridinium ions (conventional NAD^+ model compounds) may have some defect as mimics of NAD^+-dependent oxidation, and model studies of oxidations followed an unexpected course.

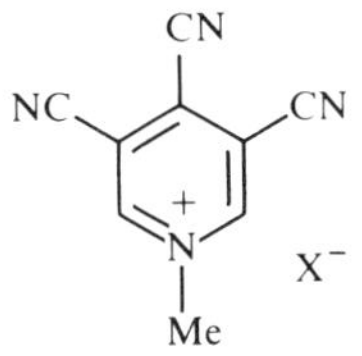

(3.48) 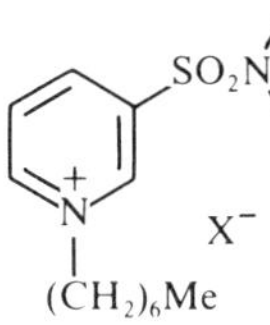   (3.49) 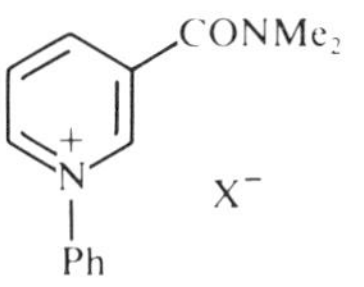   (3.50) 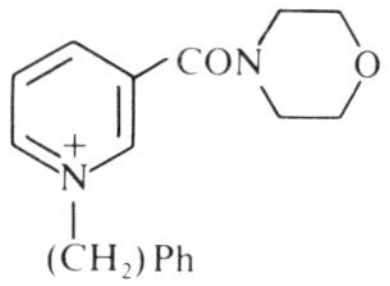   (3.51)

5-Deazaflavins (3.52) (and their reduced forms) (3.53)) were at first synthesized as analogues of flavin coenzymes (O'Brien et al., 1970; Edmondson et al., 1972). Subsequent research established, however, that the redox behaviour of 5-deazaflavins is closely akin to that of NAD^+ rather than to that of flavin coenzymes (Hemmerich et al., 1977). Surprisingly, it was later discovered that the basic skeleton of 5-deazaflavin is involved in fluorescent cofactor F_{420} of Methanobacterium (Cheeseman et al., 1972).

Yoneda et al. (1977) found that (3.54) is able to oxidize alcohols under alkaline conditions, the yields of aldehydes and ketones being 82–99%. The result indicates that the skeleton of 5-deazaflavin suitably mimics NAD^+ in the enzymes. Subsequently, they demonstrated that modified coenzyme

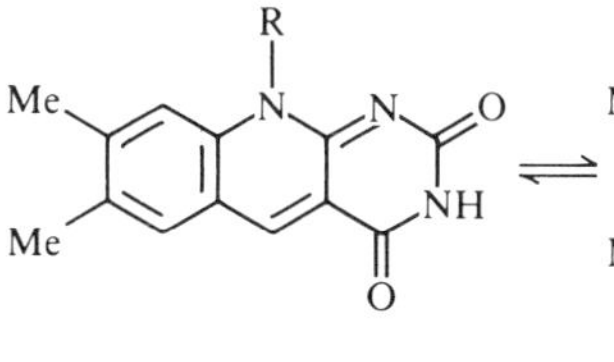

(3.52) 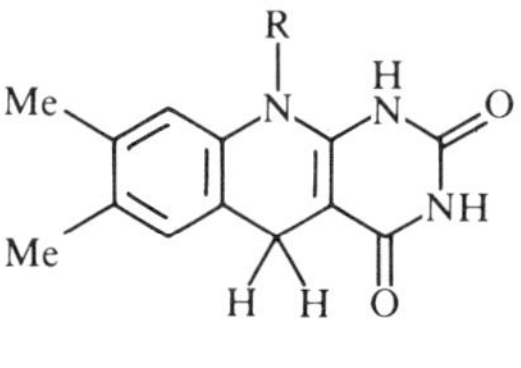   (3.53) 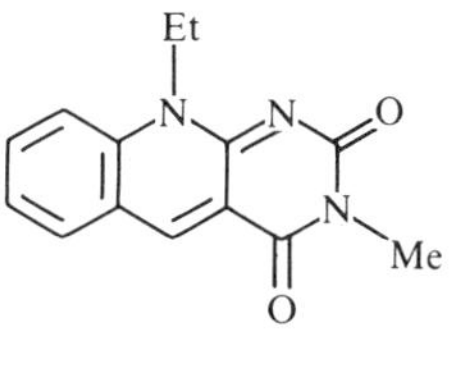   (3.54)

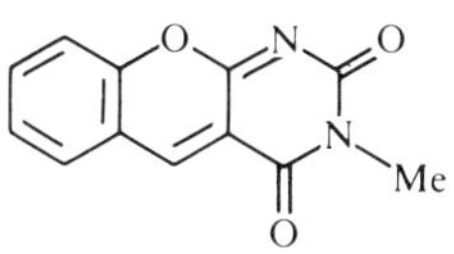

(3.55) 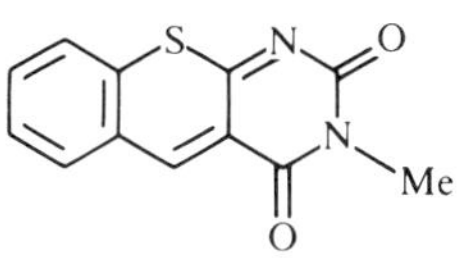   (3.56) 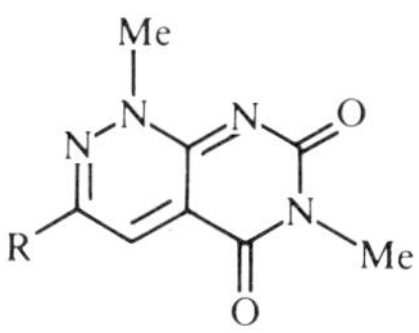   (3.57) 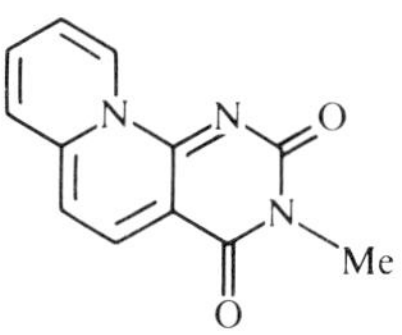   (3.58)

models (3.55)–(3.58) also oxidize alcohols (Yoneda and Nakagawa, 1980; Yoneda et al., 1978a, 1980a, b). Compound (3.55) shows strong oxidizing power owing to the electron-negative nature of the oxygen in the ring system, and oxidizes benzyl alcohol in the absence of base. The yields of the oxidation products per mole of heterocycle under aerobic conditions sometimes exceed 100% because the reduced forms are reoxidized by molecular oxygen. Yoneda et al. (1981) synthesized (3.59). As its reduced form is relatively sensitive to molecular oxygen, it acts as an excellent recycling oxidizing reagent. Shinkai et al. (1981a) designed another NAD^+ model (3.60) which is isoelectronic with 5-deazaflavin. We found that (3.60) efficiently oxidizes alcohols and the reduced form is reoxidized by molecular oxygen, the recycle number being 18–22.

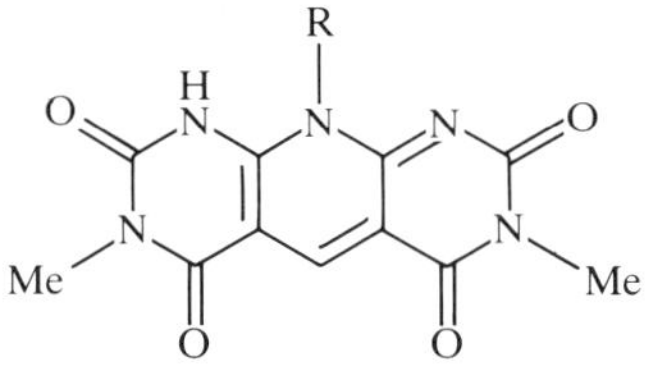

(3.59) 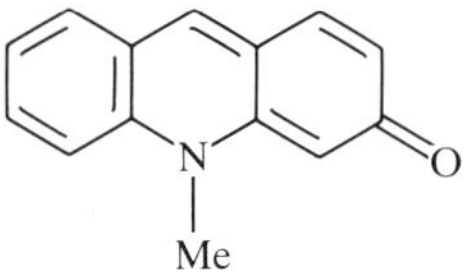   (3.60)

Successful model NAD^+ oxidants all contain the equivalent of N-alkylpyridinium rings made electrophilic by substitution with several electron-withdrawing groups in a polycyclic system. Their efficacy compared with simple NAD^+ models can be ascribed to these structural features. Some of the model oxidants also have reducing properties. For example, it is known that reduced 5-deazaflavins are able to reduce aldehydes to alcohols in the presence of acids (Shinkai and Bruice, 1973b; Yoneda et al., 1978b). It is interesting to note that the basic skeleton of an acid-stable NADH model (3.19) is involved in (3.53). This suggests that the reducing ability of (3.53) is also associated with its stability to acids.

3.3 FLAVIN CATALYSES

Flavin coenzymes (FAD, FMN and riboflavin) oxidize a wide variety of substrates, such as amino acids, oxy acids, amines, etc. in biological systems, and are themselves reoxidized by molecular oxygen (Hemmerich et al., 1970; Walsh, 1980; Bruice, 1980) (Fig. 3.8). From a chemical point of view, the oxidation ability of flavins probably stems from the electron-

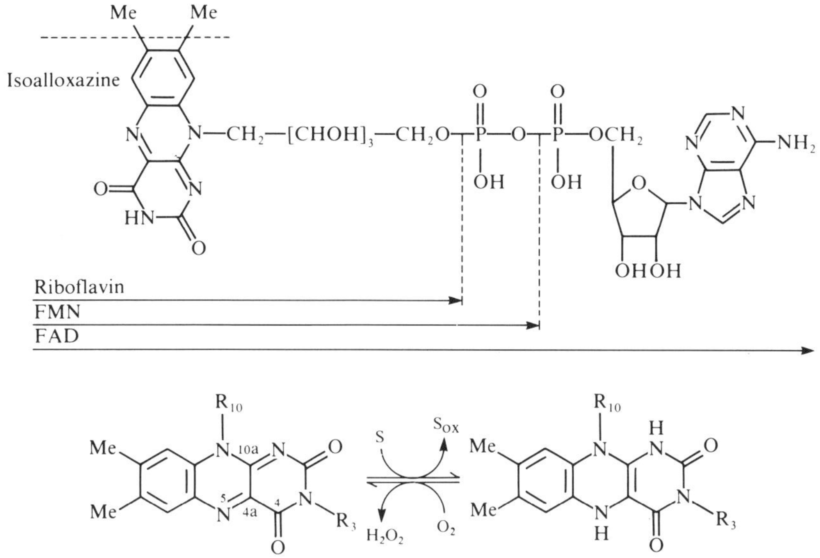

Fig. 3.8

withdrawing triketone-like structure at the 4-4a-10a linkage which is comparable with the reactive NAD^+ oxidant mimics just discussed. In simple α, β, γ -triketones such as alloxane and ninhydrin, the oxidation ability is lost in aqueous solution owing to facile hydration of the central carbonyl group. In contrast, the triketone-like structure of flavins undergoes no hydration and retains its strongly electron-deficient nature. Walsh *et al.* (1978) and Spencer *et al.* (1977) investigated the importance of nitrogen atoms by systematically replacing each nitrogen atom in the flavin with a carbon atom. They found that the 1-deaza analogue (*3.63*) is as sensitive to oxygen as the native reduced flavin (*3.62*), but the 5-deaza analogues (*3.64*) and (*3.65*) are not (or scarcely) reoxidized by oxygen. This finding suggests that the sensitivity of reduced flavin to oxygen is associated with the 5,10-enediamine structure.

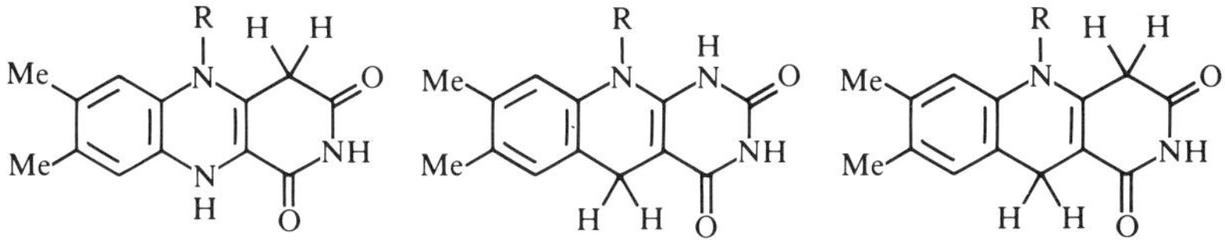

3.3.1 Flavins as oxidation catalysts

Flavin coenzymes serve as versatile oxidation catalysts in many biological systems but, like the NAD^+ model systems, flavin molecules *in vitro* have a rather small oxidation ability towards even relatively electron-rich substrates. It is known that flavin and isoalloxazines are able to oxidize dihydropyridines such as (*3.1*) and (*3.2*), aliphatic thiols and dithiols to disulphides (Gibian and Winkelman, 1969; Gascoigne and Radda, 1967; Gumbley and Main, 1976; Loechler and Hollocher, 1975, 1980), and carbanions such as (*3.66*)–(*3.69*) (Main *et al.*, 1972; Rynd and Gibian, 1970; Brown and Hamilton, 1970; Shinkai *et al.*, 1974). However, less-activated substrates such as thiophenol and nitroalkane carbanions are stable to simple flavin models.

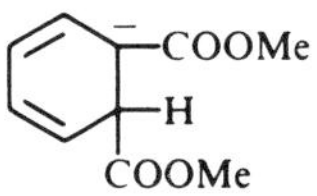

(*3.66*) 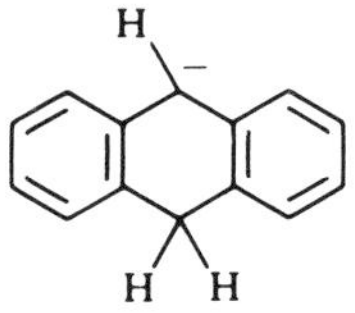   (*3.67*) 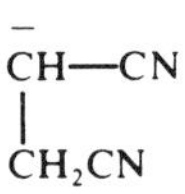   (*3.68*) 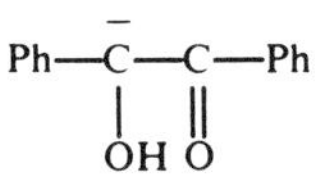   (*3.69*)

There are now two main methods recognized for 'activating' flavin molecules as oxidizing catalysts: (i) shifting the redox potential of flavins to more positive values, and (ii) converting substrates into more 'specific' ones with the aid of second cofactors or environmental effects. Method (i) is exemplified by (*3.70*)–(*3.73*) (Yokoe and Bruice, 1975; Bruice *et al.*, 1977; Knappe, 1979; Shinkai *et al.*, 1979f; Yano *et al.*, 1984), in which the electron-withdrawing substituents enhance the electron-deficiency of the isoalloxazine ring. For example, (*3.70*) is able to oxidize thiophenol and 2-nitropropane carbanion under ambient conditions (Yokoe and Bruice,

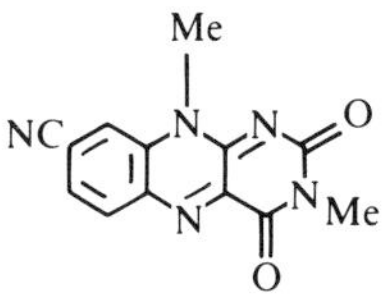

(*3.70*) 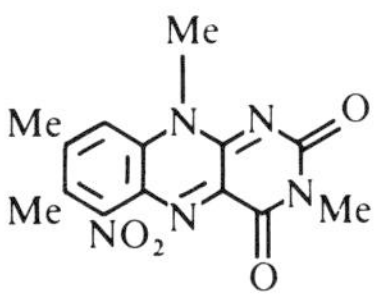   (*3.71*) 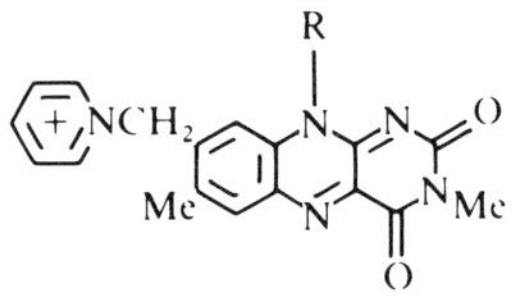   (*3.72*)

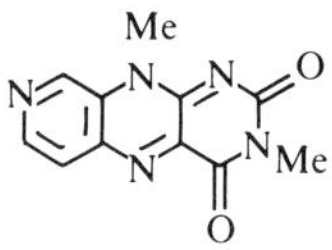

(*3.73*)

1975). Compound (*3.73*) also exhibits strong oxidizing power owing to the electron-negative nature of a nitrogen atom incorporated in the ring system (Yano *et al.*, 1984).

Flavin coenzymes serve as electron carriers to and from iron-sulphur proteins, molybdenum (xanthine oxidase) and haem proteins, and the interactions between flavins and metal ions have attracted considerable attention. However, model investigations of metal–flavin interactions have been very limited. The reason is simple: flavins have no significant affinity for metal ions except for a few such as Ag(I), Cu(I), Mo(V), Ru(II) and Fe(II) (Fritchie, 1972; Bamberg and Hemmerich, 1961; Clarke *et al.*, 1979; Selbin *et al.*, 1974). The stability of these metal–flavin complexes is probably due to a metal $\rightarrow$ flavin charge transfer. To overcome this deficiency, Shinkai *et al.* (1982a, 1983, 1986a) synthesized a metal–coordinative flavin (*3.74*) which combines the structure of a flavin with the well-known chelate 1,10-phenanthroline and binds various kinds of metal ions. The metal– (*3.74*) complex acts as a strong oxidizing agent owing to the strong electron-withdrawing nature of complexed metal ions. This system is also classified as an example of method (i) of activating a flavin.

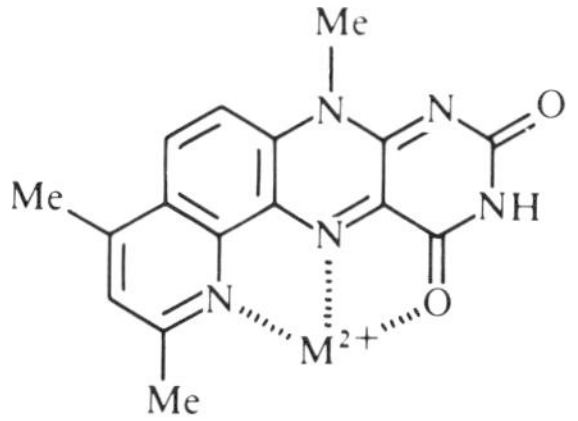

(*3.74*)

Method (ii) 'activates' substrates in the same way as apoenzymes do in the apoenzyme–coenzyme–substrate ternary complexes. The most expeditious method is to utilize the hydrophobic environments of micelles and polymers, where anions are significantly activated (Kunitake and Shinkai, 1980). In contrast to the inability of conventional flavins to oxidize thiophenol and nitroalkane carbanions in solution, hydrophobic (*3.75*) and (*3.76*) bound to a cationic micelle can oxidize these substrates under mild conditions (Shinkai *et al.*, 1980b). Probably, both micellar activation and concentration of the substrate facilitate the oxidation reaction in the micelle, suggesting the importance of the hydrophobic environments. Similarly, flavins immobilized in cationic polymers act as oxidation catalysts for thiophenol and nitroethane carbanion, and the oxidation of NADH proceeds according to Michaelis–Menten saturation kinetics owing to the electrostatic interaction between cationic polymers and anionic NADH (Shinkai *et al.*, 1978c, d, e; Spetnagel and Klotz, 1978).

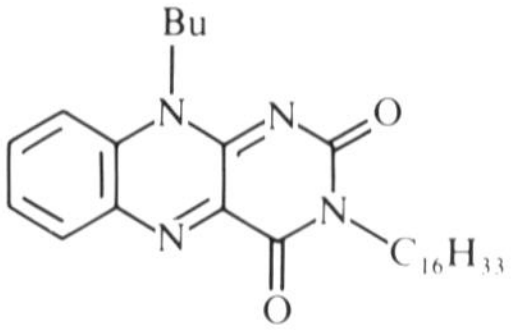

(3.75) 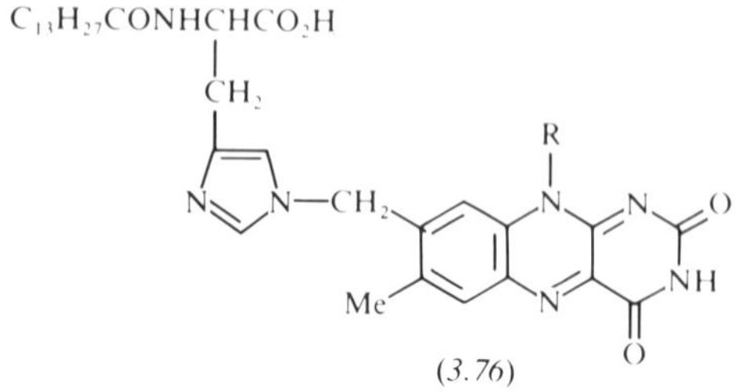   (3.76)

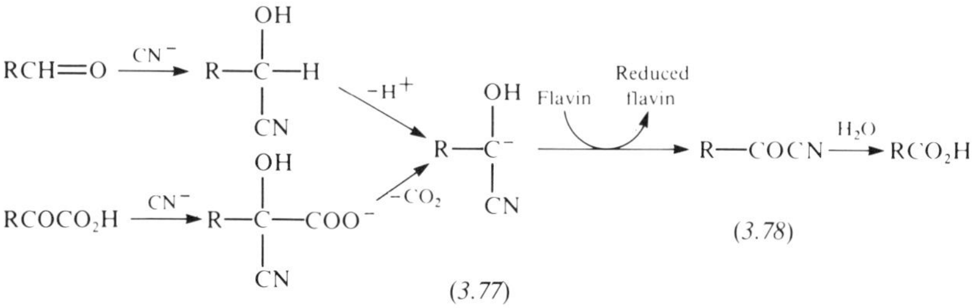

(3.77)   (3.78)

Fig. 3.9

It was proposed on the basis of the oxidation of β-chloroalanine by D-amino acid oxidase (Walsh *et al.*, 1971) that some flavoenzymes do not oxidize substrates directly but the conjugate base generated from the substrates (Kosman, 1977). The concept is very helpful for mimicking flavin oxidation in a non-enzymic system. For example, neither aldehydes nor α-ketoacids are oxidized by flavins, but in the presence of cyanide ion these substrates are converted to cyanohydrin carbanions (*3.77*) and undergo rapid flavin oxidation to give the corresponding carboxylic acids (Shinkai *et al.*, 1980c) (Fig. 3.9). Similarly, thiazolium ions, which exhibit a catalytic activity closely resembling that of cyanide ion, form 'active aldehyde' intermediates (*3.79* from aldehydes and α-ketoacids, and (*3.79*) is rapidly oxidized by flavins (Shinkai *et al.*, 1980d, e; Yano *et al.*, 1980) (Fig. 3.10). Interestingly, this reaction model mimics well the catalytic cation of pyruvate oxidase which converts pyruvic acid to acetic acid with the aid of FAD and thiamine pyrophosphate. The reactions starting from α-ketoacid are examples of the decarboxylative oxidation. Since (*3.78*) and (*3.80*) have active acyl groups, the reaction in alcohols gives the corresponding esters (Shinkai *et al.*, 1980e; Yano *et al.*, 1980). Shinkai *et al.* (1982b) synthesized a biscoenzyme (*3.81*) which has both flavin and thiazolium ion within a molecule. (*3.81*) efficiently catalyses the reaction sequence as intramolecular processes.

More recently, a new coenzyme called PQQ (or methoxatin: (*3.82*)) was

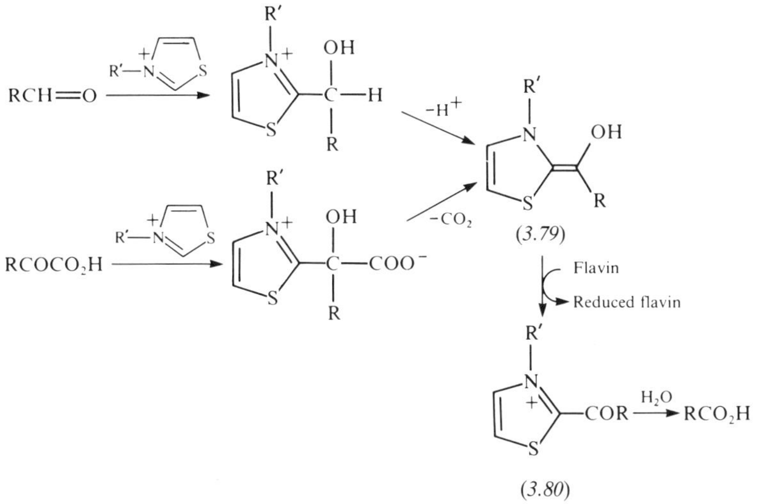

Fig. 3.10

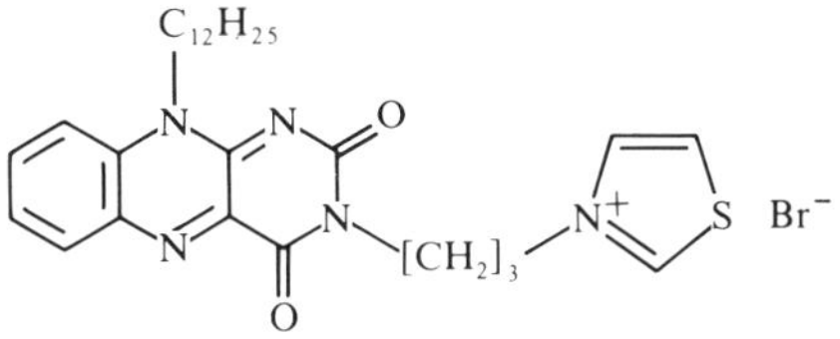

(3.79)

(3.80)

Flavin

Reduced flavin

(3.81) 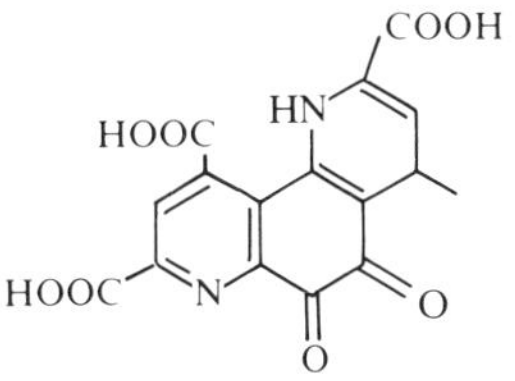   (3.82)

discovered in quinoproteins, and the structure and the reactivities have been investigated (Duine and Frank, 1981). Since the redox activity of this coenzyme stems from the *ortho*-quinone moiety, the chemical behaviour is expected to be similar to that of flavins. Evidence from enzyme inhibition studies suggests that single electron abstraction from substrates is a major pathway for PQQ-mediated oxidations (reviewed by Suckling, 1988). Indeed there are oxidases containing PQQ that catalyse the same reactions as flavins (Sherry and Abeles, 1985).

3.3.2 Flavins as photosensitizers

Massey and Hemmerich (1977, 1978) have shown that flavins and 5-deazaflavins are active catalysts for photoreduction of flavoproteins, haeme proteins, and ion-sulphur proteins by EDTA, amino acids, and oxalate as electron donors. The observation that iron–sulphur proteins of low redox potential are photoreduced by flavins and 5-deazaflavins suggests that this class of photocatalysts should also catalyse reduction of other low potential electron carriers. Krasna (1980) showed that 5-deazaflavins (*3.52*) catalyse photoproduction of hydrogen from a variety of compounds (e.g., carboxylic acids, amines, sugars, etc.) that are inactive with well-known acridines and Ru(bpy)$_3^{2+}$ and act as much more effective sensitizers for the photoreduction of methylviologen than Ru(bpy)$_3^{2+}$. It is shown, however, that continuous irradiation of (*3.52*) in the presence of oxalate anion causes photodimerization with quantum efficiencies exceeding unity (Bliese *et al.*, 1983).

    Flavins and 5-deazaflavins also serve as photosensitizers for the cleavage of *cis,syn*-thymine dimer and the production of thymine monomer (Rokita and Walsh, 1984; Jorns, 1987). The principal damage resulting from exposure of DNA to ultraviolet light is the formation of cyclobutane dimers between adjacent pyrimidine residues. Photoreactivating enzymes (DNA photolyases) repair UV-damaged DNA by splitting the dimers in a rather unusual reaction that requires visible light, and oxidized flavin (a 8-hydroxy-5-deazaflavin derivative) is present in photolyase from *S. griseus*. Thus, several characteristics of this model system mimic the *in vivo* thymine dimer repair phenomenon attributed to the action of DNA photolyases.

3.3.3   Stereochemistry and asymmetric oxidations

In contrast to a large number of examples for asymmetric reduction of carbonyl substrates by optically-active NADH models, very few precedents exist for asymmetric redox reactions mediated by flavins. To the best of our knowledge, there are only two examples of flavins with a chiral substitutent: (*3.83*) possesses an asymmetric carbon substituent at N(3) (Tanaka *et al.*, 1984) and (*3.84*) has one at N(10) (Shinkai *et al.*, 1984b).

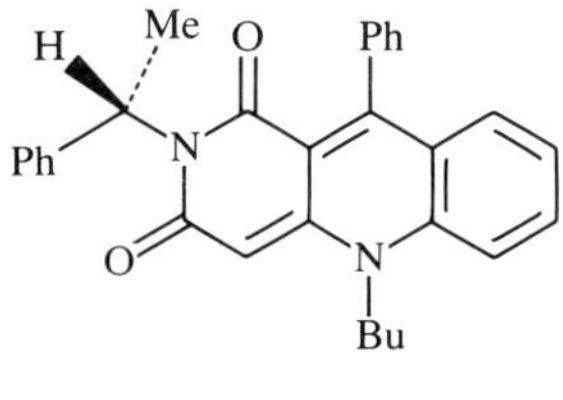

(*3.83*) 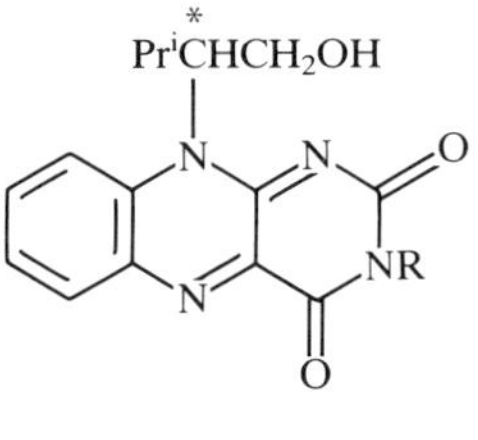   (*3.84*)

    The optical yields attained in these examples were relatively low (less than 30% e.e.). Hence, this study was extended to the synthesis of new flavins with 'large' chiral frames of reference such as axial chirality and planar chirality. (*3.85*) and (*3.86*) (X=N or CH) feature restricted rotation about the C(1′)–N(10) single bond and can be optically resolved by a liquid chromatographic method (Shinkai *et al.*, 1988a). Flavinophanes (*3.87*) (X=N or CH, $n = 6$–12) with planar chirality are also optically resolved by a liquid chromatographic method (Shinkai *et al.*, 1986b, 1987). These cyclic flavins can oxidize (*3.39*) in 60% e.e. and optically-active thiols in 43% e.e.

(*3.85*)   (*3.86*)   (*3.87*)

(*3.85*) and (*3.86*) did not racemize thermally at 70°C, but they racemized invariably when they were reduced to the 1,5-dihydro forms (except those with a 2′-substituted naphthyl group at N(10)) (Shinkai *et al.*, 1988a). The oxidized flavins and 5-deazaflavins are almost planar, whereas the reduced forms are folded like butterfly wings along N(5) (or C(5)) and N(10). Thus, the novel 'redox-induced' racemization is due to conversion of the 'planar' oxidized forms (sterically-tense) into the 'bent' reduced forms (sterically-relaxed). Shinkai *et al.* (1986b) also found that (*3.87*) with $n \geq 10$ racemize when they are reduced to the 1,5-dihydro forms but those with $n \leq 8$ do not. The racemization is due to the rope-skipping motion in the sterically-relaxed, folded reduced forms. Thus, the critical strap length for racemization is $n = 9$. These are novel examples in which the racemization is induced only by the redox reaction.

3.3.4 Flavin–oxygen complexes as oxygenase models

Reduced flavins (*3.62*) are rapidly reoxidized by molecular oxygen to yield hydrogen peroxide, but they can reduce pyruvate esters, nitrobenzene and quinones under anaerobic conditions (Williams and Bruice, 1976; Gibian and Baumstark, 1971; Gibian and Rynd, 1969). On the other hand, (*3.88*), in which N(5) of the flavin skeleton is blocked by a methyl or ethyl group, produces the radical species (*3.89*) by one-electron oxidation, and 4a-hydroperoxyflavin (*3.90*) by O_2 addition (Kemal and Bruice, 1976a; Kemal *et al.*, 1977) (Fig. 3.11). It is believed that the structure of (*3.90*) is similar to enzymically generated hydroperoxyflavin. 4a-Hydroperoxyflavin (*3.90*) transfers oxygen or dioxygen to sulphides (*3.91*) and (*3.92*), amines (*3.93*), phenolate ions (*3.94*), indoles (*3.95*), etc. (Kemal *et al.*, 1977; Miller, 1982;

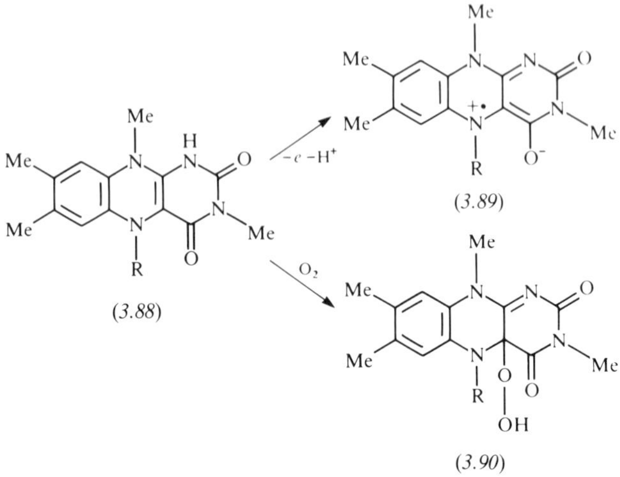

Fig. 3.11

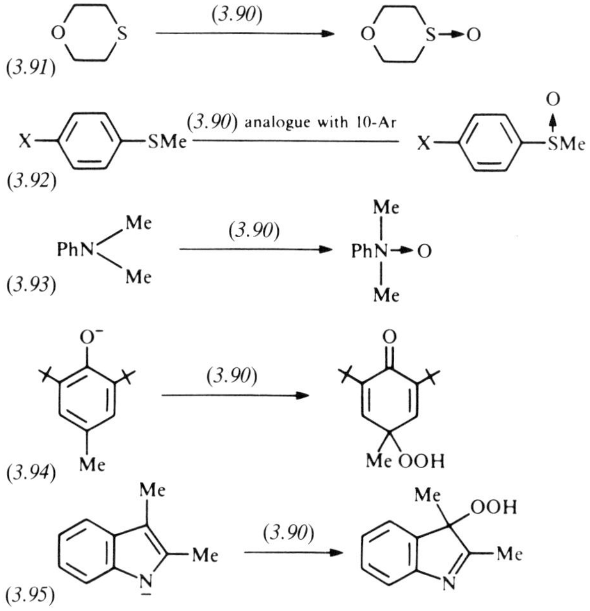

Fig. 3.12

Ball and Bruice, 1980; Kemal and Bruice, 1979; Muto and Bruice, 1980) (Fig. 3.12). Thus, these reactions become good biomimetic examples of mono- and dioxygenases.

Meanwhile, Rastetter and coworkers (1979; Frost and Rastetter, 1981) have demonstrated that a flavin N(5)-oxide (3.96) also transfers monooxygen to phenols, amines, hydroxylamines, etc. They consider that in mono-oxygenases, 4a-hydroperoxyflavin may be converted to flavin N(5)-oxide prior to oxygen transfer, and a flavin N(5)-nitroxyl radical is a viable candidate for the ultimate hydroxylation agent in the enzyme-catalysed reaction. A mechanistic understanding of flavin-mediated oxygenation is still incomplete.

A more simplified and synthetically important oxygenase model was offered by Heggs and Ganem (1979). As described previously, oxidized flavins have a triketone-like structure. In other words, the 4a-hydroperoxide in (3.90) is flanked by two electron-withdrawing groups and lies adjacent to a weakly basic, electronegative group (5-NH). They pointed out that (3.97), which is readily derived from hydrated hexafluoroacetone and hydrogen peroxide, completely satisfies these criteria; (3.97) acts as a low-cost, catalytic epoxidation reagent.

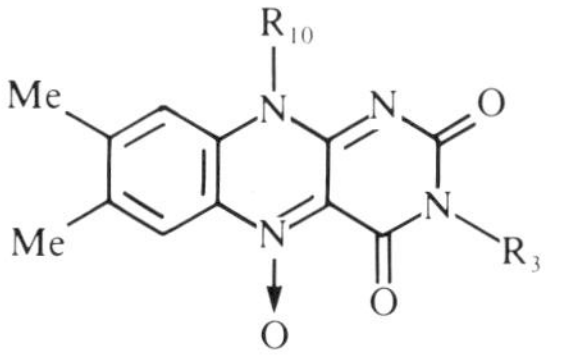

(3.96) 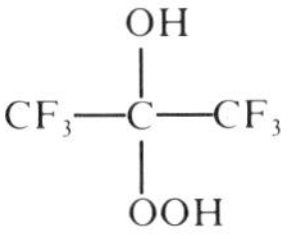   (3.97)

More recently, N(5)-ethyl derivatives of (3.87) (X=N) with planar chirality were synthesised (Shinkai et al., 1988b). In the presence of excess H_2O_2, these cyclic flaviniums act as an autorecycling catalyst for monooxygenation. Thus, sulphides (3.92) are oxidized to optically-active sulphides (65% e.e.) in 800% yield (calculated on the basis of (3.87)). This is a novel example for 'chiral multiplication' mediated by a flavin coenzyme model for oxygenases.

3.3.5 Model studies of bacterial luciferase

Another fascinating aspect of flavin chemistry is a chemiluminescence phenomenon imitating the bioluminescence of bacterial luciferase. It has been shown that the key step of bioluminescence catalysed by bacterial luciferase is the reaction of luciferase-bound 4a-hydroperoxyflavin mono-

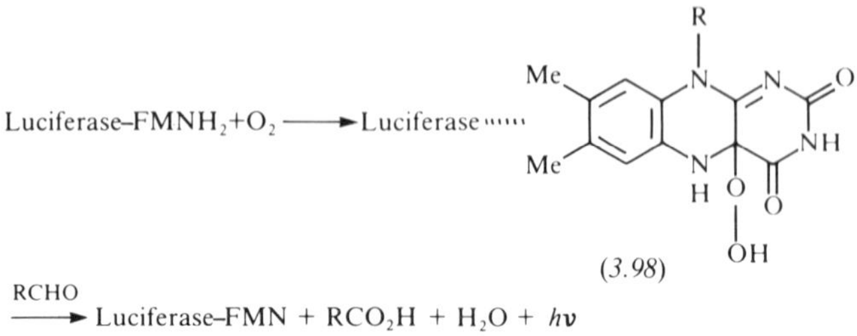
Luciferase–FMNH₂+O₂ → Luciferase
(3.98)
RCHO
→ Luciferase–FMN + RCO₂H + H₂O + hν

Fig. 3.13

nucleotide (*3.98*) with a long-chain aldehyde (Hastings *et al.*, 1973) (Fig. 3.13).

Kemal and Bruice (1976b) found that (*3.90*) undergoes a chemiluminescent reaction in the presence of an aldehyde which was visible to the dark-adapted eye. Similar light emission was observed when a flavinium salt (*3.99*) was mixed with alkyl hydroperoxides, the quantum yields being 3.0×10^{-4}–3.3×10^{-4} (Kemal and Bruice, 1977) (Fig. 3.14). They proposed a structure (*3.100*) common to chemiluminescent species.

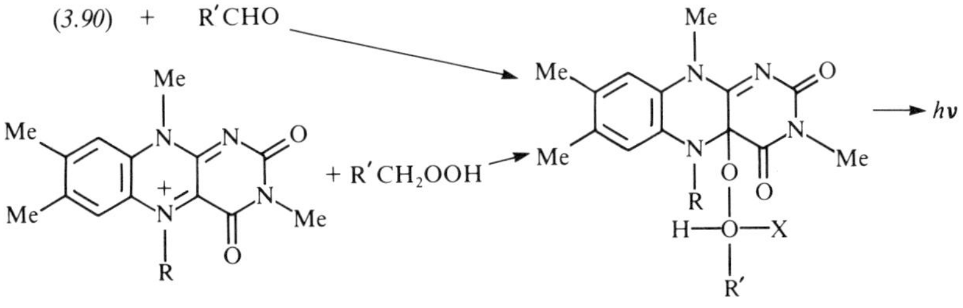
(3.90) + R′CHO
+ R′CH₂OOH
hν

(*3.99*) (*3.100*)

Fig. 3.14

McCapra and Lesson (1976) also found chemiluminescence for the reaction of (*3.101*) and peroxides (*3.102*), the quantum yields being 5×10^{-6}–2.1×10^{-5} (Fig. 3.15). They consider that, in this case, the nucleophilic addition of the peroxides occurred at position 10*a*.

These results, together with those of oxygen-transfer reactions, suggest that it is essential to substitute a labile N(5)-hydrogen with alkyl groups to imitate the flavoprotein functions. Conceivably, the N(5)-hydrogen of flavin coenzymes is tightly 'fixed' by hydrogen-bonding in the enzyme active sites.

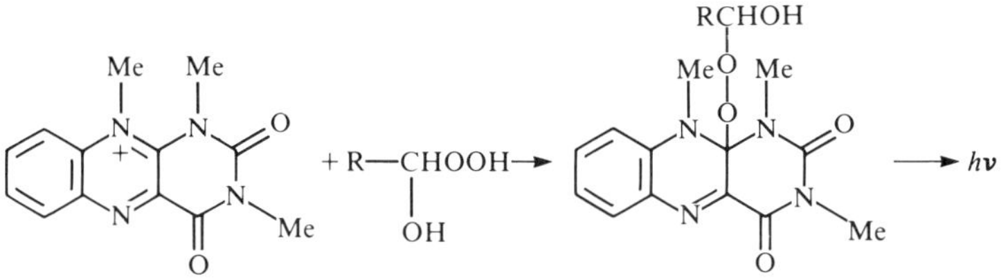

(3.101) (3.102)

Fig. 3.15

3.4 CATALYSES RELATING TO VITAMIN B_1 AND ANALOGUES

Thiamine pyrophosphate (TPP) and its analogues (*3.103*) (vitamin B_1 family) serve as catalysts for condensation, decarboxylation and reduction reactions. In 1958, Breslow showed that the catalytic role of TPP can be ascribed to the 2-H deprotonated conjugate bases of a thiazolium ion (*3.104*), which behaves analogously to cyanide ion in benzoin condensation. It is now recognized that, in the reaction sequence of thiazolium ion catalysis, the active aldehyde (*3.105*) always plays the role of a key intermediate (Fig. 3.16).

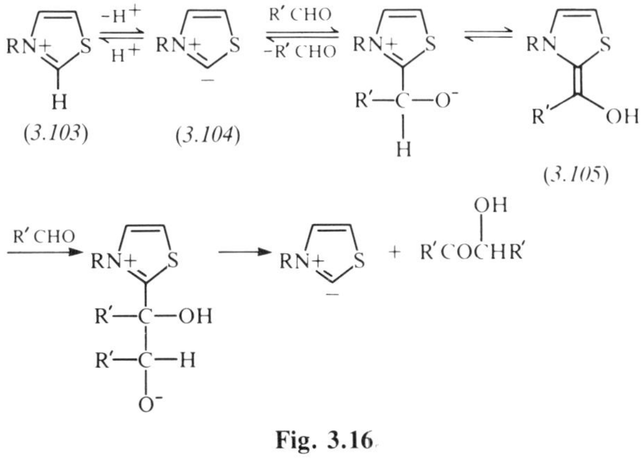

(3.103) (3.104) (3.105)

Fig. 3.16

3.4.1 Thiazolium ions as condensation catalysts

Breslow (1858) demonstrated that a variety of thiazolium ions are active in alcohol solvents in the presence of base as acyloin condensation catalysts unless position 2 is blocked by alkyl groups. The potential of the

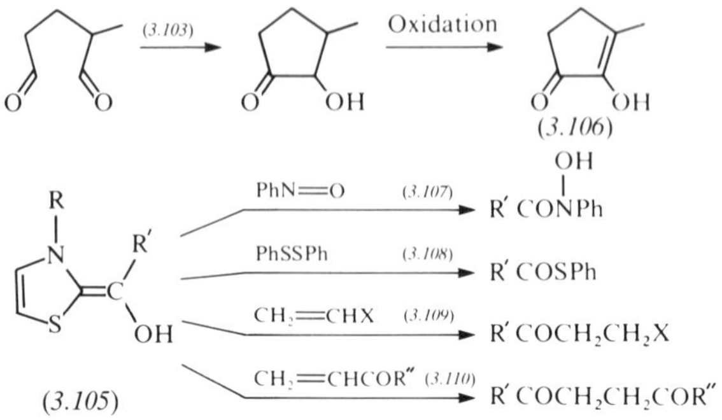

Fig. 3.17

condensation reaction stems from the strong nucleophilicity of the active aldehyde (3.105). Cookson and Lane (1976) utilized (3.103: R = benzyl) to synthesize cyclic 2-hydroxy-2-enones (3.106) from dialdehydes; this is an example of an intramolecular acyloin condensation. Compound (3.105) is able to attack not only the aldehyde group but also nitrosobenzene (3.107), disulphides (3.108), ethyl acrylate and acrylonitrile (3.109: X = COOEt, CN) and vinyl alkyl ketones (3.110) (Corbett and Chipko, 1980; Rastetter and Adams, 1981; Stetter et al., 1980; Stetter and Kuhlmann, 1976) (Fig. 3.17).

In order to use the thiazolium ions as synthetic reagents, Castells et al. (1978) and Sell and Dorman (1982) prepared a thiazolium ion immobilized in insoluble polymer supports. However, in contrast to the activity of thiazolium ions in alcohol solvents, they are relatively inactive in aqueous solution. This is probably due to OH^- catalysed decomposition of the thiazolium ring which takes place precedence over 2-H deprotonation. Tagaki and Hara (1973) found that surfactant thiazolium ions such as N-dodecylthiazolium bromide are very active in the aqueous acyloin condensation. Similarly, thiazolium ions covalently linked to cationic polymers are active in aqueous solutions (Shinkai et al., 1982c). They consider that cationic environments facilitate the deprotonation of 2-hydrogen, which leads to the enhancement in the catalytic activity.

Another interesting mimic of thiazolium ion catalysis is an asymmetric synthesis. Since acetoin and benzoin, the products of the thiazolium-catalysed condensation, have a chiral carbon, one may expect that the condensation could proceed in a stereoselective manner. Sheehan and Hara (1974) reported that thiazolium ions with chiral N-substitutents (3.111) give optically active benzoins in methanol. In particular, the optical yield in the presence of (3.111: R = p-O_2N-Ph) was significantly high (51.5%). Tagaki et al. (1980) also examined the asymmetric benzoin condensation under micellar conditions. They found on the chemical yields of benzoin that the

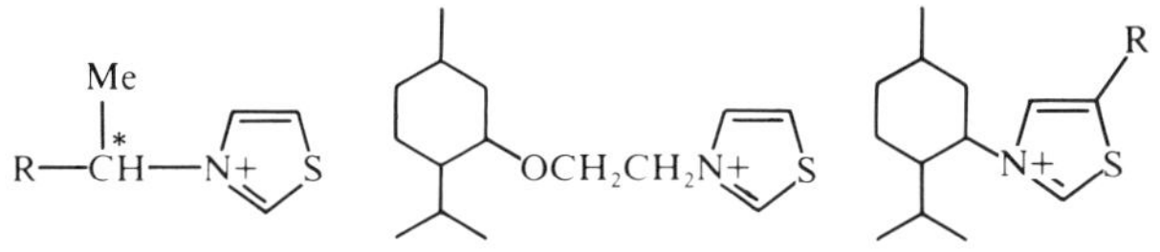

(3.111)   (3.112)   (3.113)

catalytic activity is high for (3.112) and low for (3.113), but (3.113) gave much higher optical yield (35.5%) than did (3.112) (1.8–3.5%). These results suggest that the chiral centre of the *N*-substituent should be bulky and close to the reaction centre.

The above discussion indicates that thiazolium ion catalyses are useful not only for the preparation of C–C and C–N condensation products but for asymmetric synthesis. Recently, Matsumoto *et al.* (1984) showed that in the self condensation of formaldehyde catalysed by thiazolium ions, dihydroxy-acetone (triose: $HOCH_2COCH_2OH$) is formed selectively and in high yield. This reaction, an analogue of the so-called 'formose reaction', is of much interest as a model synthesis of triose from this C_1 compound.

3.4.2   Thiazolium ions as decarboxylation catalysts

Another important TPP-mediated reaction is the decarboxylation of pyruvate to acetaldehyde which is catalysed by pyruvate decarboxylase (Crosby *et al.*, 1970; Kluger and Pike, 1979; Kluger *et al.*, 1981, 1987; see also Chapter 2). We have mentioned that thiamine catalysis in the acyloin condensation is closely akin to cyanide ion catalysis in the benzoin condensation; decarboxylation is similar. The mechanisms proposed for the thiamine-catalysed carboxylation also involve the key intermediate (3.105). One can write a reaction sequence equivalent to the cyanide ion catalysis which involves a cyanohydrin carbanion (3.77) as the key intermediate analogue (Fig. 3.18). Indeed many years ago, Franzen and Fikentscher

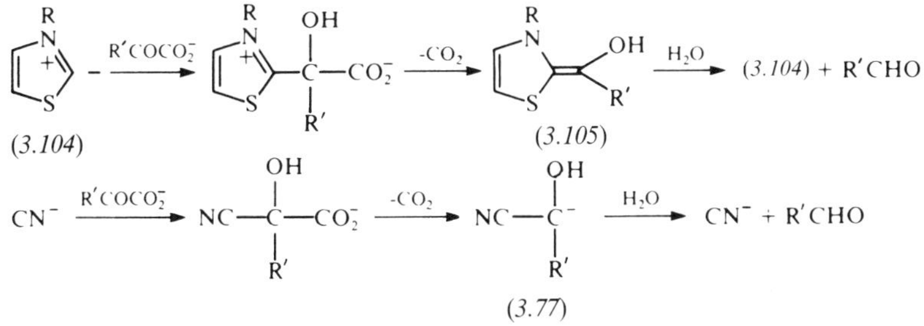

Fig. 3.18

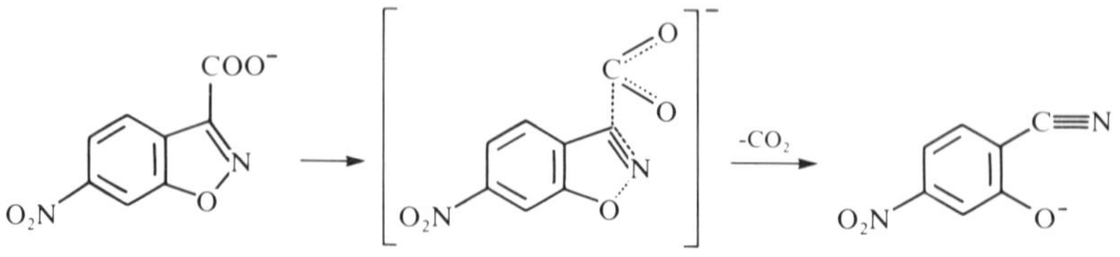

(3.114)

Fig. 3.19

(1958) found that cyanide ion acts as a good catalyst for the decarboxylation of α-keto acids.

Here, one should pay attention to one important characteristic of decarboxylation reactions, namely that they are very sensitive to solvent effects. In general, the reactions are fast in dipolar aprotic solvents and are markedly suppressed in protic solvents owing to solvation of the carboxylation anion through hydrogen-bonding (Kunitake and Shinkai, 1980) (Fig. 3.19). For example, the rate constant for the decarboxylation of (3.114) in dimethyl formamide is greater by 5.1×10^6-fold and 1.5×10^5-fold than those in water and methanol respectively (Kemp and Paul, 1970). If some part of the enzyme catalysis is reproduced by the solvent effect, this result implies that the enzymic decarboxylation would be markedly facilitated in a water-free, hydrophobic active site of an enzyme, as was discussed in Chapter 2.

As expected, thiazolium ion-mediated decarboxylation of β-ketoacids does not occur (or barely occurs) in aqueous solution (Crosby *et al.*, 1970; Shinkai *et al.*, 1980e). On the other hand, the cationic-micelle-bound thiazolium ion (3.103: $R = n\text{-}C_{16}H_{33}$) catalysed the decarboxylation of α-ketoacids under ambient conditions. This result suggests that the hydrophobic microenvironment of the micelle is favourable to the decarboxylation and is in line with the fact that the cationic micelle accelerates the decarboxylation of (3.114) by a factor of 90 (Bunton *et al.*, 1973). However, the solvent effect on the thiazolium ion-mediated decarboxylation is not fully understood.

3.4.3 Thiazolium ions as reducing agents

The key intermediate (3.105) which usually acts as a strong nucleophile becomes a reducing agent for one-electron-accepting substrates. Christen and Gasser (1980) and Cogoli-Greuter *et al.* (1979) demonstrated the irreversible inactivation of pyruvate decarboxylase and transketolase by hexacyanoferrate (III) or H_2O_2. These TPP-dependent enzymes involve the TPP analogues of (3.105) in their reaction sequences, which are oxidatively

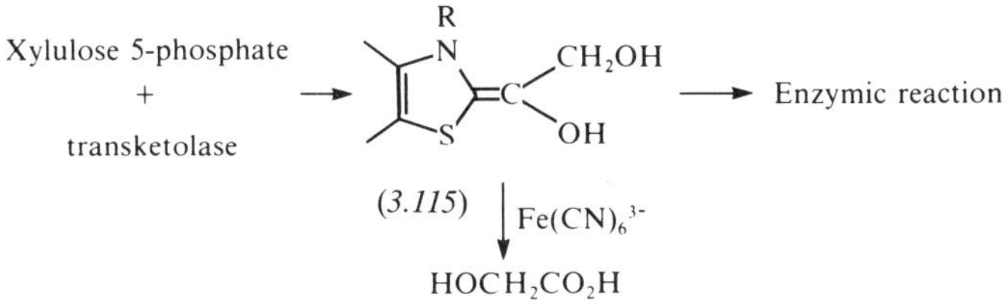

Fig. 3.20

trapped by a one-electron oxidant. For example, transketolase produces the dihydroxyethyl intermediate (3.115), which is rapidly oxidized by hexacyanoferrate (III) to glycollic acid (Fig. 3.20). The result clearly suggests that the key intermediate (3.105) can act not only as a nucleophile but also as a reducing agent.

As has been described in the section on flavin catalysis, analogues of (3.105) rapidly reduce flavins (Shinkai *et al.*, 1980d, e, 1982b; Yano *et al.*, 1980). Inoue and Higashiura (1980) found that acridine and 10-methylacridinium chloride are reduced to 9,9′-biacridan and 10,10′-dimethyl-9.9′-biacridan respectively by (3.105: R = benzyl) plus base in methanol. These examples suggest that (3.105) has the character of a one-electron donor. Castells *et al.* (1977) employed *p*-nitrobenzaldehyde as a substrate for the thiazolium ion-mediated condensation of aldehydes. Unexpectedly, they could not detect 4,4′-dinitrobenzoin but did identify the self-oxidoreduction products such as nitro-, azo- and azoxybenzoic acids and esters. This finding implies that the nitro group serves as an oxidant of (3.105). They proposed a nucleophilic attack of (3.105) on the nitro group (3.116), but the possibility of an electron-transfer mechanism cannot be rule out (Fig. 3.21).

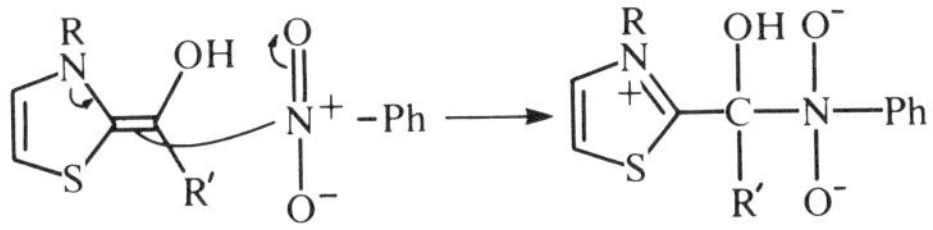

(3.116)

Fig. 3.21

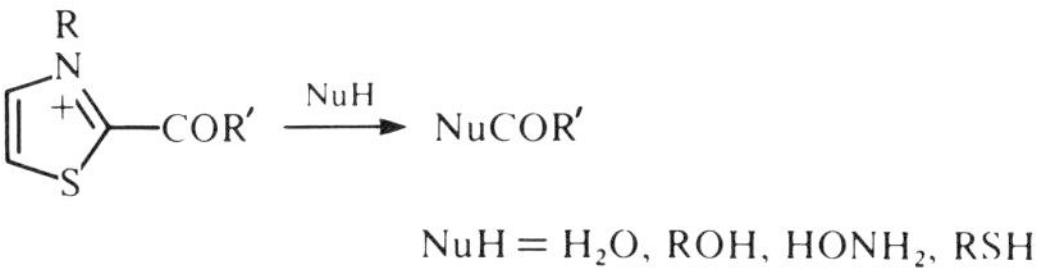

NuH = H$_2$O, ROH, HONH$_2$, RSH

(3.117)

Fig. 3.22

It has been established by Daigo and Reed (1962) and White and Ingraham (1962) that 2-acylthiazolium ions (*3.117*) are easily cleaved by nucleophiles: for example, (*3.117*) gives carboxylic acids by water, esters by alcohols, hydroxamic acid by hydroxylamine, and thioesters by mercaptans, i.e. (*3.117*) is a versatile acylation reagent (Fig. 3.22). Since (*3.117*) can be obtained from thiazolium salts + aldehydes in one step, aldehydes can easily be converted into, for example, esters (Shinkai *et al.*, 1980e; Yano *et al.*, 1980; Inoue and Higashiura, 1980).

3.5   PYRIDOXAL CATALYSES

It is well known that pyridoxal (vitamin B_6) catalyses proceed via a Schiff base and its tautomeric form (*3.120*) (Metzler *et al.*, 1954) (Fig. 3.23). Pyridoxal-dependent enzymes catalyse a broad range of reactions of amino acids such as decarboxylation, elimination, transamination. Although the mechanistic aspects of these reactions have been studied, the impact on organic chemistry is rather limited. For instance Llor and Cortijo (1977) showed that the tautomerism between (*3.119*) and (*3.120*) is useful as a measure of the solvent polarity. These two tautomeric forms have different absorption maxima (415 and 335 nm respectively), and the relative strength changes depending on the solvent polarity. They confirmed that $\triangle G$ values for the equilibrium are linearly correlated to Kosower's Z values. This model becomes an especially useful strategy for measuring the polarity of the active sites of pyridoxal-dependent enzymes.

Buckley and Rapoport (1982) have demonstrated an interesting synthetic

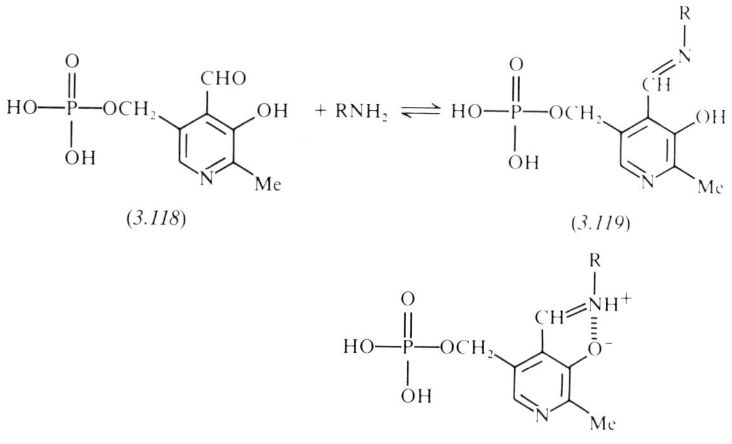

Fig. 3.23

application of a pyridoxal mimic: in the presence of base 4-formyl-1-methylpyridinium ion acts as a convenient reagent for the chemical modification of primary amines to aldehydes and ketones, the yields being relatively good (77–94%). This process is not only interesting from the viewpoint of biomimetic chemistry but it is also important as simple alternative to such transformation procedures.

Breslow *et al.* (1980) attached the pyridoxamine unit to β-cyclodextrin and examined the possibility of substrate specificity in transamination and chiral induction. While pyridoxamine itself exhibits similar reactivity for three α-ketoacids (*3.122*: R = Me, benzyl, CH$_2$-indole), (*3.121*) catalyses the transamination with indolepyruvic acid (*3.122*:R = CH$_2$-indole) about 200 times more efficiently than in the absence of the cyclodextrin (Fig. 3.24). The rate acceleration due to the neighbouring cyclodextrin was not observed for pyruvic acid (*3.122*: R = Me). Clearly, the difference is due to the binding of the indole moiety into the cavity of β-cyclodextrin. Because the cavity of β-cyclodextrin is chiral, one might expect some chiral induction in the product β-amino acids. Breslow's group reported that dinitrophenyl-tryptophan and dinitrophenylalanine have 12% and 52% excess of the L-isomer respectively. Hence, this 'artificial pyridoxal enzyme' shows significant optical induction. If one is prepared to deviate a little more from the prototype pyridoxal structure, further coenzyme analogues capable of asymmetric induction can be obtained.

Kuzuhara *et al.* (1978) synthesized a pyridoxal analogue (*3.123*) involved in an asymmetric cyclophane structure. It was found that the phane derivatives with the ring size equal to or less than fourteen members can be optically resolved into enantiomers (i.e., n = 4,5,6, in (*3.123*)) (Iwata *et al.*, 1976). The stereospecificity in the pyridoxal-like catalysis was tested in racemization of L- and D-glutamic acids with (*3.123*: n = 5, (−)-form) in the presence of cupric ions (Kuzuhara *et al.*, 1977) but the L-isomer was found to racemize only 1.3 times faster than the D-isomer. They later synthesized a cyclic pyridoxamine analogue (*3.124*) having a branched 'ansa chain' in the strap (Ando *et al.*, 1982). (*3.124*) transforms diverse α-ketoacids to the

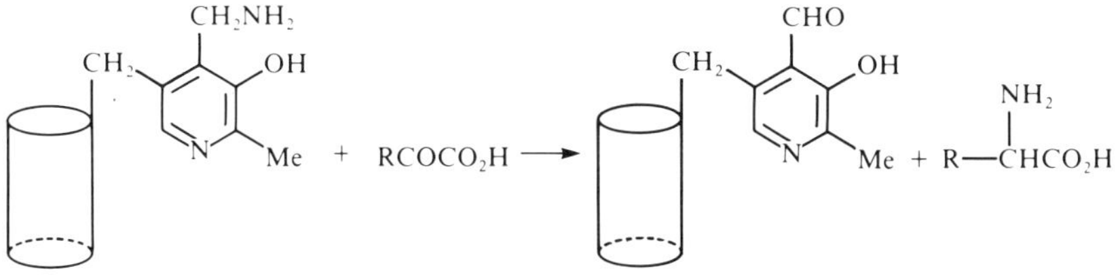

(*3.121*)   (*3.122*)

Fig. 3.24

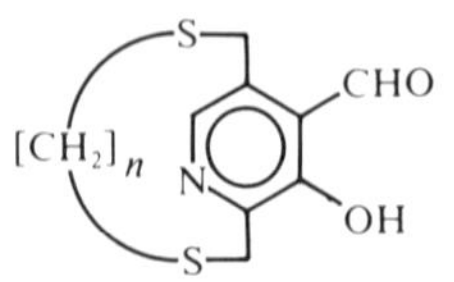

(3.123)

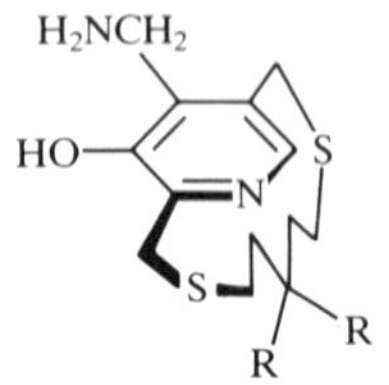

(3.124)

corresponding α-amino acids in 60–96% e.e. (Tachibana *et al.*, 1982). This suggests the importance of the steric effect on the transamination transition state.

Micellar effects on pyridoxal catalysis were examined by using α, β-elimination of *S*-phenylcysteine from the Schiff base (*3.125*) (Murakami and Kondo, 1975) (Fig. 3.25). While anionic and non-ionic micelles showed little or somewhat retarding catalytic effects, the cationic CTAB micelle promoted the elimination by a factor of 7.1. Probably, the catalytic efficiency of the CTAB micelle stems from the enhanced concentration of OH⁻ which abstracts α-hydrogen to induce the α, β-elimination. Similarly, Nakano *et al.* (1981) have found that the pyridoxal-promoted α,β-elimination of serine is facilitated by metal ion plus quaternized poly(4-vinylpyridine). The contribution of the electrostatic force between the polymer and the metal complex of the Schiff base was suggested to account for the rate stimulation. Such elimination reactions have been used as the basis for the design of enzyme inhibitors (Chapter 5).

One of the recent noteworthy developments in the design of enzyme-stimulated systems is the selective introduction of organic functionality in a stereochemically controlled manner and the maintenance of the stereochemical integrity of the functional groups through a synthesis (see also Chapter

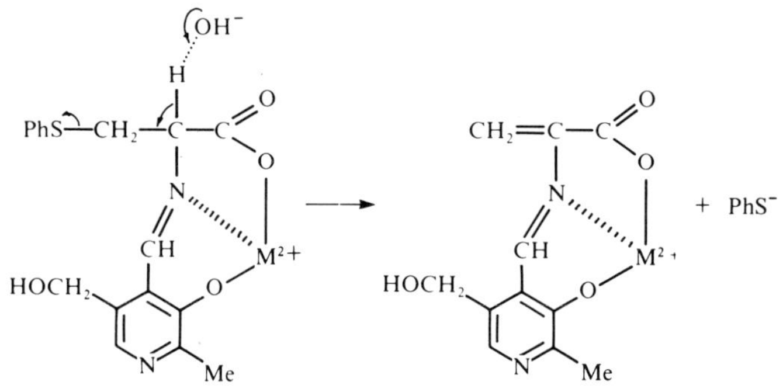

(3.125)

Fig. 3.25

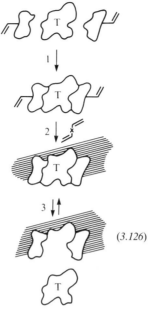

(3.126)

Fig. 3.26

4). By utilizing template synthesis of macromolecules through the insolubil-
ization process, one may give a specific 'memory' of the original template
('T') to the 'ghost' macromolecules (3.126) (Shea et al., 1980; Shinkai, 1982)
(Fig. 3.26). These investigations are still in progress (Nishide et al., 1976;
Wulff et al., 1977; Shea et al., 1980: Damen and Neckers, 1980). The scheme
reminds us of the specific binding of coenzymes to apoenzymes or substrates
to enzymes. Belokon and coworkers (1980) applied this concept to a
pyridoxal system. They prepared a polymeric gel (3.129) in which salicy-
laldehyde and lysine moieties are capable of forming an internal aldimine at
pH > 6.0 by copolymerization of (3.127) with acrylamide and (3.128) with
the subsequent removal of the copper ions by EDTA (Fig. 3.27). They
found that the equilibrium formation constants of the 'internal' aldimine are
ca. 30 times (pH 7.1) and ca. 100 times (pH 9.2) higher than that of the
model reaction. The result suggests that the 'memory' to form the aldimine
is retained in the insolubilized polymer.

Another important pyridoxal-mediated reaction is decarboxylation,
which is, as mentioned previously, highly dependent on reaction media.
One must therefore take the solvation of the carboxylate group into
consideration in order to mimic the decarboxylase catalysis. It has been
reported that pyridoxal-mediated decarboxylation of amino acids occurs

under rather drastic conditions (100°C, 4 h) (Kalyankar and Snell, 1962), but no example of a model reaction under ambient conditions exists.

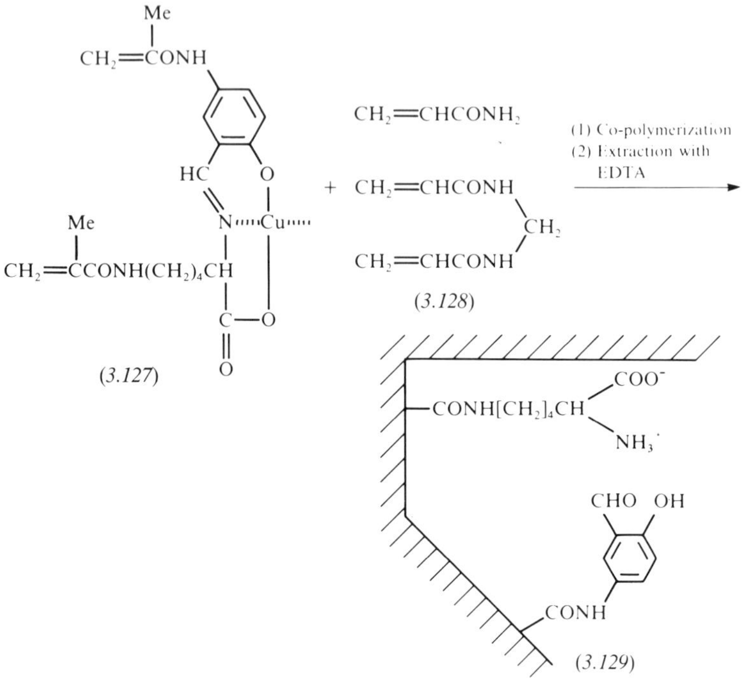

Fig. 3.27

3.6   CATALYSES OF THIOL COENZYMES

Thiol coenzymes which frequently appear in the biological systems are lipoic acid (3.130), glutathione (GSH: 3.131) and coenzyme A (CoASH: 3.132). Although these coenzymes catalyse a broad range of reactions, the catalyses may be reduced to the following classification from the viewpoint of organic chemistry: thiol coenzymes act as (i) strong nucleophiles as well as good-leaving groups, (ii) mediators of oxidoreduction reactions involving group transfers, and (iii) electron-withdrawing groups to facilitate the E1CB-type deprotonation. In this article, we do not refer to (i) (see Kunitake and Shinkai, 1980).

Lipoic acid is widely distributed among micro-organisms and is frequently coupled to TPP, i.e., it receives an acyl unit from (3.105) and transfers it to CoASH (Fig. 3.28). A model reaction has been reported by Rastetter and Adams (1981) (see Section 4.1). Takagi *et al.* (1976) utilized the redox couple between dihydrolipoic acid and lipoic acid to the activation of acyl

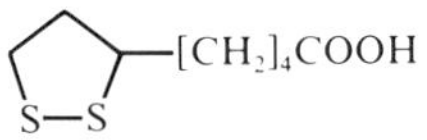

Lipoic acid (3.130)

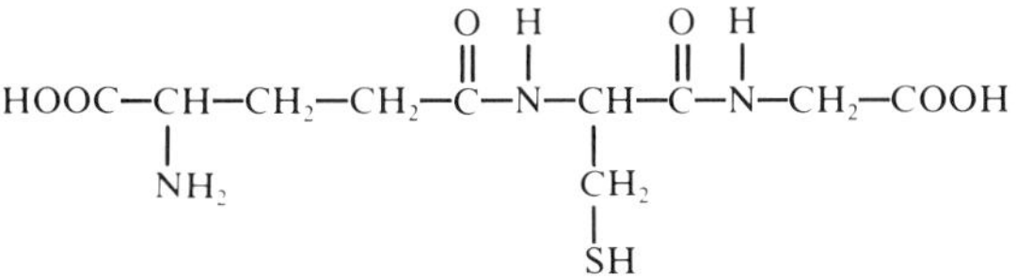

GSH   (3.131)

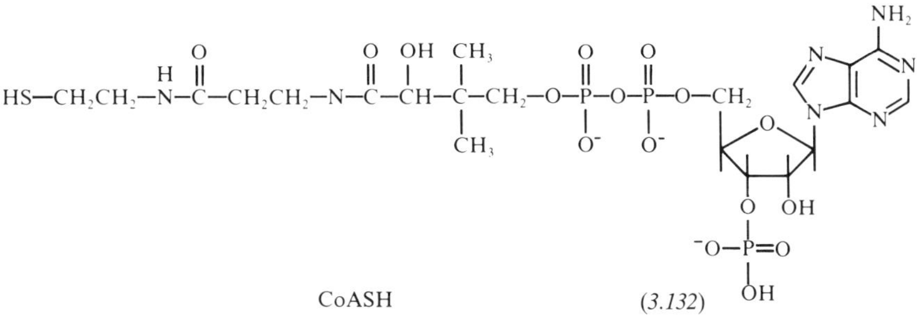

CoASH   (3.132)

groups. Structure (3.133) when oxidized in methanol gives an active acyl intermediate (3.134), which easily succumbs to the nucleophilic attack of solvent methanol (Fig. 3.29). This is an example of oxidative acyl transfer and a facile method for ester synthesis. These workers later found that (3.133) can be obtained from aldehyde (RCHO) and disulphide ($R_1R_1CCH_2SSCH_2$) under photochemical and radical conditions (Takagi et al., 1980) and proposed that the reductive fission of the S–S linkage occurs

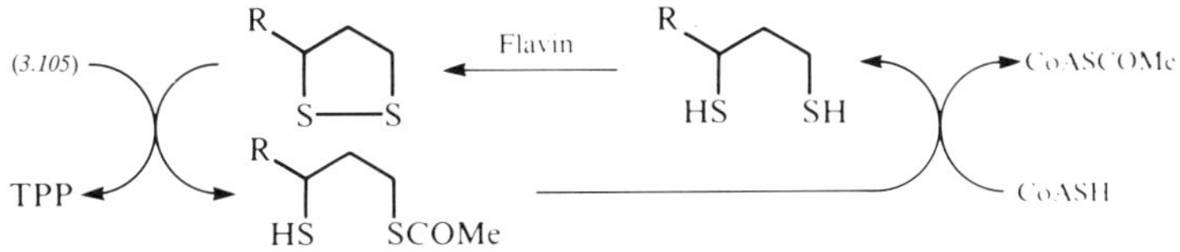

Fig. 3.28

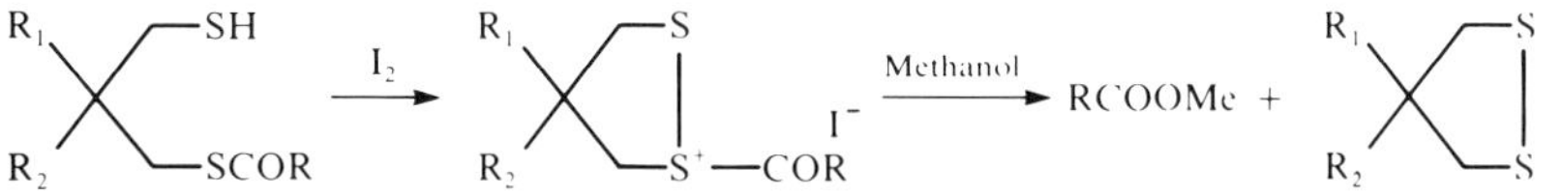

(3.133)   (3.134)

Fig. 3.29

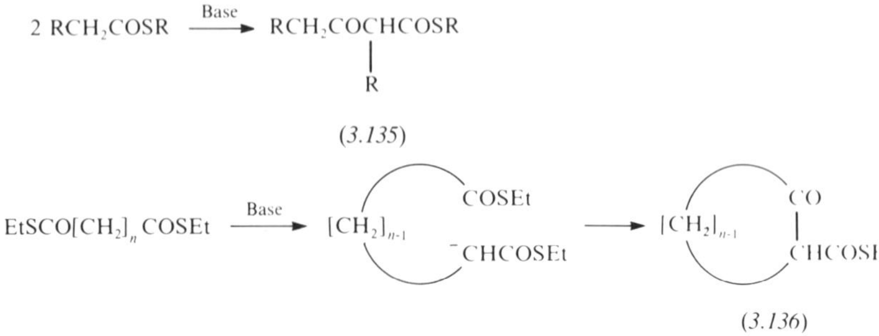

2 RCH2COSR
Base
RCH2COCHCOSR
R
(3.135)
EtSCO[CH2]n COSEt
Base
[CH2]n-1
COSEt
⁻CHCOSEt
[CH2]n-1
CO
CHCOSEt
(3.136)

Fig. 3.30

by means of the photo- or radical-initiated chain reaction. Meanwhile, Nambu *et al.* (1981) demonstrated that an analogue of (*3.133*) is produced from lipoamide and a carboxylic acid in the presence of tri-n-butylphosphine. They also prepared lipoic acid immobilized in cross-linked polystyrene beads (Nambu *et al.*, 1980).

Another interesting feature is the fact that hydrogen attached to α-carbon of thioesters is easily removed as a proton (coenzyme–$SCOCH_2R \xrightarrow{-H^+}$ coenzyme–SC$\bar{O}$HR: Yaggi and Douglas, 1977). This is a pre-equilibrium step for a subsequent condensation. The synthetic utility of thiol ester enolate anions has been explored by Wilson and Hess (1980). For example, chain elongation to (*3.135*) and cyclization to (*3.136*) are possible by the reaction of thioesters and base via the carbanionic intermediates (Fig. 3.30).

O O H Base
RC CSG
H
(3.137)
O OH
RC CHSG
OH O
RCH CSG
O⁻ OH
RC CSG
+
H Base
Flavin
O O
RC CSG
(3.138)

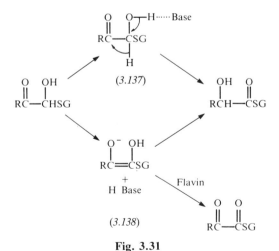

Fig. 3.31

It has been established by Oae *et al.* (1961) that the sulphur atom is able to stabilize the neighbouring carbanion, and this fact became important in the clarification of the reaction mechanism of glyoxalase I. The mechanism by which the glyoxalase I (GSH-dependent enzyme) catalyses the conversion of glyoxals to corresponding α-hydroxy acid esters has been a controversial problem. Racker (1951) and Franzen (1955) proposed the intramolecular 1,2-hydride shift (*3.137*) like the Cannizzaro reaction, but the possibility of the enediol mechanism (*3.138*) has not been excluded (Fig. 3.31). Recently, the enediol mechanism was supported by several independent experiments: (i) a solvent proton is incorporated into the product α-hydroxy acid ester (Hall *et al.*, 1978), (ii) the enediol intermediate (*3.138*) is rapidly trapped by a flavin which does not serve as an efficient oxidant for intermediates including the 1,2-hydride shift mechanism (*3.137*) (Shinkai *et al.*, 1981b) (Fig. 3.31), (iii) fluoromethylglyoxal (*3.139*) is converted to pyruvic acid via the β-elimination path (Kozarich *et al.*, 1981) (Fig. 3.32), and (iv) the substance with the enediol structure acts as a transition state analogue inhibitor (Douglas and Nadvi, 1979; Douglas and Shinkai, 1985) (see Chapter 5). In conclusion, the acidity of hemithiol acetal carbon acids is primarily due to the electron-withdrawing nature of the acyl group and, in addition, is due to the stabilization of carbanions produced by the neighbouring thioether group.

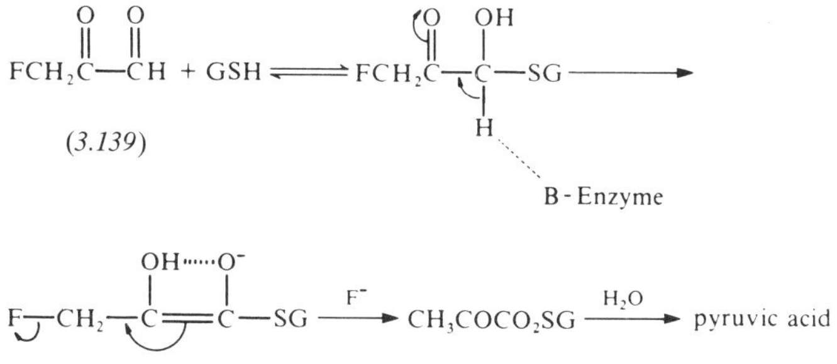

Fig. 3.32

3.7 CONCLUSION

The literature cited in this chapter consistently reveals that the reactivates of coenzymes are controlled by solvent effects, metal ions, micelles, polymers, host molecules, etc. The control of the stereoselectivities, which seems more difficult than that of the reactivities, has also been attained in some model systems. However, it is still difficult to enhance both the reactivity and the stereoselectivity of model systems at the same time, a facility that enzymes possess. Further efforts should be devoted toward this problem. It is also noteworthy that applications of coenzyme chemistry to synthesis are growing space. Nevertheless it is clear that the successes to date are closely related to the skilfully designed coenzyme skeletons, as Jencks (1969) calls pyridoxal phosphate: the best catalyst God created. Finally, space precluded discussion of biotin, folic acid, ascorbic acid, vitamin B_{12}, vitamin K, amongst others. Ascorbic acid has frequently been employed as a reducing agent, but model studies of other coenzymes and vitamins are less extensive than those described here. Haemoproteins, on the other hand, have a vast specialist literature, especially concerning hydroxylating enzymes such as cytochrome P-450. Model studies have played an important role in this field (see Suckling, 1989 for leading references) and applications of the chemisty to biomimetic synthesis have been intensely studied (Chapter 4, section 4.00). To the eyes of organic chemists, the chemical functions of these coenzymes and vitamins are fascinating as targets of biomimetic chemistry. It is to be expected that the further development of excellent coenzyme-originated catalysts will continued.

REFERENCES

Abeles, R. H., Hutton, R. F. and Westheimer, F. H. (1957) *J. Am. Chem. Soc.*, **79**, 712.

Adams, M. J., Buehner, M., Chandrasekhar, K., Ford, G. C., Hockert, M. L., LiLjas, A., Rossman, M. G., Smiley, I. E., Allison, S., Evarse, J., Kaplan, N. O. and Taylor, S. (1973) *Proc. Natl. Acad. Sci. U.S.A.*, **70**, 1968.

Ando, M., Tachibana, Y. and Kuzuhara, H. (1982) *Bull. Chem. Soc. Jpm.*, **55**, 829.

Awano, H. and Tagaki, W. (1985) *Chem. Lett.*, 669.

Ball, S. and Bruice, T. C. (1980) *J. Am. Chem. Soc.*, **102**, 6498.

Bamberg, P. and Hemmerich, P. (1961) *Helv. Chim. Acta*, **44**, 1001.

Behr, J.-P. and Lehn, J.-M. (1978) *J. Chem. Soc., Chem. Commun.*, 143.

Belokon, Y. N., Tararov, V. I., Saveleva, T. F. and Belikov, V. M. (1980) *Makromol Chem.*, **181**, 2183.

Bliese, M., Launikonis, A., Loder, J. W., Mau, A. W. and Sasse, W. H. F. (1983) *Aust. J. Chem.*, **36**, 1873.

Breslow, R. (1958) *J. Am. Chem. Soc.*, **80**, 3719.

Breslow, R., Hammond, M. and Lauer, M. (1980) *J. Am. Chem. Soc.*, **102**, 421.

Brown, L. E. and Hamilton, G. A. (1970) *J. Am. Chem. Soc.*, **92**, 7225.

Bruice, T. C. (1980) *Acc. Chem. Res.*, **13**, 256.

Bruice, T. C. and Benkovic, S. J. (1966) *Bioorganic mechanisms*, Vol. 2, Benjamin, New York, Chapter 9.

Bruice, T. C., Chan, T. W., Taulane, J. P., Yokoe, I., Elliot, D. L., Williams, R. F. and Novak, M. (1977) *J. Am. Chem. Soc.*, **99**, 6713.

Buckley, T. F. and Rapoport, H. (1982) *J. Am. Chem. Soc.*, **104**, 4446.

Bunton, C. A., Minch, M. J., Hidalgo, J. and Sepulveda, L. (1973) *J. Am. Chem. Soc.*, **95**, 3262.

Bunton, C. A., Rivera, F. and Sepulveda, L. (1978) *J. Org. Chem.*, **43**, 1166.

Castells, J., Llitjos, H. and Moreno-Manas, M. (1977) *Tetrahedron Lett.*, 205.

Castells, J., Dunach, E., Geijo, F., Pujol, F. and Segura, P. M. (1978) *Isr. J. Chem.*, **17**, 278.

Cheeseman, P., Toms-Wood, A. and Wolfe, R. S. (1972) *J. Bacteriol.*, **112**, 527.

Christen, P. and Gasser, A. (1980) *Eur. J. Biochem.*, **107**, 73.

Clarke, M. J., Dowling, M. G., Garafalo, A. R. and Brennan, T. F. (1979) *J. Am. Chem. Soc.*, **101**, 223.

Cogoli-Greuter, M., Hausner, U. and Christen, P. (1979) *Eur. J. Biochem.*, **100**, 295.

Cookson, R. C. and Lane, R. M. (1976) *J. Chem. Soc., Chem. Commun.*, 804.

Corbett, M. D. and Chipko, B. R. (1980) *Bioorg. Chem.*, **9**, 273.

Creighton, D. J. and Sigman, D. S. (1971) *J. Am. Chem. Soc.*, **93**, 6314.

Crosby, J., Stone, R. and Lienhard, G. E. (1970) *J. Am. Chem. Soc.*, **92**, 2891.

Daigo, K. and Reed, L. J. (1962) *J. Am. Chem. Soc.*, **84**, 659.

Damen, J. and Neckers, D. C. (1980) *Tetrahedron Lett.*, 1913.

de Kok, P. M. T. and Buck, H. M. (1985) *J. Chem. Soc., Chem. Commun.*, 1009.

de Vries, J. G. and Kellog, R. M. (1979) *J. Am. Chem. Soc.*, **101**, 2759.

Dittmer, D. C. and Fouty, R. A. (1964) *J. Am. Chem. Soc.*, **86**, 91.

Douglas, K. T. and Nadvi, I. N. (1979) *FEBS Lett.*, **106**, 393.

Douglas, K. T. and Shinkai, S. (1985) *Angew. Chem.*, 5.

Duine, J. A. and Frank, J. (1981) *Trends Biochem. Sci.*, **6**, 278.

Dupas, G., Bourguignon, J., Ruffin, C. and Queguiner, G. (1982) *Tetrahedron Lett.*, **23**, 5141.

Edmondson, D. E., Barman, B. and Tollin, G. (1972) *Biochemistry*, **11**, 1133.

Endo, T., Hayashi, Y. and Okawara, M. (1977) *Chem. Lett.*, 392.

Franzen, V. (1955) *Chem. Ber.*, **88**, 1361.

Franzen, V. and Fikentscher, L. (1958) *Liebigs, Ann. Chem.*, **613**, 1.

Fritchie, C. J., Jr (1972) *J. Biol. Chem.*, **247**, 7459.

Frost, J. W. and Rastetter, W. H. (1981) *J. Am. Chem. Soc.*, **103**, 5242.

Fukuzumi, S., Ishikawa, M. and Tanaka, T. (1985) *J. Chem. Soc., Chem. Commun.*, 1069.

Gascoigne, I. M. and Radda, G. K. (1967) *Biochim. Biophys. Acta*, **131**, 498.

Gase, R. A. and Pandit, U. K. (1977) *J. Chem. Soc., Chem. Commun.*, 480.

Gase, R. A., Boxhoon, G. and Pandit, U. K. (1976) *Tetrahedron Lett.*, 2889.

Gibian, M. J. and Baumstark, A. L. (1971) *J. Org. Chem.*, **36**, 1389.

Gibian, M. J. and Rynd, J. A. (1969) *Biochem. Biphys. Res. Commun.*, **34**, 594.

Gibian, M. J. and Winkelman, D. W. (1969) *Tetrahedron Lett.*, 3901.

Gumbley, S. J. and Main, L. (1976) Tetrahedron Lett., 3209.

Hadju, J. and Sigman, D. S. (1975) *J. Am. Chem. Soc.*, **97**, 3524.

Hadju, J. and Sigman, D. S. (1977) *Biochemistry*, **16**, 2841.

Hall, S. S., Doweyko, A. M. and Jordan, F. (1978) *J. Am. Chem. Soc.*, **100**, 5934.

Hastings, J. W., Balny, C., Le Peuch, C. and Douzou, P. (1973) *Proc. Natl. Acad. Sci. U.S.A.*, **70**, 3468.

Heggs, R. P. and Ganem, B. (1979) *J. Am. Chem. Soc.*, **101**, 2484.

Hemmerich, P., Nagelschneider, G. and Veeger, C. (1970) *FEBS Lett.*, **8**, 69.

Hemmerich, P., Massey, V. and Fenner, H. (1977) *FEBS Lett.*, **84**, 5.

Inoue, H. and Higashiura, K. (1980) *J. Chem. Soc., Chem. Commun.*, 549.

Inoue, H., Aoki, R. and Imoto, E. (1974) *Chem. Lett.*, 1157.

Ishitani, O., Pac, C. and Sakurai, H. (1983) *J. Org. Chem.*, **48**, 2941.

Ishitani, O., Ihama, M., Miyauchi, Y. and Pac, C. (1985) *J. Chem. Soc., Perkin Trans. 1*, 1527.

Iwata, M., Kuzuhara, H. and Emoto, S. (1976) *Chem. Lett.*, 983.

Jencks, W. P. (1969) *Catalysis in Chemistry and Enzymology*, McGraw-Hill, New York.
Jones, J. B. and Beck, J. F. (1976) in *Application of Biochemical Systems in Organic Chemistry* (eds J. B. Jones, C. J. Sih and D. Perlman), John Wiley & Sons, New York, pp. 107–401.
Jorns, M. S. (1987) *J. Am. Chem. Soc.*, **109**, 3133.
Kalyankar, G. D. and Snell, E. E. (1962) *Biochemistry*, **1**, 594.
Kemal, C. and Bruice, T. C. (1976a) *J. Am. Chem. Soc.*, **98**, 3955.
Kemal, C. and Bruice, T. C. (1976b) *Proc. Natl. Acad. Sci. U.S.A.*, **73**, 995.
Kemal, C. and Bruice, T. C. (1977) *J. Am. Chem. Soc.*, **99**, 7064.
Kemal, C. and Bruice, T. C. (1979) *J. Am. Chem. Soc.*, **101**, 1635.
Kemal, C., Chan, T. W. and Bruice, T. C. (1977) *J. Am. Chem. Soc.*, **99**, 7272.
Kemp, D. S. and Paul, K. G. (1970) *J. Am. Chem. Soc.*, **92**, 2533.
Kill, R. J. and Widdowson, D. A. (1976) *J. Chem. Soc., Chem. Commun.*, 755.
Kill, R. J. and Widdowson, D. A. (1978) in *Bioorganic Chemistry*, Vol. 4 (ed. E. E. van Tamelen), Academic Press, New York, pp. 239–275.
Kluger, R. and Pike, D. C. (1979) *J. Am. Chem. Soc.*, **101**, 6425.
Kluger, R., Chin, J. and Smyth, T. (1981) *J. Am. Chem. Soc.*, **103**, 884.
Kluger, R., Karimian, K. and Kitamuya, K. (1987) *J. Am. Chem. Soc.*, **109**, 6368.
Knappe, W.-R. (1979) *Liebigs Ann. Chem.*, 1067.
Kojima, M., Toda, F. and Hattori, K. (1981) *J. Am. Chem. Soc., Perkin Trans 1*, 1647.
Kosman, D. J. (1977) in *Bioorganic Chemistry*, Vol. 2 (ed. E. E. van Tamelen), Academic Press, New York, pp. 175–195.
Kozarich, J. W., Chari, R. V. J., Wu, J. C. and Lawrence, T. L. (1981) *J. Am. Chem. Soc.*, **103**, 4953.
Krasna, A. I. (1980) *Photochem. Photobiol.*, **31**, 75.
Kunitake, T. and Shinkai, S. (1980) *Adv. Phys. Org. Chem.*, **17**, 435.
Kuzuhara, H., Iwata, M. and Emoto, S. (1977) *J. Am. Chem. Soc.*, **99**, 4173.
Kuzuhara, H., Komatsu, T. and Emoto, S. (1978) *Tetrahedron Lett.*, 3563.
Llor, J. and Cortijo, M. (1977) *J. Chem. Soc., Perkin Trans 2*, 1111.
Loechler, E. L. and Hollocher, T. C. (1975) *J. Am. Chem. Soc.*, **97**, 3235.
Loechler, E. L. and Hollocher, T. C. (1980) *J. Am. Chem. Soc.*, **102**, 7312, 7322, 7328.
MacInnes, I., Nenhebel, D. C., Orszulik, S. T. and Suckling, C. J (1982) *J. Chem. Soc., Chem. Commun.*, **121**, 1146.
Main, L., Dasperek, G. J. and Bruice, T. C. (1972) *Biochemistry*, **11**, 3991.
Makino, T., Nunozawa, T., Baba, N., Oda, J. and Inoue, Y. (1979) *Tetrahedron Lett.*, 1683.
Massey, V. and Hemmerich, P. (1977) *J. Biol. Chem.*, **252**, 5612.
Massey, V. and Hemmerich, P. (1978) *Biochemistry*, **17**, 9.
Matsumoto, T., Yamanoto, H. and Inoue, S. (1984) *J. Am. Chem. Soc.*, **106**, 4829.
McCapra, F. and Lesson, P. (1976) *J. Chem. Soc., Chem. Commun.*, 1037.
Metzler, D. E., Ikawa, M. and Snell, E. E. (1954) *J. Am. Chem. Soc.*, **76**, 648.
Meyers, A. I. and Oppenlaender, T. (1986) *J. Am. Chem. Soc.*, **108**, 1989.
Miller, A. (1982) *Tetrahedron Lett.*, **23**, 753.
Moras, D., Olsen, K. W., Sabesan, M. N., Buehner, M., Ford, G. C. and Rossman, M. G. (1975) *J. Biol. Chem.*, **250**, 9137.
Murakami, Y. and Kondo, H. (1975) *Bull. Chem. Soc. Jpn*, **48**, 541.
Murakami, Y., Aoyama, Y. and Kikuchi, J. (1981a) *J. Chem. Soc., Chem. Commun.*, 444.

Murakami, Y., Aoyama, Y. and Kikuchi, J. (1981b) *J. Chem. Soc., Perkin Trans 1*, 2809.

Muto, S. and Bruice, T. C. (1980) *J. Am. Chem. Soc.*, **102**, 7559.

Nakano, H., Nishioka, M., Sangen, O. and Yamamoto, Y. (1981) *J. Polym. Sci., Polym. Chem. Ed.*, **19**, 2919.

Nambu, Y., Endo, T. and Okawara, M. (1980) *J. Polym. Sci., Polym, Chem. Ed.*, **18**, 2793.

Nambu, Y., Endo, T. and Okawara, M. (1981) *J. Polym. Sci., Polym. Chem. Ed.*, **19**, 1937.

Nishide, H., Deguchi, J. and Tsuchida, E. (1976) *Chem. Lett.*, 169.

Nishiyama, K., Baba, N., Oda J. and Inoue, Y. (1976) *J. Chem. Soc., Chem. Commun.*, 101.

Oae, S., Tagaki, W. and Ohno, A. (1961) *J. Am. Chem. Soc.*, **83**, 5036.

O'Brien, D. E., Weinstock, L. T. and Cheng, C. C. (1970) *J. Heterocyclic Chem.*, **7**, 99.

Ohnishi, Y. and Kitami, M. (1978) *Tetrahedron Lett.*, 4035.

Ohnishi, Y., Kagami, M. and Ohno, A. (1975a) *Tetrahedron Lett.*, 2437.

Ohnishi, Y., Kagami, M. and Ohno, A. (1975b) *Chem. Lett.*, 125.

Ohnishi, Y., Kagami, M. and Ohno, A. (1975c) *J. Am. Chem. Soc.*, **97**, 4766.

Ohnishi, Y., Kagami, M., Numakunai, T. and Ohno, A. (1976) *Chem. Lett.*, 915.

Ohno, A., Ikeguchi, M., Kimura, T. and Oka, S. (1979) *J. Am. Chem. Soc.*, **101**, 7063.

Ohno, A., Yasui, S., Gase, R. A., Oka, S. and Pandit, U. K. (1980) *Bioorg. Chem.*, **9**, 199.

Ono, N., Tamura, R. and Kaji, A. (1980) *J. Am. Chem. Soc.*, **102**, 2851.

Ono, N., Tamura, R., Tanikaga, R. and Kaji, A. (1981) *J. Chem. Soc., Chem. Commun.*, 71.

Oppenheimer, N. J., Arnold, L. J. and Kaplan, N. O. (1971) *Proc. Natl. Acad. Sci., U.S.A.*, **68**, 3200.

Pac, C., Ihama, M., Yasuda, M., Miyauchi, Y. and Sakurai, H. (1981) *J. Am. Chem. Soc.*, **103**, 5495.

Pandit, U. K. and Mas Cabré, F. R. (1971) *J. Chem. Soc., Chem. Commun.*, 552.

Pattison, S. E. and Dunn, M. F. (1976) *Biochemistry*, **15**, 3691.

Racker, E. (1051) *J. Biol. Chem.*, **190**, 685.

Rastetter, W. H. and Adams, J. (1981) *J. Org. Chem.*, **46**, 1882.

Rastetter, W. H., Gadek, T. R., Tane, J. P. and Frost, J. W. (1979) *J. Am. Chem. Soc.*, **101**, 2228.

Rob, F., van Ramesdonk, H. J., van Gerresheim, W., Bosma, P., Scheele, J. J. and Verhoeven, J. W. (1984) *J. Am. Chem. Soc.*, **106**, 3826.

Rokita, S. E. and Walsh, C. T. (1984) *J. Am. Chem. Soc.*, **106**, 4589.

Rynd, J. A. and Gibian, M. J. (1970) *Biochem. Biophys. Res. Commun.*, **41**, 1097.

Sarma, R. H. and Kaplan, N. O. (1970) *Biochemistry*, **9**, 539.

Seki, M., Baba, N., Oda, J. and Inoue, Y. (1981) *J. Am. Chem. Soc.*, **103**, 4613.

Selbin, J., Sherrill, J. and Bigger, C. H. (1974) *Inorg. Chem.*, **13**, 2544.

Sell, C. S. and Dorman, L. A. (1982) *J. Chem. Soc., Chem. Commun.*, 629.

Shea, K. J., Thompson, E. A., Pandey, S. D. and Beauchamp, P. S. (1980) *J. Am. Chem. Soc.*, **102**, 3149.

Sheehan, J. C. and Hara, T. (1974) *J. Org. Chem.*, **39**, 1196.

Sherry, B. and Abeles, R. H. (1985) *Biochemistry*, **24**, 2594.

Shinkai, S. (1982) *Prog. Polym. Sci.*, **8**, 1.

Shinkai, S. and Bruice, T. C. (1973a) *Biochemistry*, **12**, 1750.

Shinkai, S. and Bruice, T. C. (1973b) *J. Am. Chem. Soc.*, **95**, 7526.
Shinkai, S. and Kunitake, T. (1977) *Chem. Lett.*, 297.
Shinkai, S., Kunitake, T. and Bruice, T. C. (1974) *J. Am. Chem. Soc.*, **96**, 7140.
Shinkai, S., Ando, R. and Kunitake, T. (1975) *Bull. Chem. Soc. Jpn*, **48**, 1914.
Shinkai, S., Shiraishi, S. and Kunitake, T. (1976a) *Bull. Chem. Soc. Jpn*, **49**, 3656.
Shinkai, S., Ando, R. and Kunitake, T. (1976b) *Bull. Chem. Soc. Jpn*, **49**, 3652.
Shinkai, S., Hamada, H., Ide, T. and Manabe, O. (1978a) *Chem. Lett.*, 685.
Shinkai, S., Ide, T. and Manabe, O. (1978b) *Bull. Chem. Soc. Jpn*, **51**, 3655.
Shinkai, S., Yamada, S. and Kunitake, T. (1978c) *J. Polym. Sci., Polym, Lett. Ed.*, **16**, 137.
Shinkai, S., Ando, R. and Kunitake, T. (1978d) *Biopolymers*, **17**, 2757.
Shinkai, S., Yamada, S. and Kunitake, T. (1978e) *Macromolecules*, **11**, 65.
Shinkai, S., Hamada, H., Kusano, Y. and Manabe, O. (1979a) *Bull. Chem. Soc. Jpn*, **52**, 2600.
Shinkai, S., Hamada, H., Kusano, Y. and Manabe, O. (1979b) *J. Chem. Soc., Perkin Trans*, 699.
Shinkai, S., Hamada, H. and Manabe, O. (1979c) *Tetrahedron Lett.*, 1397.
Shinkai, S., Hamada, H., Kusano, Y. and Manabe, O. (1979d) *Tetrahedron Lett.*, 3511.
Shinkai, S., Nakano, T., Hamada, H., Kusano, Y., and Manabe, O. (1979e) *Chem. Lett.*, 229.
Shinkai, S., Mori, K., Kusano, Y. and Manabe, O. (1979f) *Bull. Chem. Soc. Jpn*, **52**, 3606.
Shinkai, S., Hamada, H., Dohyama, A. and Manabe, O. (1980a) *Tetrahedron Lett.*, 1661.
Shinkai, S., Kusano, Y., Manabe, O. and Yoneda, F. (1980b) *J. Chem. Soc., Perkin Trans 2*, 1111.
Shinkai, S., Yamashita, T., Kusano, Y., Ide, T. and Manabe, O. (1980c) *J. Am. Chem. Soc.*, **102**, 2335.
Shinkai, S., Yamashita, T., Kusano, Y. and Manabe, O. (1980d) *Tetrahedron Lett.*, **21**, 2543.
Shinkai, S., Yamashita, T., Kusano, Y. and Manabe, O. (1980e) *J. Org. Chem.*, **45**, 4947.
Shinkai, S., Hamada, H., Kuroda, H. and Manabe, O. (1981a) *J. Org. Chem.*, **46**, 2333.
Shinkai, S., Yamashita, T., Kusano, Y. and Manabe, O. (1981b) *J. Am. Chem. Soc.*, **103**, 2070.
Shinkai, S., Ishikawa, Y. and Manabe, O. (1982a) *Chem. Lett.*, 809.
Shinkai, S., Yamashita, T., Kusano, Y. and Manabe, O. (1982b) *J. Am. Chem. Soc.*, **104**, 563.
Shinkai, S., Hara, Y. and Manabe, O. (1982c) *J. Polym. Sci., Polym. Chem. Ed.*, **20**, 1097.
Shinkai, S., Ishikawa, Y. and Manabe, O. (1983) *Bull. Chem. Soc. Jpn*, **56**, 1964.
Shinkai, S., Era, H., Tsuno, T. and Manabe, O. (1984a) *Bull. Chem. Soc. Jpn.*, **57**, 1435.
Shinkai, S., Nakao, H., Tsuno, T., Manabe, O. and Ohno, A. (1984b) *J. Chem. Soc., Chem. Commun.*, 849.
Shinkai, S., Nakao, H., Honda, N., Manabe, O. and Müller, F. (1986a) *J. Chem. Soc., Perkin Trans. 1*, 1825.
Shinkai, S., Yamaguchi, T., Nakao, H. and Manabe, O. (1986b) *Tetrahedron Lett.*, **27**, 1611.

Shinkai, S., Yamaguchi, T., Kawase, A., Kitamura, A. and Manabe, O. (1987) *J. Chem. Soc., Chem. Commun.*, 1506.

Shinkai, S., Nakao, H., Kuwahara, I., Miyamoto, M., Yamaguchi, T. and Manabe, O. (1988a) *J. Chem. Soc., Perkin Trans. 1*, 313.

Shinkai, S., Yamaguchi, T., Manabe, O. and Toda, F. (1988b) *J. Chem. Soc., Chem. Commun.*, 1399.

Shirra, A. and Suckling, C. J. (1977) *J. Chem. Soc., Perkin Trans, 2*, 759.

Sigman, D. S., Hadju, J. and Creighton, D. J. (1978) in *Bioorganic Chemistry*, Vol. 4 (ed. E. E. van Tamelen) Academic Press, New York, pp. 385–407.

Singh, S., Trehan, A. K. and Sharma, V. K. (1978) *Tetrahedron Lett.*, 5029.

Singh, S., Chhina, S., Sharma, V. K. and Sachdev, S. S. (1982) *J. Chem. Soc., Chem. Commun.*, 453.

Singh, S., Sharma, V. K., Gill, S. and Sahota, R. I. K. (1985) *J. Chem. Soc., Perkin Trans. 1*, 437.

Singh, S., Nagrath, S. and Chanana, M. (1986) *J. Chem. Soc., Chem Commun.*, 282.

Spencer, R., Fisher, J. and Walsh, C. (1977) *Biochemistry*, **16**, 3586.

Spetnagel, W. J. and Klotz, I. M. (1978) *Biopolymers*, **17**, 1657.

Steffens, J. J. and Chipman, D. M. (1971) *J. Am. Chem. Soc.*, **93**, 6694.

Stetter, H. and Kuhlmann (1976) *Chem. Ber.*, **109**, 2890.

Stetter, H., Basse, W. and Nienhaus, J. (1980) *Chem. Ber.*, 113, 690.

Suckling, C. J. (1988) *Angew. Chem. Int. Edn., Engl.*, **27**, 537.

Tachibana, Y., Ando, M. and Kuzuhara, H. (1982) *Chem. Lett.*, 1765.

Tagaki, W. and Hara, H. (1973) *J. Chem. Soc., Chem. Commun.*, 891.

Tagaki, W., Tamura, T. and Yano, Y. (1980) *Bull. Chem. Soc. Jpn*, **53**, 478.

Takagi, M., Goto, S, Isihara, R. and Matsuda, T. (1976) *J. Chem. Soc., Chem. Commun.*, 993.

Takagi, M., Goto, S., Tazaki, M. and Matsuda, T. (1980) *Bull. Chem. Soc. Jpn*, **53**, 1982.

Talma, A. G., Jouin, P., De Vries, J. G., Troostwijk, C. B., Buning, G. H. W., Waninge, J. K., Visscher, J. and Kellog, R. M. (1985) *J. Am. Chem. Soc.*, **107**, 3981.

Tanaka, K., Okada, T. and Yoneda, F. (1984) *Tetrahedron Lett.*, **25**, 1941.

Tintillier, P., Dupas, G., Bourguignon, J. and Queguiner, G. (1986) *Tetrahedron Lett.*, **27**, 2357.

van Bergen, T. J. and Kellog, R. M. (1977) *J. Am. Chem. Soc.*, **99**, 3882.

van Eikeren, P. and Grier, D. L. (1976) *J. Am. Chem. Soc.*, **98**, 4655.

van Ramesdonk, H. J., Verhoeven, J. W., Pandit, U. K. and de Boer, Th. J. (1978) *Rec. Trav. Chim. Pays-Bas.*, **97**, 195.

Wallenfels, K. and Hanstein, W. (1965) *Angew. Chem. Int. Ed. Engl.*, 869.

Wallenfels, K., Ertel, W. and Friedrich, K. (1973) *Leibigs Ann. Chem.*, 1663.

Walsh, C. (1980) *Acc. Chem. Res.*, **13**, 148.

Walsh, C., Schonbrunn, A. and Abeles, R. H. (1971) *J. Biol. Chem.*, **246**, 6855.

Walsh, C., Fisher, J., Spencer, R., Graham, D. W., Ashton, W. T., Brown, J. E., Brown, R. D. and Rogers, E. F. (1978) *Biochemistry*, **17**, 1942.

White, F. G. and Ingraham, L. L. (1962) *J. Am. Chem. Soc.*, **84**, 3109.

Williams, R. F. and Bruice, T. C. (1976) *J. Am. Chem. Soc.*, **98**, 7752.

Wilson, G. E., Jr and Hess, A. (1980) *J. Org. Chem.*, **45**, 2766.

Wulff, G., Vesper, W., Grode-Einsler, R. and Sarhan, A. (1977) *Makromol. Chem.*, **178**, 2799.

Yaggi, N. F. and Douglas, K. T. (1977) *J. Am. Chem. Soc.*, **99**, 4844.

Yano, Y., Hoshino, Y. and Tagaki, W. (1980) *Chem. Lett.*, 749.

Yano, Y., Ohshima, M., Sutch, S. and Nakazato, M. (1984) *J. Chem. Soc., Chem. Commun.*, 1031.

Yokoe, I. and Bruice, T. C. (1975) *J. Am. Chem. Soc.*, **97**, 450.

Yoneda, F. and Nakagawa, K. (1980) *J. Chem. Soc., Chem. Commun.*, 878.

Yoneda, F., Sakuma, Y. and Hemmerich, P. (1977) *J. Chem. Soc., Chem. Commun.*, 825.

Yoneda, F., Kawazoe, M. and Sakuma, Y. (1978a) *Tetrahedron Lett.*, 2803.

Yoneda, F., Sakuma, Y. and Nitta, Y. (1978b) *Chem. Lett.*, 1177.

Yoneda, F., Hirayama, R. and Yamashita, M. (1980a) *Chem. Lett.*, 1157.

Yoneda, F., Ono, M., Kira, K., Tanaka, H., Sakuma, Y. and Koshiro, A. (1980b) *Chem Lett.*, 817.

Yoneda, F., Yamato, H. and Ono, M. (1981) *J. Am. Chem. Soc.*, **103**, 5943.

4 | Selectivity in synthesis – chemicals or enzymes?

Colin J. Suckling

4.1 INTRODUCTION

A characteristic feature of recent developments in organic synthesis is an emphasis upon selectivity. In chemical reactions, selectivity takes many forms. At the simplest level, the chemist is concerned with ensuring that, as far as possible, only the desired reaction takes place (chemoselectivity), and one of the first areas to reach a high state of development in this context was peptide synthesis. The selective synthesis of peptides has been possible because protecting or blocking groups were devised to permit the sequential construction of the required peptide. Today, an experienced peptide chemist can make near-optimal choices of protecting groups. Further developments of protecting-group chemistry have led to efficient procedures for the specific synthesis of oligo-nucleotides.

At the next level of control, the chemist wishes to determine at which of two or more positions of similar reactivity a reaction will occur (regioselectivity). There are venerable chemical strategies for controlling regioselectivity. For instance, it has usually been found that the less reactive a reagent, the more selective its action; also, it is sometimes possible to control a reaction by choosing the conditions such that either the kinetically or the thermodynamically favoured product predominates. An example of the former is the greater selectivity of bromine atoms in alkane halogenation compared with the more reactive chlorine, and the latter may be exemplified by the regiospecific formation of ketone enolates according to the reaction conditions. Perhaps the most subtle control required of the organic chemist is the control of stereochemistry and in particular the formation at will of only one optical isomer (stereoselectivity). In recent years, the control of stereochemistry has taken pride of place in most synthetic investigations. As we shall see, enzyme chemistry can contribute greatly both directly and indirectly to control in synthesis. For convenience in the discussion, I shall refer to the use of enzymes and micro-organisms in synthesis as biochemical methods and conventional reactions involving neither as chemical methods.

There are several reasons for the modern chemist's pre-occupation with selectivity. He has the academic aim of controlling the regio- and stereoselectivity of a synthetic reaction as completely as possible; this endeavour has its origin not only in striving for elegance in the execution of a synthesis, but also in practical reasons. On the one hand, many naturally occurring compounds, which are common synthetic targets, are biologically active in only one stereoisomeric form and on the other hand, the economic pressure in industry to avoid wasteful and possibly harmful by-products daily demands more attention from the chemist.

The classic case of failure to take account of chirality concerned the drug thalidomide, a mild analgesic prescribed for pregnant women in the 1960s (Fig. 4.1(a)). This relatively simple compound contains only one chiral centre yet the chirality proved crucial to the biological effects of the drug. Since the working environment of drugs in nature is chiral, enantiomers are not identical chemical entities. With respect to their receptor complexes they have a diastereoisomeric relationship. In the case of thalidomide, no heed was taken of the potential consequences of this relationship and the drug was manufactured and sold as a racemic mixture. Tragically, women who had taken the drug gave birth to deformed offspring. Further research showed that only the S isomer was teratogenic and the R isomer was safe.

The skills of chemists to prepare enantiomerically pure compounds in quantity are now immensely greater than those of the 1960s and it is reasonable to expect that new drugs and agrochemicals should be sold enantiomerically pure. As more examples emerge, it becomes clearer that it is the rule to expect a difference in biological activity between enantiomers and not the exception. Indeed it has been argued that the inclusion of additional chiral centres will lead to increased selectivity of action of drugs. Regulatory authorities are beginning to require that compounds be sold optically pure; in Switzerland and Sweden, racemic phenoxypropanoate herbicides (Fig. 4.1(b)) will be banned in 1990 and the countries of the European Community are expected to follow with controls. These compounds are manufactured and sold on a scale of more than 20 000 t.p.a. and on this scale, very efficient chemistry is essential to produce pure enantiomers. One effective route is to employ enzymic catalysis. A further two

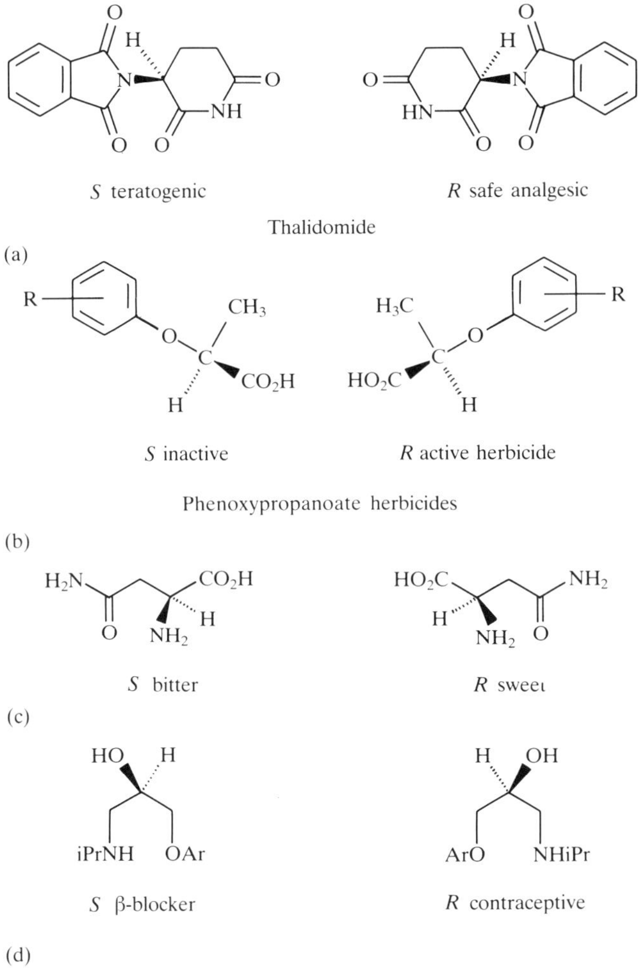

S teratogenic   R safe analgesic

Thalidomide

(a)

S inactive   R active herbicide

Phenoxypropanoate herbicides

(b)

S bitter   R sweet

(c)

S β-blocker   R contraceptive

(d)

Fig. 4.1 Enantiomers and biological activity.

examples of enantiomers with very different properties are shown in Fig. 4.1(c) and (d).

The word 'selectivity' features in the titles of many current papers and Norman (1982) had identified several strategies open to the chemist to control selectivity. He considered established chemical devices such as neighbouring-group effects and the reactivity controlling methods mentioned above but added for the first time in such a broad discussion the challenge of enzymic selectivity by including some biomimetic examples. However, with the exception of polynucleotide synthesis, chemists have been slow to make use of enzymes. It is probably not generally realized that the number of synthetic reactions to which enzyme chemistry can be applied is very large. Perhaps it is difficult to see how the use of an enzyme might aid a synthesis or at what stage in a synthesis enzyme chemistry can be of most value. This chapter explores these avenues by comparing the current ability of selected chemical reagents to control synthetic transformations with the properties of closely analogous enzyme-catalysed reactions. With an eye to the future, the achievements and potential of biomimetic systems and of new biologically based strategies in synthesis will also be discussed. Recently, a topical series of four articles directly relevant to this discussion has appeared from outstanding scientists at the Oxford school (Davies, 1989; Brown, 1989; Pratt, 1989; Fleet, 1989).

Since the first edition of this book there has been an explosive growth not only in reports of applications of enzymes in organic synthesis but also of new opportunities for the creative discovery of selective biological catalysts. On the one hand, the former has resulted largely from the ingenuity of chemists in capitalizing upon the opportunities presented by enzymes. Of particular note is the introduction of organic solvents as reaction media (see section 4.7). On the other hand, the latter has been made possible by innovations of technology in biological sciences, namely cloning and gene manipulation in molecular and cell biology, and monoclonal antibody production in animals. The biological background to these fields is presented in Chapter 9 (see section 4.8). Although the technology is essentially biological, much of it was made possible by skilful organic chemistry. Moreover the potential of the new technologies for the chemical industry cannot be underestimated. For the the first time it becomes realistic for a chemist seriously to consider designing a protein-based catalyst. This can be approached either by genetic engineering or monoclonal antibody techniques. It is even possible to modify biosynthetic pathways in cells to produce improved quantities of desirable natural products. Examples of these new opportunities will be found later in this chapter.

Until the early 1980s, comparatively little attention had been paid to the harnessing of enzymes in general organic synthesis. An illustration of the prevailing view is provided by a few lines from Fleming's valuable teaching book *Selected Organic Synthesis* (1973).

'Actually, this . . . synthesis, using a microbiological oxidation, is not a total synthesis, even in principle; the use of micro-organisms which have not themselves been synthesised disqualifies the route as a total synthesis in the usually accepted academic sense.'

And later

'They (enzymes) have the inestimable advantage of being able to . . .

carry out highly specific reactions, whether in a polyfunctional molecule or in a completely unfunctionalised one. . . . They are not convenient laboratory reagents, although some enzymes, such as ribonuclease, which have rather predictable properties certainly are.'

These lines are suggestive of a rigorous, but now dated, academic viewpoint and they highlight accurately the situation in the early 1970s. Interestingly, only two examples of biochemical synthetic steps appear in the book and both refer to industrial syntheses. Fleming's comment 'From the point of view of industrial practice, this was simply the best way to do the reaction' is surely sound advice to any modern synthetic chemist interested in reaching a target. Nevertheless, even fifteen years ago the methodology for using enzymes and micro-organisms was not sufficiently well advanced to attract the attention of synthetic chemists at large. The potential was clearly there, but so were the problems (Suckling and Suckling, 1974). Principal amongst these were the stability of enzymes in use and, most significantly, the recycling of coenzymes. The nature of coenzymes as essentially enzyme-bound reagents will have become obvious from the preceding chapter. Such great strides have been made in solving these problems (Chenault *et al.*, 1988) that it is now possible to compare enzymes and conventional procedures on equal terms. However, the third arm of this chapter's comparison, biomimetic chemistry, is currently as enzymes were fifteen years ago, promising but problematical. On the other hand the importance of the new opportunities opened by biological technologies mentioned above will also become apparent.

It is important to distinguish between the use of purified enzyme preparations, which contain essentially only one catalyic activity, and microbiological fermentations, which contain many. Obviously a synthetic chemist requires predictability in his reactions, and a microbiological method will not usually be attractive although, as we shall see, it can be most effective. On the other hand, many enzyme-catalysed reactions are today highly predictable and have earned their place in the synthetic organic chemist's repertoire. Regrettably, most standard reference works to organic synthesis omit enzymes. Fieser's *Reagents* series lists none, but Theilheimer is a useful source. An extensive compilation of relevant early information is in Weissberger's *Techniques of Organic Chemistry* Series (Jones *et al.*, 1976). In view of this situation, it might pay one of the enzyme suppliers to promote a commercial catalogue of synthetically useful enzymes. Their sales slogan might be 'An accessible source of controlled selectivity'. Several reviews chronicle the development of synthetic applications of enzymes (Whitesides and Wong, 1985; Jones, 1986; Yamada and Shimizu, 1988).

4.2 PROBLEMS OVERCOME

To capitalize upon the selective advantages of enzymes, it was essential that practical problems including coenzyme recycling, enzyme stability, and compatibility with organic substrates should be solved. These three topics illustrate well the great strides made in recent years.

4.2.1 Coenzyme recycling

It has often been argued that the carbonyl group is the most important functionality in synthesis. The ability to reduce aldehydes and ketones to alcohols with control of chirality of the product is obviously significant. The conventional chemical solution is to use bulky groups either in the substrate or the reagent to hinder approach from one of the two enantiotopic faces of the carbonyl group. An enzyme has a structure that will control the direction of attack. However, such redox systems (dehydrogenases) require the assistance of the coenzyme NAD in its reduced form, NADH, and in stoichiometric quantities (Fig. 4.2). This coenzyme is much more costly (£2000 per mole) than the most expensive competitive selective chemical reducing agent. It is therefore essential that they be recycled if the enzyme-catalysed reduction of carbonyl groups is to be practical in synthesis. The first extensively used method employed *coupled substrates*

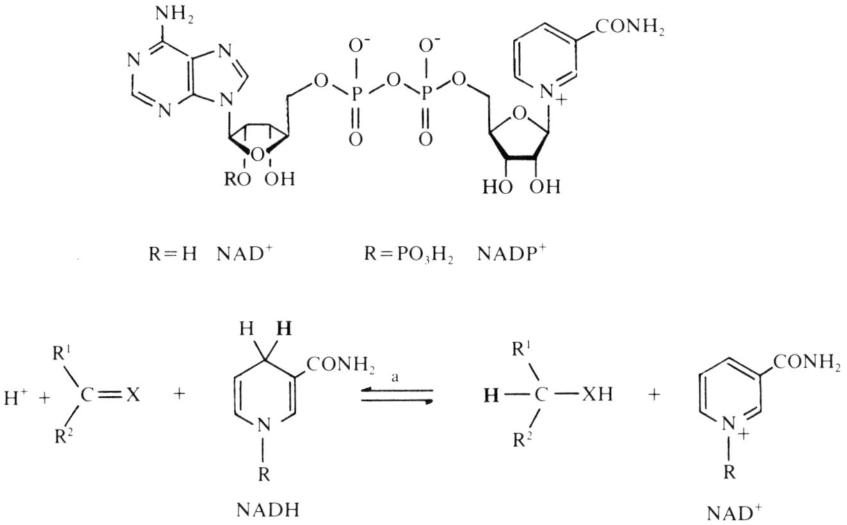

Fig. 4.2 Nicotinamide–adenine dinucleotide and some reactions mediated by dehydrogenases, e.g. a = horse liver alcohol dehydrogenase (HLADH), R^1 = alkyl, R^2 = H, X = O; lactate deyhdrogenase (LDH), R^1 = CH_3, R^2 = CO_2H, X = O.

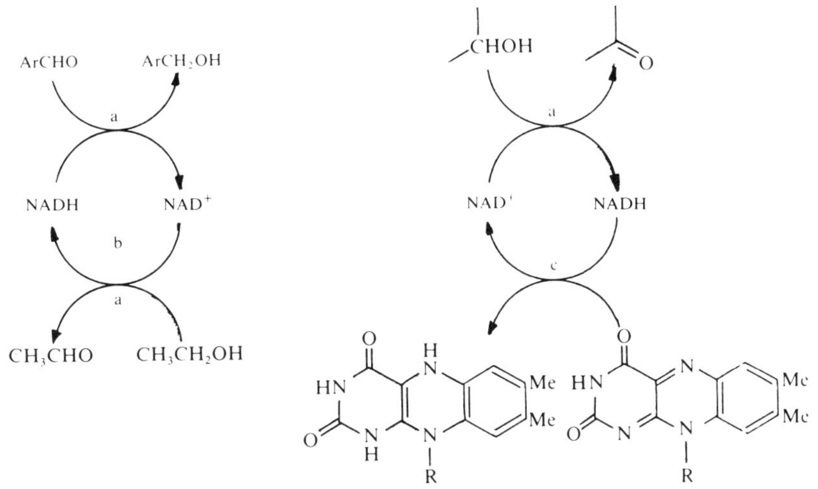

Fig. 4.3 Some recycling methods for NAD⁺/NADH. a, HLADH; b, Na₂S₂O₄; c, chemical oxidation using flavin mononucleotide.

(e.g. Battersby *et al.*, 1975) and illustrates elegantly a further ramification of enzymes in synthesis, reversible catalysis (Fig. 4.3). In this reduction of an aldehyde, an excess of an alcohol was used: the oxidized NAD would then act with the enzyme to oxidize the cosubstrate alcohol and thereby regenerate the required reducing agent, NADH. This method had been widely applied in the synthesis of chirally labelled benzylic alcohols. Shortly afterwards, Jones and Beck (1976) and Jones and Taylor (1976) showed that coenzyme recycling could also be accomplished using added chemical oxidizing or reducing agents. If NAD(P)H was required for a reduction, sodium dithionite served to regenerate the coenzyme and, if an oxidation was needed, flavin mononucleotide (FMN) was effective in recycling NAD(P).

However, none of these methods was applied on quantities larger than a few grams of substrate, and it was left to Whitesides to give a virtuoso exposition of recycling methods for NAD(P)H on up to a mole scale. The success of Whitesides' work is based upon the fact that it is possible to use a mixture of several enzymes, each with different catalytic activities, in the same reaction flask without one interfering with the other. Chemical reagents are scarcely ever as compatible. Using a *coupled enzyme*, NAD(P)⁺ can be reduced to NAD(P)H with concomitant oxidation of a substrate for the second enzyme. Experience has shown that the best enzyme for relaying the reduction of NAD(P)⁺ is glucose 6-phosphate dehydrogenase; thus glucose 6-phosphate becomes the terminal reductant

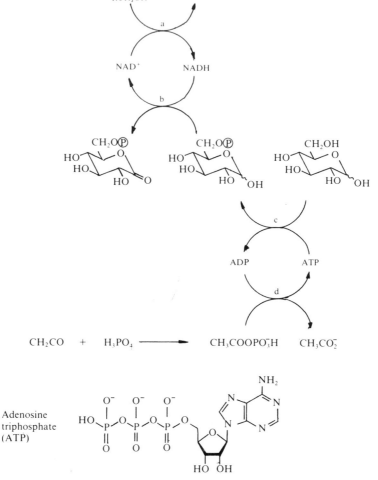

Fig. 4.4 ATP and NAD⁺ recycling coupled together. a, HLADH; b, glucose 6-phosphate dehydrogenase; c, hexokinase; d, acetate kinase.

or fuel for the reduction (Fig. 4.4). However, since glucose 6-phosphate is not cheap, it would be advantageous if it too could be regenerated enzymically. This can be achieved using hexokinase, the enzyme that catalyses the phosphorylation of glucose by ATP. The enzyme chain has now relayed the task of driving the whole reaction sequence to ATP which can fortunately be produced cheaply either chemically or enzymically (Lewis *et al.*, 1979).

Such chains of coupled enzyme catalysed reactions are efficient when used on up to a few moles of reactants but on larger scales still, the number and

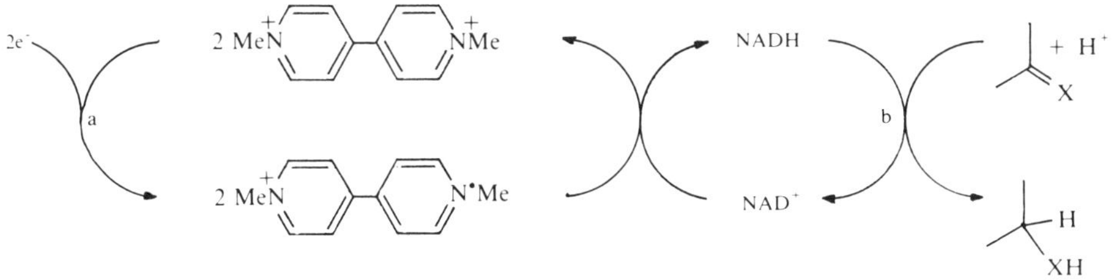

Fig. 4.5 Electrochemical recycling of NADH. a, Reduction of Methyl Viologen at a cathode (W or Hg); b, an enzymic reduction, e.g. HLADH or enoate reductase.

quantity of co-operating enzymes and relay substrates become prohibitively large. It then becomes appropriate to use electrochemical methods or hydrogen gas as the terminal reductant. In the former case, a coupled system is still used in which Methyl Viologen acts as an electron carrier from the cathode to the enzyme that reduces $NAD(P)^+$ (Fig. 4.5; DiCosimo et al., 1981). Such procedures are not restricted to purified enzymes as Simon and his colleague have shown. Reductions by micro-organisms can also be driven by electrochemical energy (Simon et al., 1985; Thanos and Simon, 1986). When hydrogen is the terminal reductant, the relay component is an enzyme, a hydrogenase, that reduces NAD^+ to NADH.

Significant as the virtuosic expositions of cofactor recycling are, a major stride was still required to demonstrate the practicality of dehydrogenases in industrial scale applications. As described below, great improvements in enzyme behaviour on a large scale can often be achieved by immobilization. Because the whole of the cofactor molecule is not involved in the chemical transformation and the non-reactive part binds close to the surface of the enzyme, it is possible to prepare an immobilized NAD derivative with high coenzyme activity. This was achieved using polyethylene glycol (PEG 20 000) by coupling to an adenine-modified NAD (Fig. 4.6). The derivatized NAD was thus converted into a water-insoluble macromolecule that could be retained on one side of a membrane together with the recycling enzymes whilst the substrate and product can diffuse freely across. Under optimum conditions, 80 000 molecules of product can be processed per molecule of NAD using formate dehydrogenase as regeneration enzyme. In this way, the cofactor costs for aminoacid production by ammonia:ketoacid dehydrogenases have been reduced to less than 50 cents per kg. (Wandrey and Wichmann, 1985; Wandrey et al., 1984).

At this stage in the discussion, it is worth making the first direct comparison between a conventional chemical reagent and an enzyme-mediated synthetic step. The reduction of deuteriated benzaldehydes has been particularly well documented for both approaches. Midland et al.

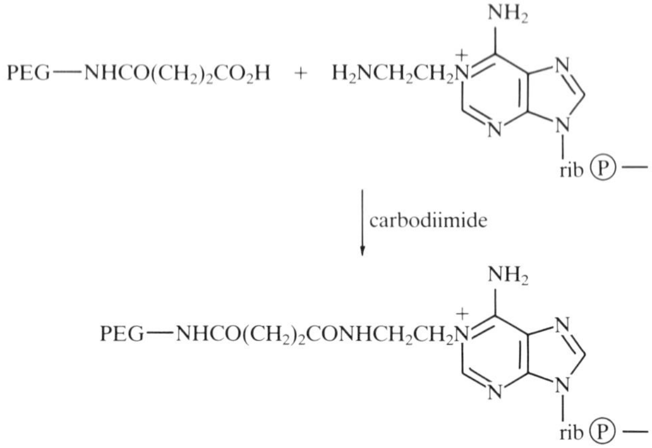

PEG—NHCO(CH$_2$)$_2$CO$_2$H   +   H$_2$NCH$_2$CH$_2$N

NH$_2$

rib Ⓟ —

| carbodiimide

PEG—NHCO(CH$_2$)$_2$CONHCH$_2$CH$_2$N

NH$_2$

rib Ⓟ —

Fig. 4.6 NAD linked to a synthetic macromolecule. PEG = polyethylene glycol, mol. wt. not less than 20 000.

(1979) described the synthesis of chiral deuteriobenzyl alcohols using chiral boranes derived from pinene. His group achieved an enantiomeric excess of greater than 90% for the S-isomer on a small laboratory scale. The analogous enzymic reaction was catalysed by horse liver alcohol dehydrogenase (HLADH) and was carried out on 0.4 mol (Wong and Whitesides, 1981). A product with 95% enantiometric excess was obtained. Significantly, the same technology using lactate dehydrogenase was used to prepare D-lactate in similar optical purity. It cannot be doubted from these examples that both chemical and enzymic techniques are practical. In Fleming's terms, the purist might discount the enzymic method from the point of view of total synthesis. However, since the source of chirality in the chemical procedure is pinene, a natural product of enzymic asymmetric induction in a tree, the only distinction between the two methods is the location at which an enzyme acts. Objections on philosophical grounds to the use of enzymes in synthesis therefore appear to be weakly founded. It's not cheating to use an enzyme in synthesis; if it does the job, it's useful, if it doesn't, try something else.

4.2.2 Enzyme stability

In his preparative reductions just described, Whitesides made extensive use of *immobilized enzymes*. Just as immobilized chemical reagents offer advantages in each stage of work-up, so immobilized enzymes permit the rapid separation of the expensive catalyst from the solution of reactants and products (Chibata, 1978; Suckling, 1977). This is valuable but more importantly, it has been generally found that immobilization enhances the stability of the enzyme. In this way it is possible to solve the second practical problem of enzymes in synthesis. Of the many methods available for immobilizing an enzyme, a specially synthesized polymer of acrylamide and the N-hydroxysuccinimide ester of acrylic acid has proved widely applicable (Pollak *et al.*, 1977, 1978). The risk of damaging the enzyme's active site through attack by the immobilizing reagents can be minimized by carrying out immobilization in the presence of the substrate or an inhibitor, either of which will bind at the active site and protect it from attack. In this way, a remarkably large number of enzymes has been immobilized, many in quantities sufficient for chemical synthesis on a mole scale. In industrial applications, the ability to obtain an active, stable enzyme preparation makes the difference between viability and otherwise of a process. There are several clear examples of this including the use of columns of immobilized fungal cells to catalyse specifically 11-β-hydroxylation of steroids and a subsequent 1,2-dehydrogenation (Mosbach and Larsson, 1970; Chibata, 1978) leading to corticosteroids: these reactions are related to those described by Fleming (1973). Notably, the microbial transformations using immobilized catalyst carry out firstly a regio- and stereoselective hydroxylation at a non-activated carbon atom and secondly a specific dehydrogenation. Both of these transformations are impossible using conventional chemical reagents. Later in this chapter (section 4.6), however, we shall see how biomimetic experiments can approach such problems.

A further aspect of enzyme stability is the sensitivity of many enzymes to oxygen. Usually such enzymes contain important thiol groups that are readily oxidized by air to catalytically inactive disulphides. To prevent oxidative inactivation, the enzyme-catalysed reaction can be carried out under an inert atmosphere, preferably argon, in the presence of an added thiol such as dithiothreitol or mercaptoethanol; these thiols are oxidized by adventitious oxygen in place of the enzyme's sensitive functional group. With characteristic thoroughness, Whitesides' group has also studied this problem (Szajewski and Whitesides, 1980). The experimental technique is no more difficult than handling modern air-sensitive reagents. It is now no longer the case that enzymes maximally active at 40°C only are available for synthesis. Enzymes isolated from organisms known as *thermophiles* that live at temperatures from 40°C up to 120°C, for example in hot springs in New Zealand, have significantly increased stability over those that live at normal amibient conditions. The ability to work at higher temperatures does not lead to a great increase in overall rate of reaction because thermophilic enzymes are less catalytically active, a price that has to be paid for the greater stability of the enzyme which may be related in turn to a more rigid protein structure. Examples of the application of thermophilic organisms can be found below (section 4.3) and in chapter 9 (thermophilic glucose oxidase). A further extension to the general applicability of enzymes in organic synthesis is shown by the ability of enzymes to tolerate organic solvents. An old conventional wisdom held that enzymes and organic solvents are incompatible but there is now ample data to show that this is not the case (Zaks and Klibanov, 1985, 1986). Indeed enzymes surrounded by non-polar media have been shown to retain catalytic activity even at temperatures as high as 100°C for many minutes (Zaks and Klibanov, 1984). It is believed that the organic solvent inhibits pathways that lead to inactivation by denaturation. More significant than the enhanced stability of enzymes under these conditions, however, is the fact that a new dimension in control of synthetic reactions using enzymes has been defined. Examples of efficient acyl transfer reactions including newly discovered selectivities will be given in section 4.7.

4.3   LOGIC AND ANALOGY IN THE SYNTHETIC USES OF ENZYMES AND MICRO-ORGANISMS

Every modern organic chemist learns his craft with the aid of structural and mechanistic analogy. If biochemical methods are to be readily acceptable, it is not unreasonable for the chemist to expect to be able to incorporate enzyme-catalysed reactions into the same system. Even micro-organisms show recognizable reactivity patterns towards organic compounds in the view of Sih and Rosazza (1976) and Perlman (1976). They pointed out that yeasts readily transform carbonyl groups, fungi cause hydroxylations, but bacteria are generally too destructive. Twenty years on, such a clear distinction is less tenable, especially with regard to bacteria. Aerobic micro-organisms are especially simple to use in synthesis because, unlike purified enzymes, they can recycle their own coenzymes; all they need is a growth medium containing suitable nutrients. The major practical problem, which can be serious on a large scale, is the isolation of the product. However attractive microbiological methods may appear in their simplicity, most chemists will not show interest unless telling examples are available. Fortunately, some recent work using yeast to prepare chiral synthons as the basis of complex natural product syntheses has been described (Fronza et al., 1980; Fuganti and Graselli, 1979, 1982). It has been known for many years that yeast will reduce ketones to chiral secondary alcohols; the reaction is even listed in *Organic Syntheses*. Fuganti has ingeniously extended this utility by capitalizing upon the observation that cinnamaldehydes, and homologues, undergo not only carbonyl group reduction, but also C=C reduction and aldol condensation (Fig. 4.7). The benzylic fragment is redundant after it has channelled the substrate into yeast metabolism; as a synthon, it acts as a latent carbonyl function. The target

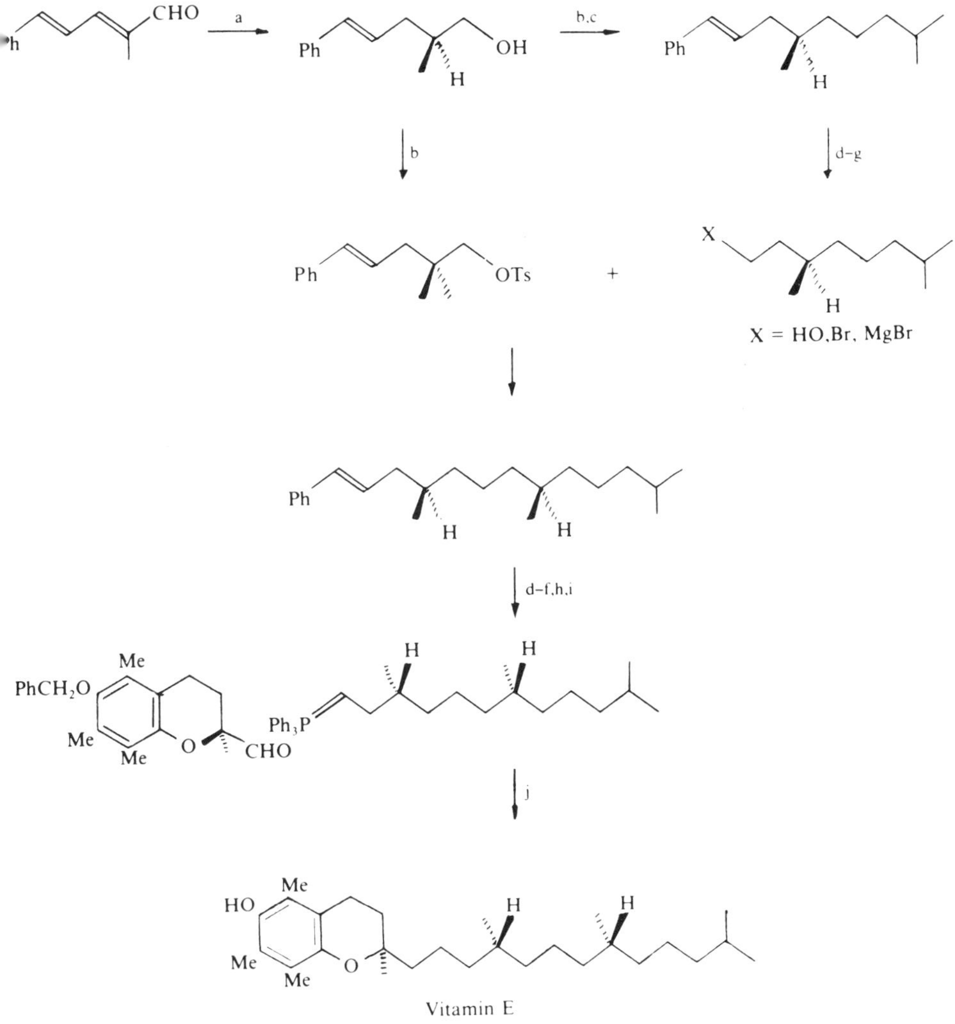

Fig. 4.7 Multiple use of a yeast-generated chiral building block in a synthesis of vitamin E. (a) Fermenting yeast; (b) TsCl/py; (c) BrMg(CH₂)₂CHMe₂ LiCuCl₄; (d) O₃; (e) LiAlH₄; (f) NBS Ph₃P; (g) Mg; (h) Ph₃P; (i) base; (j) H₂/cat.

molecules were important natural products, namely tocopherol, a natural antioxidant found in lipids, and daunosamine, a component of the antitumour antibiotic daunomycin. The repeated chiral structures of tocopherol's side chain make this synthetic strategy ideal. The readily availability of yeast and the clear opportunity for logical extension of these reactions should attract further synthetic applications. For comparison with chemical methods, the use of the stereospecific addition of carbanions to vinyl lactones controlled by palladium(0) complexes as templates illustrates an alternative strategy for the construction of the tocopherol side chain (Trost and Klun, 1979, 1981). Template strategies figure in biomimetic chemistry discussed below (section 4.6.1).

Extending further from the purely chemical domain, a new development eminently amenable to logical approaches has been made possible through genetic engineering. Consider the synthesis of ascorbic acid, vitamin C biosynthesised from glucose in micro-organisms. Ascorbate has been synthesized since the 1930s industrially from glucose by non-enzymic methods and a key intermediate is 2-ketolaevogluconic acid (Fig. 4.8, 2-KLG) which is easily converted into ascorbic acid. This intermediate can be produced by stereospecific reduction of 2,5-diketogluconic acid (2,5-DKG) in *Corynebacterium* spp. and *Erwinia herbicola* manufactures 2,5-DKG from glucose. The logical question then arises whether a fermentation process could be constructed in an organism containing both enzymes so that the entire conversion of glucose into 2-KLG can be carried out in a single fermentation. The gene that codes for 2-KLG reductase was cloned from *Corynebacterium* and successfully introduced into *E. herbicola*. After some problems with side reactions caused by unexpected enzymic activities were circumvented, the required conversion was shown to be possible (Estell *et al.*, 1984). This is a good example of biological process design for a commodity chemical (vitamin C). In the future, then, the synthetic chemist who can synthesize organisms may evolve.

Many enzymes have been so thoroughly studied that their behaviour in synthesis can be predicted reliably. These include the peptidase, chymotrypsin, for which an active-site model has been deduced from a combination of specificity studies and X-ray crystallography on the enzyme (Jones and Beck, 1976). Undoubtedly the classic example of the development of predictability concerns the enzyme horse liver alcohol dehydrogenase,

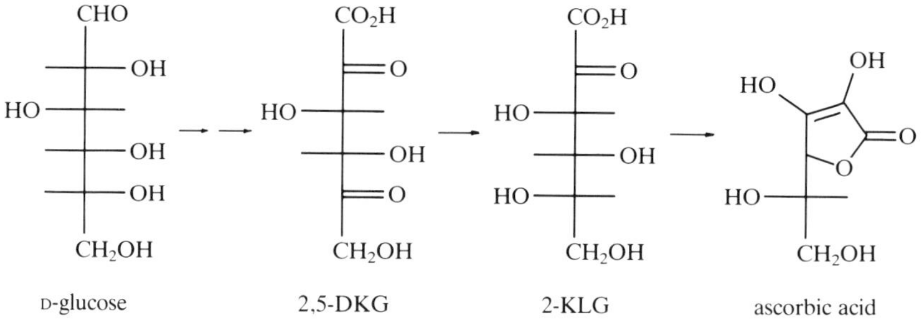

Fig. 4.8   Ascorbic acid synthesis from D-glucose.

which has been mentioned above (Fig. 4.2). The early studies of this enzyme in the laboratories of Prelog in Zurich have recently been spectacularly extended by Jones' group in Toronto (Jones, 1980). HLADH is a relatively inexpensive enzyme. Although specific with regard to the chirality of the redox reaction it catalyses at the carbonyl group of the substrate, it will accept an unusually wide range of organic substrates containing typically 2 to 15 carbon atoms. The broad specificity of this enzyme reflects the biological source, liver, which is the organ principally responsible for detoxification of organic compounds foreign to the metabolism of the animal. It may well be that other synthetically useful enzymes are present in the same source. The structural origin of the broad specificity of HLADH has been identified as a large non-polar pocket adjacent to the catalytically active site; this binding site is especially suited to alicyclic molecules. What the synthetic chemist wants to know is whether his substrate will fit the binding site and will be reduced at a useful rate by HLADH. Prelog sought an answer to this question by offering the enzyme a series of substituted cyclohexanones and decalinones. Some compounds were found to be reduced rapidly, some slowly, and some not at all. It was then possible to assess where substituents favoured binding and reduction, and where they hindered it. Since cycloalkanones are chiefly built from tetrahedral carbon atoms, Prelog constructed a reactivity map in the form of a so-called 'diamond lattice' (Fig. 4.9(a); Prelog, 1964). The chemist could then fit his substrate to the lattice and see whether any unfavourable interactions were present.

Undoubtedly, Prelog's work established the predictability of enzyme-catalysed reactions in synthesis and it was natural that when beginning his studies, Jones should build upon Prelog's foundation. At that time, the only recognized substrates for HLADH were ethanol, benzaldehyde and substituted cyclic alkanones. There were some notable gaps in possible substrates. For instance, aldehydes were easily reduced by the enzyme; if labelled substrate or coenzyme was used, a chiral product resulted. However, acyclic ketones were poor substrates and also, no attention had been paid to stereochemical control in oxidative reactions. The newly discovered NAD recycling technique made it possible for Jones to attempt oxidations of achiral compounds and use the enzyme's natural asymmetry to induce chirality in the product (Jones and Lok, 1979). Figure 4.10 illustrates the oxidation of achiral alcohols to afford good yields of optically pure lactones. Two enzyme-catalysed reactions occur: firstly, the enzyme selects one arm of the substrate for oxidation, the pro-R in this case, and secondly, the hemiacetal which is in equilibrium with the aldol is oxidized to the lactone.

In order to convert an acyclic ketone into a substrate, Jones constrained the ketone into a ring using sulphur which could readily be removed with Raney nickel after chiral reduction. Thus the synthetic equivalent of an acyclic ketone became, from the enzyme's point of view, a welcome

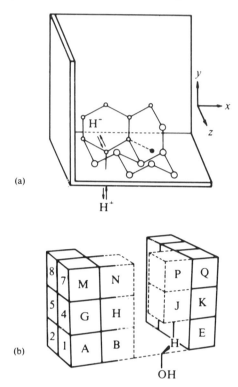

(a)

(b)

Fig. 4.9 (a) Prelog's diamond lattice model. (b) Jones' cubic model. Jones' model is depicted viewed from a left-front viewpoint. The forbidden regions are shown bounded by solid (———) lines and the limited regions (–––) lines. The open spaces are the allowed areas favouring substrate binding. For specificity analyses, the substrate model is built to the same scale. Its $C = O$ or $CH(OH)$ group is first positioned in the required orientation at the oxidation site located at the bottom of the Cl,Dl intersection. Its various orientations in the cubic lattice section are then compared in order to identify which, if any, will permit the formation of a productive ES complex. Both figures reproduced with permission; Fig. 4.9(a) by courtesy of *Pure Appl. Chem.*, IUPAC.

cycloalkyl substrate (Fig. 4.11; Jones and Davies, 1979; Jones and Schwartz, 1981).

The extended utility of HLADH illustrated by the above examples nevertheless exposed a limitation in the predictive value of the diamond lattice model. Five-membered rings and compounds containing sulphur do not readily fit the points of the diamond lattice. Clearly what was required was a model that specified regions where substituents could be tolerated and where they could not. Jones has constructed a convenient model based upon

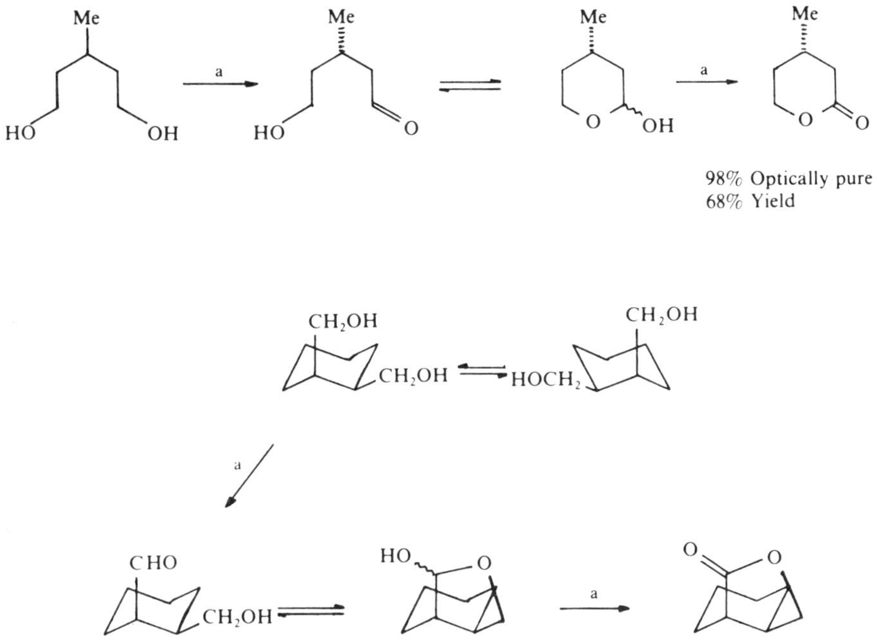

Fig. 4.10 Generation of chiral lactones using (a) HLADH with NAD$^+$ recycling.

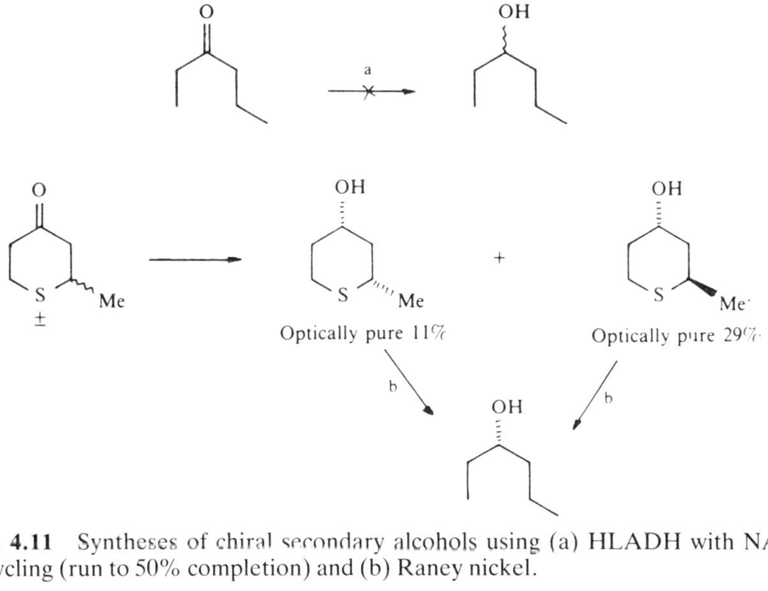

Fig. 4.11 Syntheses of chiral secondary alcohols using (a) HLADH with NAD$^+$ recycling (run to 50% completion) and (b) Raney nickel.

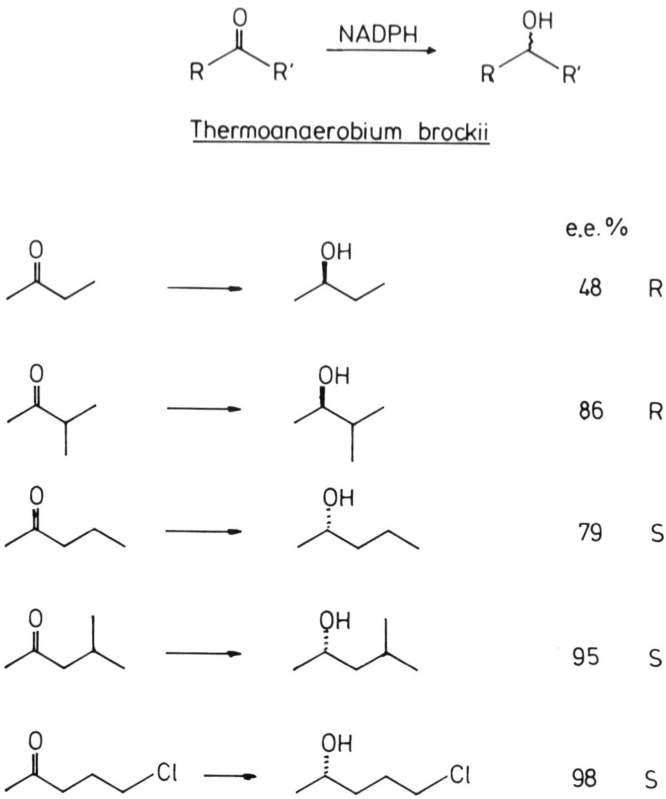

Fig. 4.12 Enantioselective reduction with a dehydrogenase from a thermophilic organism.

cubic spatial sections which allows the chemist to assess easily the likely success of an oxidation or reduction (Fig. 4.9(b); Jones and Jakovac, 1982).

HLADH is by no means the only alcohol dehydrogenase to be useful in synthesis. Dehydrogenases were at the centre of the 2-KLG fermentation described above and recently two isolated dehydrogenases have come into prominence for laboratory synthesis. The properties of the alcohol dehydrogenase from *Thermoanaerobium brockii* are of interest. Firstly, *T. brockii* is a thermophile and its alcohol dehydrogenase (TBADH) can be used at temperatures of up to 65°C. Secondly, it complements HLADH in that it accepts acyclic ketones and secondary alcohols readily (Fig. 4.12; Keinan *et al.*, 1986) and thirdly there are some interesting examples of chemoselectivity in its reactions. Alkyl halides, for example do not interfere with the reaction.

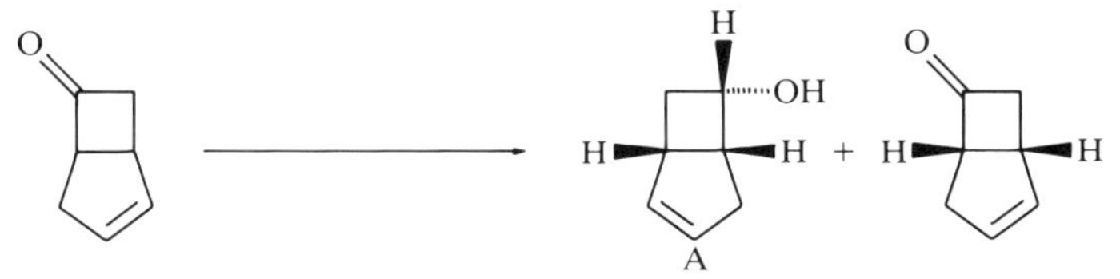

(a)

Horse liver alcohol dehydrogenase   ⟶ A <10% e.e.
(CH₃CH₂OH recycling)

Hydroxysteroid dehydrogenase   ⟶ A <10% e.e.
(glucose: glucose dehydrogenase recycling)

Thermoanaerobium brockii dehydrogenase   ⟶ A >95% e.e.
iPrOH recycling)

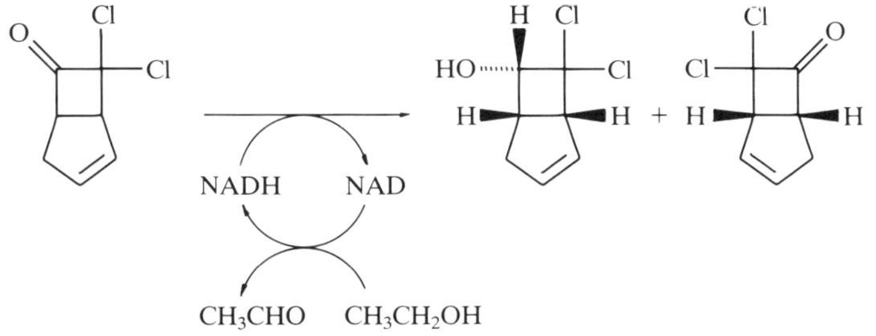

(b)   yeast alcohol dehydrogenase

Fig. 4.13 Dehydrogenases for reduction of a prostaglandin precursor.

This extended capability was helpful to Roberts who, in his synthesis of prostaglandins, required to reduce the bicyclic ketone (Fig. 4.13(a); Baxter and Roberts, 1986). TBADH succeeded where HLADH failed. However in an analogous sequence aimed at leucotrienes, logic and analogy broke down because neither HLADH nor TBADH was effective (Fig. 4.13(b)). It was necessary to screen a range of enzymes for the required activity and hydroxysteroid dehydrogenase was found to be suitable. This opened the way to a stereocontrolled synthesis of several prostaglandins.

4.4 ENZYMES AND CHEMICAL REAGENTS IN 'COMPETITION'

From the foregoing discussion there can be no doubt that biochemical synthetic methods are amenable to the normal logical processes of organic chemistry. It is now appropriate to examine how biochemical methods fare in competition with chemical reagents in the tough testing ground of natural product synthesis and the related synthesis of labelled precursors for biosynthetic studies. There are three areas to consider; two of these are synthon preparation and functional group modification, but first I shall discuss the biosynthetically important area of chiral isotopically labelled compounds.

4.4.1   Chiral isotopic labels

The synthesis of chiral isotopically labelled compounds was one of the first fields in which the significance of enzyme stereoselectivity was realized. Indeed the important initial demonstration of enzymic selectivity at prochiral centres demanded the use of isotopic labels. One example, the reduction of benzaldehyde, has already been described. Related reactions with substituted benzaldehydes have been especially valuable in the synthesis of precursors for biosynthetic studies of alkaloids (Battersby and Staunton, 1974). The use of tritiated and deuteriated phenethylamines has permitted the elucidation of the stereochemical course of benzylic hydroxylation in alkaloid biosynthesis. However, a less frequently discussed example concerns the synthesis of $[^{13}C]$-valine for the study of the biosynthesis of β-lactam antibiotics, a problem considered from a different perspective in Chapter 7.

There are two nicely contrasting syntheses of chiral isotopically labelled valine. The first uses a chemical reaction of known stereochemical course in combination with a traditional resolution of a diasteroisomeric salt to incorporate the chirality (Baldwin *et al.*, 1973) and the second introduces the chirality by means of an enzyme-catalysed step (Kluender *et al.*, 1973). Both syntheses are conceptually quite short but require many functional group manipulations to obtain the required product, and the steps significant for the generation of chirality are shown in Fig. 4.14. In both cases, the introduction of chirality involved carboxylic acids which were subsequently reduced to methyl groups. Baldwin's chemical route relied upon the stereospecific reduction of *trans*-2-methylcyclopropanecarboxylate (step g) to generate the chiral isopropyl group: the stereochemical course of ring-opening reactions is usually well defined and is a reliable strategy for the control of stereochemistry in an organic synthesis. Although this synthesis afforded a chiral isopropyl group (100% optically pure, 95% $[^{13}C]$-labelled), the formation of the other chiral centre at C-2 was not controlled and a mixture of diastereoisomers was obtained. The use of a mixture is, however, of little consequence to the subsequent biosynthetic experiment because the enzymes that synthesize penicillins will accept only the natural 2S isomer as substrate; the other isomer may bind to the enzymes, but it will not be transformed. There is still the challenge of synthesizing (2S)-valine contain-

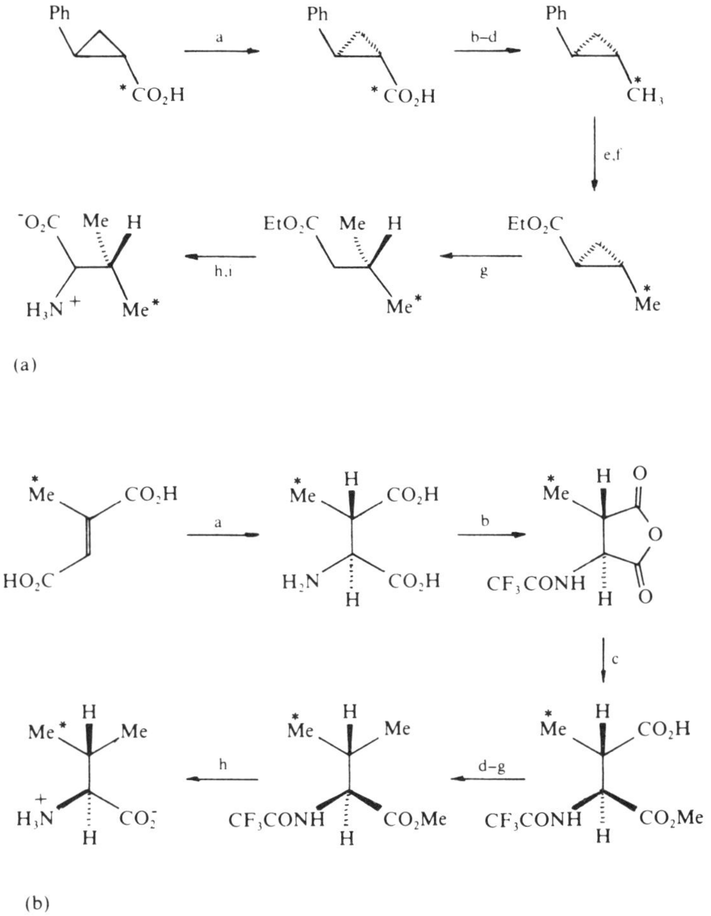

Fig. 4.14 Chemical and enzyme-based syntheses of chiral [^{13}C]-labelled valine. (a) Baldwin's synthesis. Reagents: a, resolve-quinine salt; b, LiAlH$_4$; c, MsCl; d, LiAlH$_4$; e, O$_3$; f, CH$_3$CHN$_2$; g, Li/NH$_3$; h, PCl$_3$ Br$_2$; i, NH$_3$. (b) Sih's synthesis. Reagents: a, β-methylaspartase with recycling of unreacted starting material; b, CF$_3$CO$_2$H; c, dry MeOH – the major product (80%) is shown; d, B$_2$H$_6$; e, MsCl/py; f, NaI/acetone; g, Pt/H$_2$; h, aq. HCl.

ing a chiral isopropyl group and this challenge has been met by Sih using the enzyme β-methylaspartase. This enzyme catalyses the addition of ammonia to the olefinic bond of methylfumaric acid (step a). If the methyl group is [^{13}C]-labelled, than *trans*-addition of ammonia from one enantiotopic face of the substrate as controlled by the enzyme generates the *S*-configuration at both centres. Subsequent chemical modifications then led to a specimen of chiral isotopically labelled valine, 90% [^{13}C]-enriched and enantiomerically pure. Interestingly, a relative of this enzyme, aspartase, is used industrially for the synthesis of aspartic acid, which has value for enriching foodstuffs and animal feeds (Chibata, 1978). The enzyme preparation used industrially occurs intracellularly in *Escherichia coli*, and the cost of isolating it on a large scale led the Japanese scientists to immobilize whole *E. coli* cells. In this way, the cost of producing aspartic acid enzymically was reduced to 60% of that for batch fermentation.

In most syntheses of chiral compounds, a single synthetic step can be identified as the key to controlling the ultimate stereochemical outcome of the synthesis. Sometimes in a chemically based synthesis, a ring opening or closing reaction which can take only one course, either for stereoelectronic reasons or because of geometrical constraints, is crucial. An alternative device for controlling chirality is to use a severely hindered chiral reagent or substrate (see below). The synthetic chemist will plan his strategy around this key step. There is no alteration to this strategy if a biochemical method is used to introduce chirality; the enzyme-mediated step simply becomes the focus of the synthesis and dictates the subsequent strategy. It is not uncommon for two centres of chirality to be generated by an enzyme, as was the case with valine and the synthetic use of yeast described above. In this respect, biochemical methods can again compete with stereochemically controlled ring-opening reactions widely used in conventional syntheses. For the preparation of isotopically labelled compounds, chemical and biochemical methods are therefore valuable alternatives. Further examples have been discussed in a more extensive review of this topic by Whitlock (1976).

4.4.2 Chiral intermediate preparation

I have just alluded to the significance of steps involving the generation of chiral centres in a synthesis and the design around a key chiral intermediate. There is, of course, no need for the chirality to be introduced by isotopic labels and in most cases, labelling is not wanted.

Because of the importance of the control of chirality in both academic synthetic endeavours and in fine chemical manufacture, the preparation of chiral synthons is perhaps the most studied aspect of enzymes in organic synthesis. Two classes of enzymes have been the mainstays of this work namely dehydrogenases and hydrolases. Although the exploitation of dehydrogenases preceded that of hydrolases, the perceived problems of cofactor recycling led to an upsurge of interest in hydrolases which, of course, require only water as a coreactant when operating in the hydrolytic direction. On the other hand, resolution of a racemic mixture by enzyme-catalysed hydrolysis leads to a maximum yield of only 50% because one isomer does not react. Unless both isomers are required or the unwanted one can be epimerized, a substantial cost must be borne for the waste. Both classes of enzyme are often able to operate with polyfunctionality without need for protecting groups, a potential advantage over chemical reagents.

A summary of typical reductions effected by dehydrogenases in yeast is given in Table 4.1 together, where appropriate, with a note of the target molecule. Other examples have already been cited. In the context of reductions of β-dicarbonyl compounds, some chemical systems as effective as yeasts have been devised (Kawano *et al.*, 1988). Interestingly, these are typically phosphine derivatives of chiral binaphthyls, a class of compound encountered in several subsequent examples.

As a further example of the application of chiral building blocks to natural products, Jones' synthesis of grandisol is interesting because it can be approached using either HLADH as in his original work (Jones *et al.*, 1982) or using hydrolases (Sabbioni and Jones, 1987). Figure 4.15 shows the relevant conversions of a chiral lactone that was available from either route.

This example conveniently introduces hydrolases, many of which have applications in synthesis. Several enzymes have been extensively investigated especially pig liver esterase (PLE) and many lipases including that from porcine liver (PPL) and from *Candida cylindraceae* (CCL). The specificity of PLE has been carefully investigated to open the way to logical use (Fig. 4.16; Sabbioni and Jones, 1987) and in the course of this study, some surprises were encountered. Experience with most enzymes had suggested that the stereochemical course of a reaction was consistent within a series of similar substrates such as a homologous series. It was most unexpected when it was found that this trend was broken in the series of cycloalkyl-1,2-dicarboxylates. As Fig. 4.16 shows, the behaviour of 3- and 4-membered rings was the same giving high enantiomeric excesses of the *S*-isomer. However, stereoselectivity returns, but with the opposite chirality when cyclohexane-1,2-dioic esters were substrates. Although such abrupt changes in stereoselectivity as this are rare, it is important to be prepared to define the chirality of a product rigorously when investigating a new substrate.

The range of substrates that can be transformed by hydrolases is impressive and examples that have been the basis of natural product synthesis are shown in Table 4.2. Peptidases as well as esterases are applicable. Just as dehydrogenase-catalysed reactions are reversible with the appropriate coenzyme recycling system, so hydrolases can be made to catalyse ester or amide synthesis under suitable conditions. The conditions required the avoidance of water and this was one of the primary stimuli for the development of the use of enzymes in organic solvents. Table 4.2 shows some examples and the topic is discussed in more detail in section 4.7.

The use of acylases in the industrial production of L-amino acids is one of the classics of industrial enzyme chemistry. As long ago as 1953, Japanese

Table 4.1 Yeast dehydrogenases in chiral synthon preparation (see also Figs 4.10–4.13)

Substrate	Product	Target	Reference
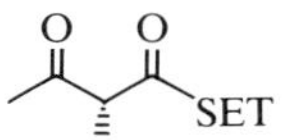	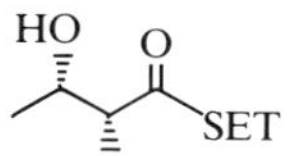	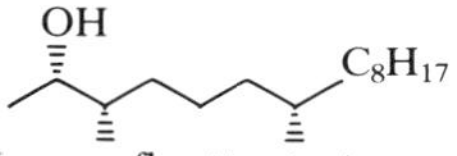	Itoh et al, 1986
		pine saw fly attractant	Christen and Crout, 1987
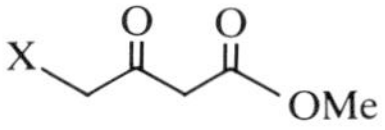	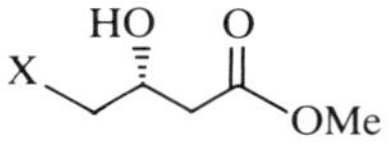	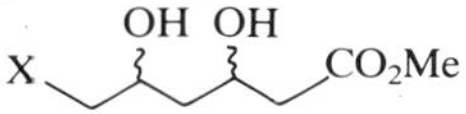 all 4 1,3-dihydroxyisomers	
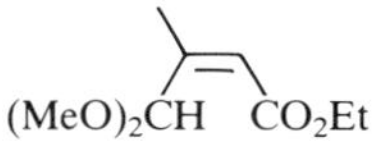	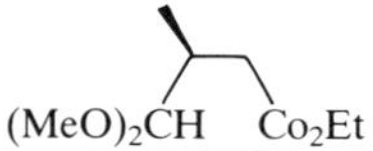	Steroid side chain	Fernaboschi et al, 1987
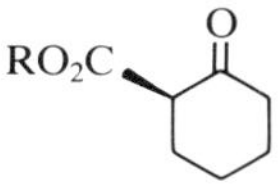	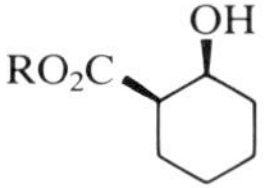		Seebach and Herradon 1987
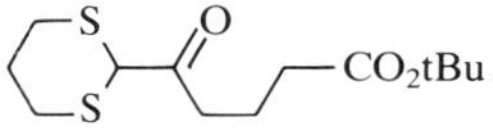	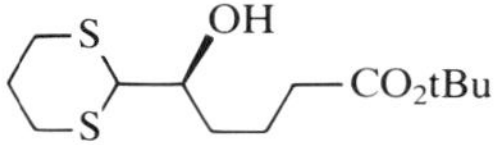	leucotriene B_4	Han et al, 1986
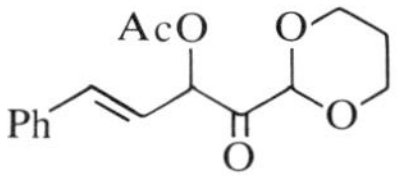	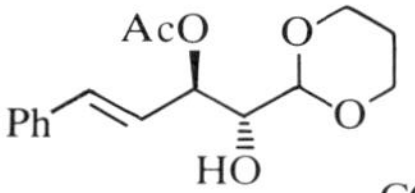	non-natural carbohydrates	Fronza et al, 1985
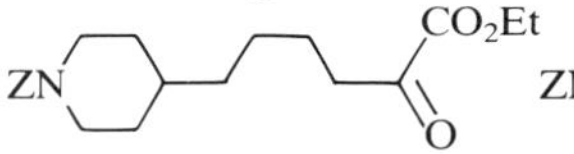	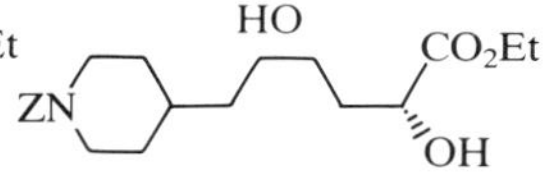	angiotensin converting enzyme inhibitor	Kori et al, 1987

In examples marked * the enantiomer was obtained using another organism or enzyme.

Table 4.2 Hydrolases in natural product synthesis (see also Figs 4.15 and 4.16)

Substrate	Enzyme or organism	Product	Target	Reference
 MeO_2C—epoxide—CO_2Me	Pig liver esterase	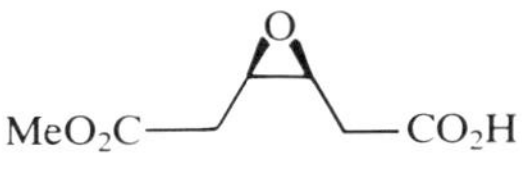 MeO_2C—epoxide—CO_2H	β-hydroxy-GABA nonactic acid	Mohr et al. 1987
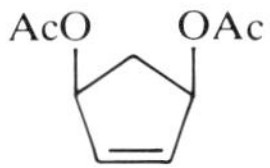 AcO—OAc (cyclopentene)	Acetylcholine esterase	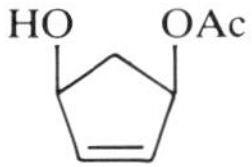 HO—OAc (cyclopentene)	prostaglandins	Deardoff et al. 1986
 MeO_2C—CO_2Me (cyclohexene)	Pig liver esterase	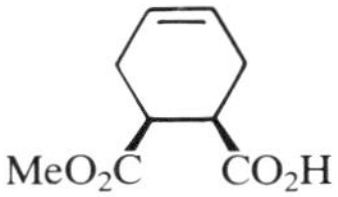 MeO_2C—CO_2H (cyclohexene)	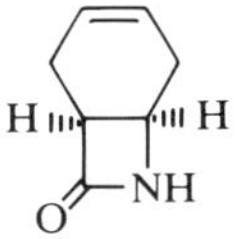 β-lactams	Kurihara et al, 1985
 $C_{11}H_{23}$—CO_2H, H NHCOCH$_2$Cl	Aspergillus spp. amino acylase	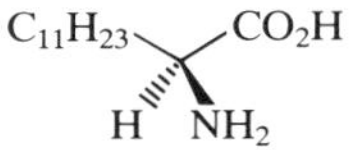 $C_{11}H_{23}$—CO_2H, H NH_2	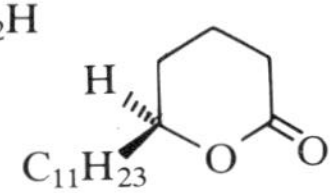 oriental hornet pheromone	Mori and Otsuka, 1985
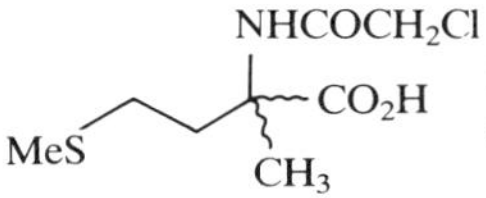 MeS—NHCOCH$_2$Cl—CO_2H CH$_3$	Pig kidney amino acylase	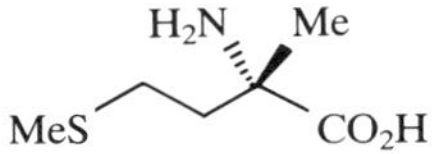 H_2N Me—CO_2H MeS	unnatural amino acids	Bednarski et al, 1987
 MeCONH—CO_2Me (cyclopentene)	Pig liver esterase	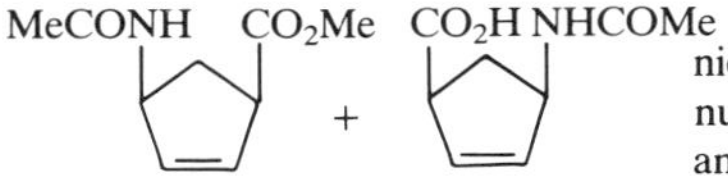 MeS CO_2Me + MeCONH CO_2H NHCOMe	nicotinamide nucleoside analogues	Sicsic et al, 1987

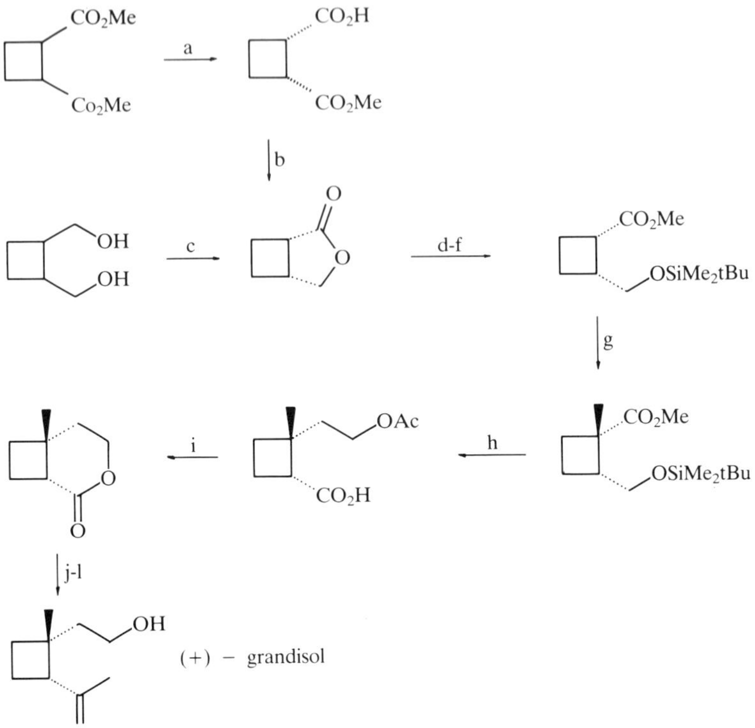

Fig. 4.15 Jones' syntheses of grandisol. a, pig liver esterase; b, LiBH₄; c, HLADH; d, KOH; e, CH₂N₂; f, tBuMe₂SiCl; g, LDA, MeI; h, several steps modifying upper side chain via Wittig reaction and hydroboration, also modifying lower substitutent by deprotection and oxidation; i, 5% H₂SO₄; j, MeLi; k, Ac₂O; l, LiAlH₄.

scientists were hydrolysing acyl DL-amino acids with an aminoacylase from the mould *Aspergillus oryzae* (Chibata, 1978). This enzyme combines high catalytic activity with broad substrate specificity, properties that are ideal for synthetic application. The early work was carried out in batch processes but the labour involved in isolating enzymes and in purifying products severely hampered operations. This difficulty stimulated the initial development of immobilized enzymes. Not only were chemical factors such as enzyme stability and efficiency investigated but also the new engineering problems of handling enzymes on a large scale were tackled. One of the main streams of modern biotechnology thus has its source in this work. As with immobilized *E. coli* cells mentioned earlier, the saving in operating costs using immobilized enzymes was more than 40%. Industrial processes also use enzymes to hydrolyse amides in more complex structures; the selective cleavage of the side chain in penicillins leaving the important β-lactam intact

Table 4.3 Chiral building blocks obtainable using enzymes (see also Tables 4.1 and 4.2, Figs 4.10–4.16)

Substrate	Enzyme	Product	Comments	Reference
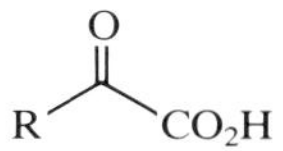 (R–CO–CO$_2$H, keto acid)	Lactate dehydrogenase	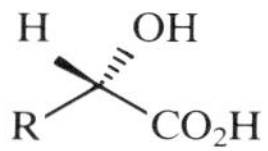 (H, OH, R, CO$_2$H)	>20 keto acids studied	Kim and Whitesides, 1988
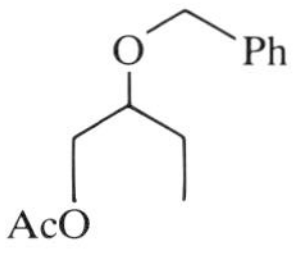 (PhCH$_2$O-substituted diacetate, AcO)	Pig kidney lipase	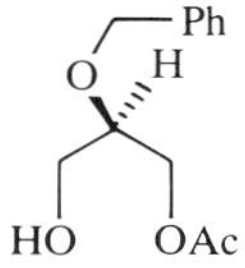 (PhCH$_2$O, H, HO, OAc)	benzyl ether gives good substrate characteristics	Breitgoff et al, 1986
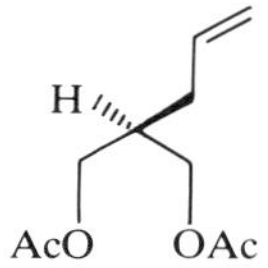 (allyl-substituted, AcO, H, OAc)	Pig kidney lipase	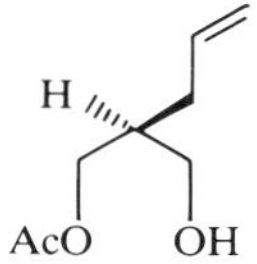 (allyl, H, AcO, OH)	in aqueous solution, S-isomer obtained by hydrolysis	Tomuto et al, 1986
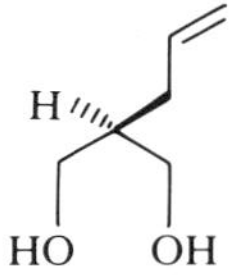 (allyl-substituted, HO, H, OH)		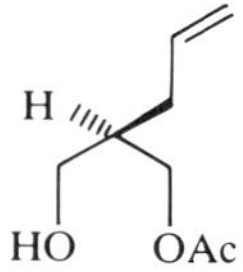 (allyl, H, HO, OAc)	in organic solvent, R-isomer obtained by esterification	

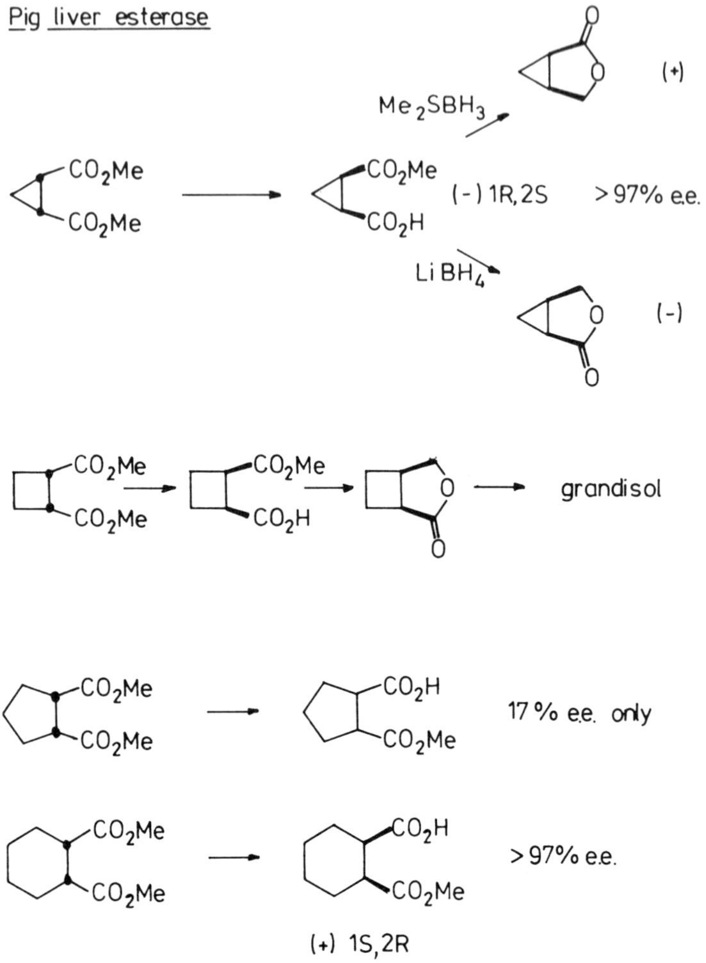

Pig liver esterase

Fig. 4.16   Enantioselective hydrolysis with pig liver esterase.

is perhaps the best-known example. Enzyme-catalysed reactions are also very useful for introducing chirality (Table 4.3), not only into amino acids as we have already seen, but also into hydroxy acids and notably benzaldehyde cyanohydrin. The last example has been mimicked by a chemical model (see section 4.6.6), but in practical synthetic terms, the enzyme is superior. Notably, this enzyme has also been used on a large scale in immobilized form and will accept a wide range of aliphatic, aromatic and heterocyclic aldehydes.

4.4.3   New reagents

The success of enzyme-catalysed reactions in preparing chiral synthons has posed a continuous challenge to organic chemists to match the selectivity of enzymes. As will be described later, one focus of this activity has been the study of biomimetic chemistry (section 4.6), but by evolving and designing compounds with very demanding steric requirements, impressive and important advances in synthetic technology have been achieved. Two fields provide especially good examples, the synthesis of chiral epoxides and aldol condensations.

Chiral epoxides can be prepared in two main ways, either by resolution of a racemic mixture or by stereospecific oxidation of an asymmetrical alkene. An effective enzymic approach uses the former and an efficient new reagent system uses the latter. The particular interest in chiral epoxides arises from the fact that two chiral centres are established in a correlated manner in a reactive functional group. In order to effect an enzymic resolution, a third functional group, an ester, must be present, a group that could in principle be distant from the epoxide or adjacent to it for a successful resolution. In the latter case, the epoxide is that derived from an allylic alcohol; this substrate is also that for the epoxidation reagent devised by Sharpless (Gao *et al.*, 1987).

Deriving from his studies of selective epoxidation of terpenes (see section 4.6.5), Sharpless discovered in 1980 that titanium isopropoxide in the presence of a tartaric acid, which is one of the purest chiral natural products available, and a hydroperoxide (t-butyl or cumene) mediated the epoxidation of allylic alcohols in extremely high optical purity (Fig. 4.17). The reaction was found to be general and has become one of the classics of modern organic chemistry. The competition between enzymic and Sharpless methods was most acutely demonstrated in the synthesis of the epoxide of allyl alcohol, which is a key intermediate in a synthesis of a class of drug

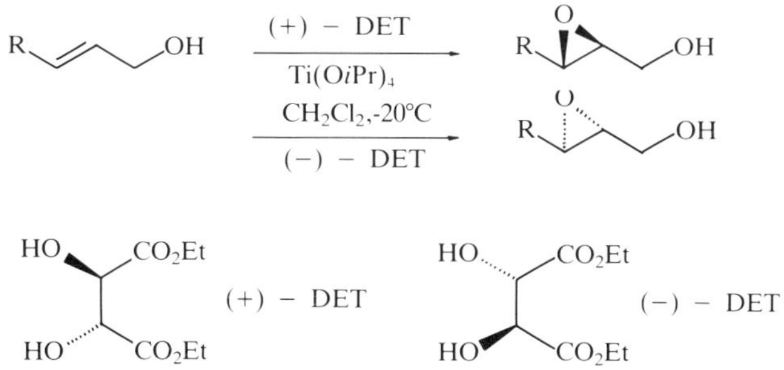

Fig. 4.17   Sharpless epoxidation.

known as β-blockers, for example propranolol (Fig. 4.18; Klunder *et al.*, 1986). These drugs are an excellent example of the developing significance of the control of chirality in pharmaceutical compounds. Although the racemic drugs have been very successful in alleviating the symptoms of angina by reducing stimulation to the heart muscles, it is known that the *S*-isomer has the desired activity on the heart. In the case of propranolol, its

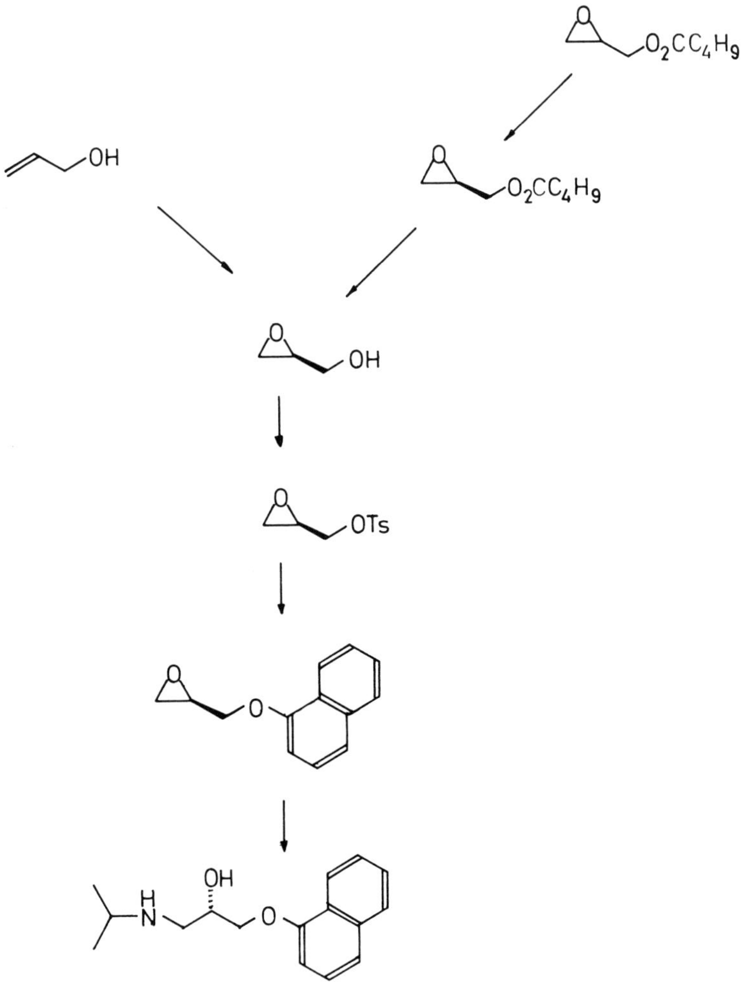

Propranolol

Fig. 4.18   β-Blocker synthesis using chiral synthons generated by non-enzymic (Sharpless) and enzymic (esterase) methods.

enantiomer does not have the catastrophic side effects of the wrong isomer of thalidomide but it does have some bizarre side effects; the *R*-isomer has contraceptive properties. Newer β-blockers with improved selectivity of action are now being prescribed and a contribution to their synthesis comes from stereospecific synthesis.

Epoxidation of allyl alcohol itself proved to be one of the most difficult to achieve in good yield and in high enantiomeric purity. After much experimentation it was found that cumene hydroperoxide was a suitable oxidizing agent and the reaction was capable of development on an industrial scale. The sequence of reactions leading to β-blockers is shown in Fig. 4.18. Obviously, by use of the opposite enantiomer of tartaric acid, the enantiomeric epoxide could be obtained. In the non-enzymic method, only one isomer of the epoxide is formed in a given reaction. On the other hand, the enzymic method affords both isomers, one as the unhydrolysed ester, and one as the alcohol. Whitesides (Bednarski *et al.*, 1987) showed that PPL was able to hydrolyse the butyric ester of glycidol efficiently and the required isomer for β-blocker synthesis was obtained from the unhydrolysed ester. The hydrolysis has also been examined commercially.

Both approaches are thus demonstrably successful. One may have a personal prejudice about the intrinsic scientific elegance of each approach but in an industrial context, the successful method is chosen for efficiency and cost. The competition cannot be successfully resolved by a cursory inspection; costs for enzyme, reagents, and chiral auxiliaries must all be evaluated. Isolation costs are often significant in biotransformations. Indeed a commercial choice may depend not so much upon the intrinsic properties of the chiral reaction but upon the engineering skills and facilities available to the company. The beauty of modern stereoselective synthesis in industry is that a genuine choice exists.

The central act of organic synthesis has always been the formation of C–C bonds and for the modern organic chemist, this must be achieved with control of chirality where necessary. Addition reactions to carbonyl groups are probably the most important class of reactions in this context. In view of the synthetic challenge posed by polyhydroxy antibiotics such as erythromycin, much recent attention has been directed to aldol reactions (for example Short and Masamune, 1987). Spectacular successes have been obtained in controlling the stereochemical course of aldol reactions by taking careful account of the conformation of transition states in addition of an enolate to an aldehyde. Careful temperature control and choice of enolate (boronate, silyl, or lithium, for example) can lead to high diastereofacial selectivity. Such methods can now be used routinely. It is noticeable, however, that the substrates carry a heavy burden in protecting groups.

One of the potential advantages of enzymes is their ability to catalyse chemo-, regio-, and stereoselective reactions without need for protecting groups and with regard to stereospecific C–C bond formation, there is an important class of enzymes, the aldolases, that are beginning to be exploited in synthesis. Aldolases are especially common in carbohydrate transformations. There are a great many compounds that act as the aldehyde component of an aldolase-catalysed reaction, but only four compounds from which the enzyme-bound enolate is received: dihydroxyacetone phosphate (DHAP), pyruvate, pyruvate enol phosphate (PEP) and acetaldehyde. A limitation of aldolases is their high specificity for their carbanion component but their ability to handle polyfunctional carbohydrates more than compensates.

There are many complex oligosaccharides that are important as compounds that define blood groups (blood group determinants) and others that provide recognition sites on cell surfaces (for example, sialic acids). The synthesis of such compounds by conventional polysaccharide chemistry is an immense labour involving much protection and deprotection. On the other hand, enzymes make it possible to prepare important oligosaccharides and their analogues directly. For example, Wong has shown that sialic acid aldolase will accept a very wide range of β-hydroxyaldehydes (Durrwachter and Wong, 1988; Kim *et al.*, 1988). If a way could be found to loosen the specificity of the carbanion-providing partner in aldolase-catalysed reactions, the openings for the synthesis of polyfunctional molecules with high enantioselectivity would be immensely wide.

Many chemists, mindful of the significance of addition reactions to carbonyl groups in synthesis, have wondered whether a general chiral addition reaction could be developed in the absence of enzymes. The requirement to achieve this is a chiral reagent with very high steric demands that distinguish the enantiotopic reaction paths. Such chiral mediators have been prepared from organic materials (Oppolzer *et al.*, 1987) but perhaps the most successful are those built from organometallic complexes exploited magnificently by Davies (for example, Brown *et al.*, 1986, Davies *et al.*, 1984, 1985, Bashiardes and Davies, 1987).

The versatility of Davies' reactions is outstanding (Fig. 4.19). The key reagent is a chiral iron acetyl complex in which access to the acetyl group is possible from one direction only. In the simplest case, the anion can be added to aldehydes to give, after oxidative cleavage of the organometallic auxiliary, a chiral aldol adduct (Fig. 4.19(a)). As with many enzyme-catalysed reactions, it is possible to control two chiral centres simultaneously (Fig. 4.19(b)). These chiral additions have served as the basis of syntheses of many natural products and of drugs (Fig. 4.19(c)). Such reagents are not cheap because the majority of the material is not the substrate of interest, it is simply the chiral auxiliary. Nevertheless the power of this type of approach cannot be overlooked especially as developments of generally applicable catalytic systems are to be expected.

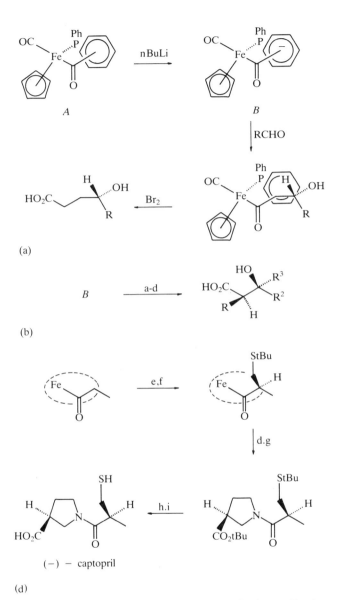

(a)

(b)

(d)

Fig. 4.19 Davies' chiral auxiliaries and some synthetic applications a, R¹Br;
b, nBuLi; c, R²COR³; d, Br₂; e, nBuLi; f, tBuSCH₂Br; g, proline t-butylester;
h, CF₃CO₂H; i, Hg(OAc)₂.

4.5 LATE-STAGE FUNCTIONAL-GROUP MODIFICATION

If the overall success of a synthesis depends crucially upon the strategy, especially where chirality is involved, then the bringing of an elegant strategy to fruition relies upon the chemist's ability to modify selectively functional groups in the molecule. At this late stage in a synthesis, difficulties often centre around the introduction of isolated chiral centres remote from control elements such as rings; differentiation of similarly reacting groups should have been controlled by the use of suitable protecting groups. However, there is no reason why enzymes cannot also be useful in this phase of a synthesis and there are cases where they have been more effective than chiral chemical reagents. For example, in prostaglandin synthesis, Sih et al., (1975) (Fig. 4.20) found that it was impossible to reduce the cyclopentatrione selectively with a chiral rhodium phosphine complex that had proved successful elsewhere. His experience with micro-organisms led him to expect that biochemical methods would be successful and, by choice of a suitable micro-organism, it was possible to produce either enantiomer at will.

The reader will by now be familiar with the notion that enzymes control selectivity by bringing the reactants together in the required juxtaposition. Perhaps the most successful but remote chemical relative of this device is the use of steric hindrance to block the attack of a reagent upon a particular part of a substrate as was illustrated in Davies' chiral auxiliaries in the previous

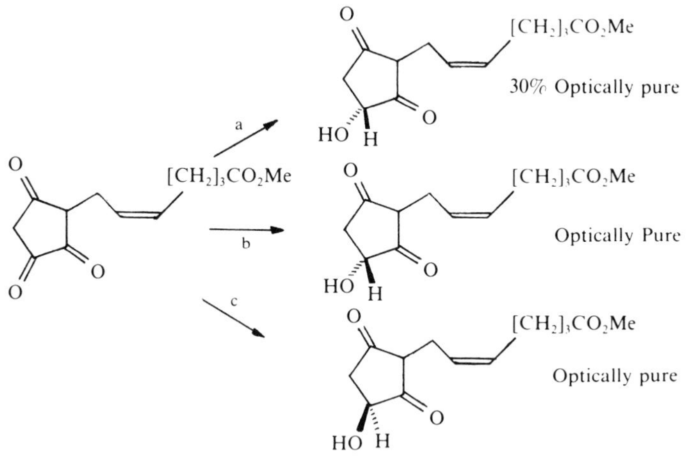

Fig. 4.20 Enantioselective reductions of prostaglandin precursors using chemical and microbiological methods. a, Chiral rhodium complex; b, *Dipodascus uninucleatus*; c, *Mucor rammanianus*.

tBu   →   tBu
                                        H

O   OH

Fig. 4.21 Enzymic and chemical reductions of conformationally restricted cycloal-kanones. a, HLADH, NADH 95% *cis*; b, LiHB(*s*-Bu)₃ 96.5% *cis*.

section. The unreactivity of the β-face of steroids was one of the first general examples of this phenomenon to be exploited, but chemists have developed this strategy to attain chiral reductions with selectivity rivalling enzymes (Krishnamurthy, 1974). Simple substrates such as cycloalkanes can be reduced with high selectivity by complex borohydrides such as lithium or potassium tris-*s*-butyl borohydride; clearly attack by such a bulky reagent will take place more readily from the less hindered side of a molecule to give, for instance, the axial alcohol (Fig. 4.21). As an alternative to letting a bulky reagent generate selectivity, it is possible to insert a large group into the substrate that will temporarily block attack from one direction. Both approaches have been exploited in controlling the stereochemical course of reduction at C-15 in the prostaglandin system (Fig. 4.22). In Corey's method (Corey *et al.*, 1972) the side-chain ketone was reduced by a chiral borane with high selectivity from the rear to yield the *S*-isomer (92% optically pure). Anyone who takes the trouble to read Corey's paper will appreciate not only how much experimental work is required to achieve such selectivity but also how unpredictable the choice of reagent can be. It could be argued that an enzyme, if a suitable one existed, would be more predictable. However, these difficulties have not deterred others from trying to equal or better Corey's selectivity. Using the bulky-reagent strategy, Yamamoto achieved a similar result reducing a close relative of Corey's blocked intermediate (Fig. 4.22; Iguchi *et al.*, 1979), and Noyori *et al.* (1979) claimed a still better result using an extremely hindered chiral binaphthyl aluminate (Fig. 4.22). We shall meet chiral binaphthyls again later in discussion of biomimetic systems. An experimental feature common to all of these strategies depending upon steric hindrance is that it is essential to conduct reactions at low temperatures, usually less than −78°C otherwise thermal energy will be sufficient to overcome the hindrance introduced by the bulky group. Since these pioneering experiments, many other sophisticated reagents have been devised, especially based upon pyrrolidines, for reactions at carbonyl groups (Corey *et al.*, 1987; Fujisawa *et al.*, 1984).

It might be claimed that enzymes are superior to chemical reagents in generating chiral intermediates but that chemical reagents are more effective in late-stage functional-group modification. Such a clear-cut distinction may have some limited value as a crude generalization, but things are rarely so simple. The cold non-polar conditions required for the use of

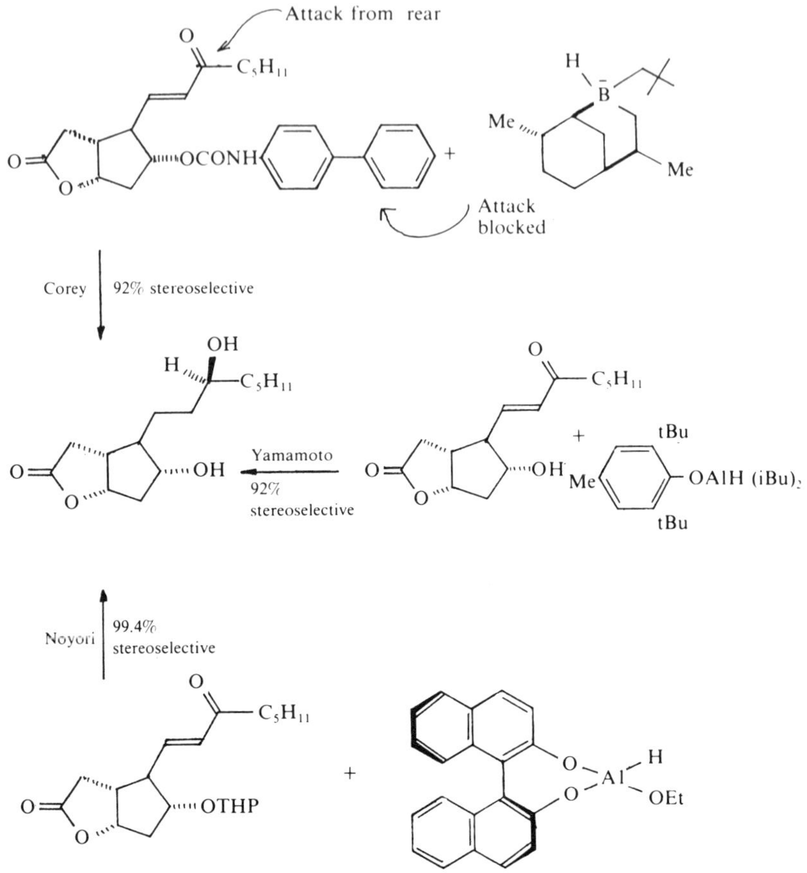

Fig. 4.22 Chemical strategies in reduction of enones in prostaglandin synthesis.

chiral boranes and hindered compounds will not suit all substrates. Although derivatization can often overcome solubility problems, it is best to treat each case on its merits. After all, there are examples of enzyme-catalysed reactions for which chemical reagents have no competitor, for instance Simon's enoate reductase. Equally there are cases when chemical selectivity has the advantage. Krishnamurthy (1981) has shown that the very hindered aluminate, lithium tris-3-ethylpentyloxy aluminium hydride is too bulky to reduce even unhindered cycloalkanones at low temperature, but will reduce aldehydes (Fig. 4.23).

An interesting example that highlights the intelligent coupling of chemical and enzymic methods in a commercial synthesis of a PGE analogue whilst avoiding the difficulty of effecting selective late stage modifications has been

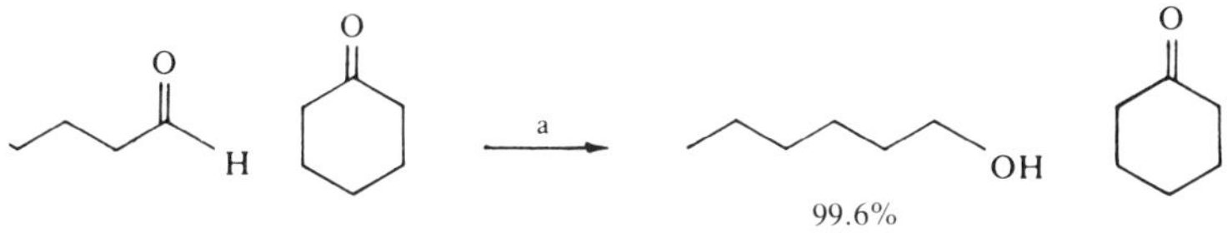

99.6%

Fig. 4.23   Reduction by a hindered aluminium hydride in $(Et_3CO)_3$ AlH, $-78°C$.

described by scientists from Merrell Dow (Kolb *et al.*, 1988; Fig. 4.24). Advantage was taken of enzymic resolution to introduce the chiral alcohol into the cyclopentane ring (Fig. 4.24(a)); Laumen and Schneider, 1984) and control of the stereochemistry of the side chain was effected using Sharpless epoxidation (Fig. 4.24(b)). Many steps in the synthesis involve protection and functional group modification to obtain the required reactivities for coupling the components. The key step to establishing the overall stereochemistry of the target molecule is a three-component coupling in which an organocuprate adds from the less hindered face of cyclopentenone (Fig. 4.24(c)). The resulting enolate is trapped by addition of an aldehyde. Subsequent dehydration and reduction leads to the required *trans* stereochemistry of the long chains of the prostaglandin analogue. In addition to its strategic interest, this synthesis incorporates many significant modern features such as Swern epoxidation, silyl protecting groups, and organocuprate chemistry.

The solubility and physical characteristics of the substrate may determine the choice of method. In our work on the synthesis of chiral tetrahydrofolate derivatives we investigated the use of a wide range of borate complexes for the reduction of 7,8-dihydrofolic acid (DHF) to 5,6,7,8-tetrahydrofolate (THF) (Fig. 4.25; Rees *et al.*, 1986). Tetrahydrofolate derivatives such as 5-formyl THF (leucovorin) are important in cancer chemotherapy as rescue agents and it is known that the 6S-isomer has the desired biological activity. The drug is, however, sold as a mixture of the 6R and 6S-diastereoisomers (an additional chiral centre exists in the glutamate side chain) and an opportunity for a chiral synthesis is apparent. Since folic acid is cheaply available as a natural product, a late-stage reduction is clearly the most economical approach. However none of the reagents that we tried was satisfactory, chiefly for solubility reasons, and the cue was given to investigate the enzyme-catalysed alternative.

Dihydrofolate reductase (DHFR) is the enzyme that catalyses the reduction of DHF to THF using NADPH as reductant. It stereospecifically produces the 6-S diastereoisomer. The major problem to be solved in adopting DHFR to preparative use is that of cofactor recycling. Through an extensive seies of coupled enzyme-catalysed reactions (Fig. 4.26) which also involved the regeneration of ATP, we were able to prepare up to gram

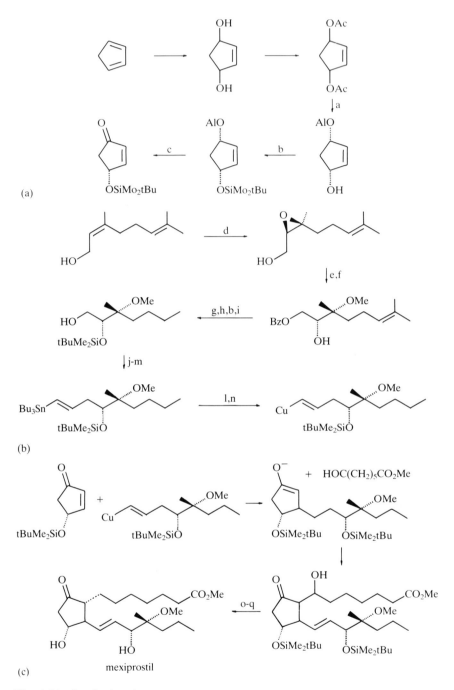

Fig. 4.24 Synthesis of mexiprostil, a PGE₁ analogue. a, pig liver esterase; b, tBuMe₂SiCl; c, pyridinium chlorochromate; d, (-)DET, Ti(OiPr)₄, tBuOOH; e, BzBr, KOBuᵗ; f, MeOH, H⁺ resin; g, O₃; h, Ph₃P⁺CH₃Br, KOBut; i, H₂ Pd; j, COCl₂, DMSO then Et₃N; k, PPh₃, CBr₄ then Et₃N, l, nBuLi; m, nBu₃SnH, AIBN; n, CuI, Ph₃P; o, MsCl, DMAP; p, nBu₃SnH, tBuOOH; q, AcOH, THF.

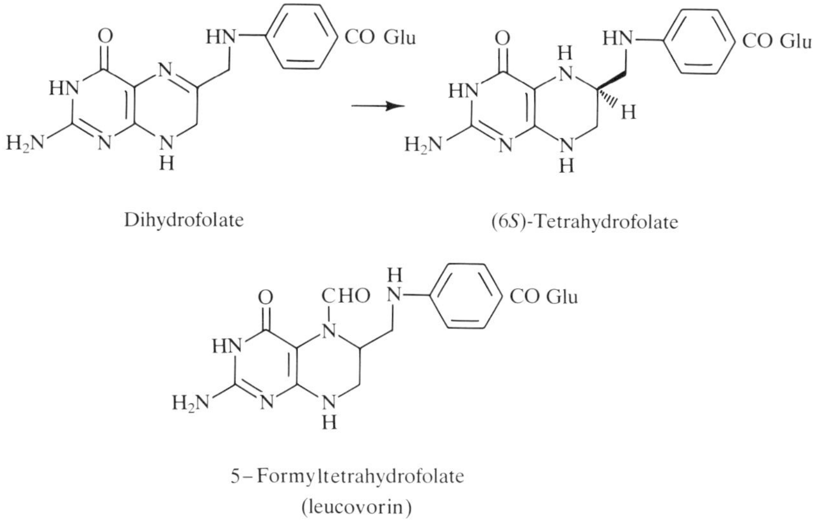

Dihydrofolate   (6S)-Tetrahydrofolate

5–Formyltetrahydrofolate
(leucovorin)

Fig. 4.25 Reduction of dihydro- to tetrahydrofolate.

quantities of diastereoisomerically pure 6-S-THF and hence leucovorin in the natural configuration. The optical purity of the product was demonstrated by NMR and HPLC of a derivative prepared using 1-naphthyl-1-isocyanatoethane. This detail is significant because it led to the discovery of an alternative route to the required diastereoisomerically pure leucovorin.

We had in mind backing up the enzyme-based synthesis with a chromatographic separation of isomers. During a survey of potential chiral auxiliaries we found that menthyl chloroformate affords a derivative of THF (Fig. 4.27) that was resolvable by simple solvent extraction with n-butanol (Rees *et al.*, 1987). Reactions were then developed to convert the menthyl urethane into each isomer of leucovorin in good optical purity. Because of the low catalytic activity of DHFR and the cost of recycling, it is likely that a commerical route to leucovorin and its relatives would be based upon separation methods such as the above. It is ironic that, having promoted strongly the virtues of enzymes in organic synthesis, our most-studied example should appear to be better approached by classical methods of resolution.

4.6 BIOMIMETIC CHEMISTRY IN SYNTHESIS

A study may be termed 'biomimetic' when the chemist consciously designs his programme based upon a feature of the chemistry of a natural system. The system could be a transport mechanism, a biosynthetic pathway, a group of receptors, or a nucleic acid as well as an enzyme. For the purposes of this chapter, we are concerned with synthetically applicable biomimetic

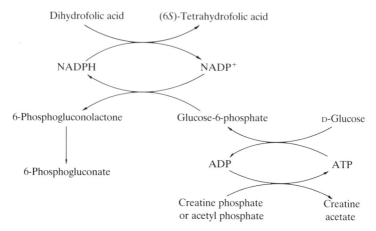

Fig. 4.26 Coupled multienzyme system for the preparation of diastereoisomerically pure tetrahydrofolate.

chemistry. Much has already been written concerning the modelling of synthetic strategy upon biosynthetic pathways, the so-called biogenetic type synthesis (Scott, 1976; Suckling *et al.*, 1978), but the significant comparison for this discussion is between enzymes used in single synthetic steps and biomimetic reactions that attempt in some way to profit from our understanding of enzyme mechanisms. Serious attempts by chemists to gain selectivity by means of enzyme mimicking is a relatively recent development despite Emil Fischer's optimism (see Chapter 1) and a great deal of credit for its development goes to Ronald Breslow. His two reviews of his work (Breslow, 1972 and 1980) are major signposts in the field.

There are several approaches to biomimetic systems. In many cases the objective has been to match the catalytic efficiency of enzymes. Although such work has had a major influence upon the subject by highlighting important concepts, the emphasis upon hydrolysis reactions of *p*-nitrophenyl esters has minimized the synthetic value. Exceptions to this limitation may be noted in those cases where stereoselective reactions have been attempted. Indeed the essence of synthetic biomimetic chemistry is the control of selectivity by some form of binding of the substrate to the enzyme mimic. As in the preceding sections in this chapter, comparisons will be made where appropriate with closely related enzyme-catalysed reactions. However, in biomimetic chemistry, a new degree of flexibility is introduced. Although the reaction is based upon enzyme chemistry, there need not be a known enzyme that catalyses the reaction in question. Hence a more meaningful comparison will be between the biomimetic system and a conventionally designed chemical reaction.

Several key systems capable of providing useful binding for biomimetic synthesis have been discovered. The first to be investigated extensively was

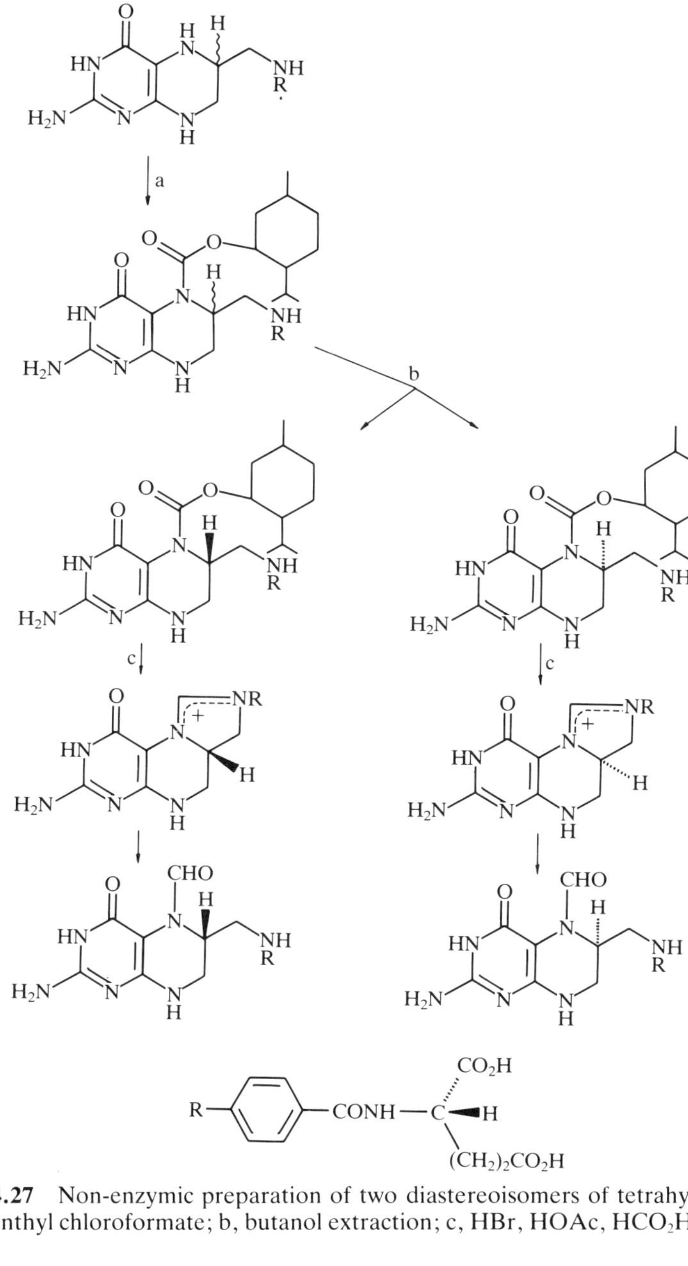

Fig. 4.27 Non-enzymic preparation of two diastereoisomers of tetrahydrofolate.
a, menthyl chloroformate; b, butanol extraction; c, HBr, HOAc, HCO₂H.

the cyclodextrin system. *Cyclodextrins* are cyclic oligomers of glucose,
β-1,4-linked, containing six, seven or eight glucose molecules. They possess
a toroidal structure with an outer coat of hydroxyl groups and a largely
non-polar inner cavity (Fig. 4.28). It is the inclusion of non-polar substrates

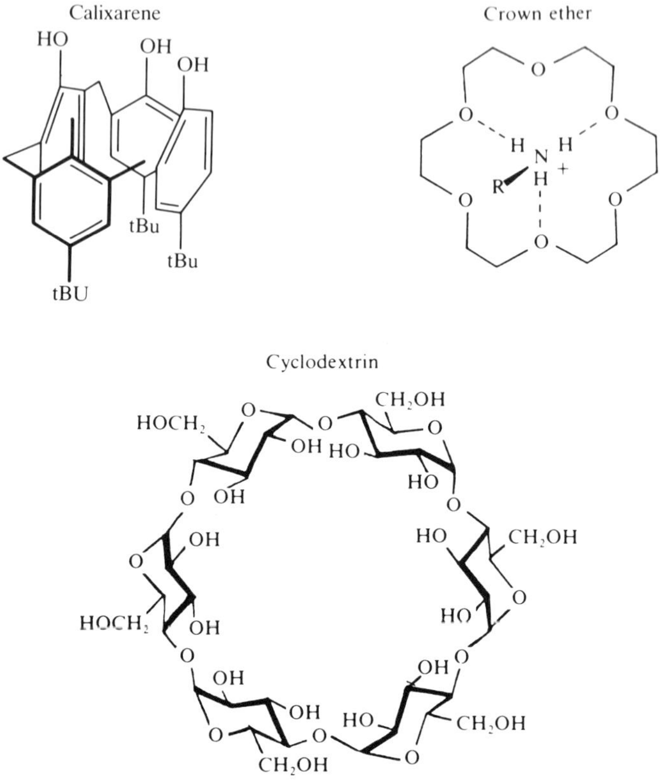

Calixarene

Crown ether

Cyclodextrin

Fig. 4.28   Some building blocks for biomimetic systems.

within this cavity that gives cyclodextrins their useful biomimetic properties. Secondly, *crown ethers* have developed a substantial chemistry. The first of these molecules to be described were macrocyclic derivatives of ethylene oxide (Fig. 4.28). Unlike cyclodextrins, which bind non-polar substrates, crown ethers complex with ions; originally they were significant for their ability to bind metal cations but the discovery that primary ammonium cations will also bind has opened up an organic chemistry. More recently, analogues have been described that will bind anions. The third major class of biomimetic systems is less well defined and concerns *micelles*. The structure of these multimolecular aggregates of detergent-like molecules is still a matter of debate but they have demonstrably interesting biomimetic properties as we shall see.

In addition to these systems, several new cavities have been constructed with properties suitable for binding host molecules (Fig. 4.28). *Calixarenes* have a variable cavity size and, by choice of substitutent X, can be made organic solvent or water soluble (Gutsche and Alam, 1988). Other aromatic and cycloalkyl systems can be combined to generate binding cavities (Tabushi *et al.*, 1981). The possibility that assemblies of such molecules might have important properties akin to some functions of living cells or be valuable in communications devices had prompted developments in a new field known as *supramolecular chemistry* (Lehn, 1988). Most of the host molecules exemplified above have relatively non-specific binding capabilities and recently the construction of molecules designed with precise molecular recognition characteristics has been demonstrated (Rebek *et al.*, 1987). However, so far relatively few hosts have been induced to take part in chemical reactions.

The essence of control of selectivity by any of the systems outlined above is ordered binding of the substrate. All types of chemical forces can be used to control binding. In some of the first successful biomimetic syntheses, covalent bonding was used to attach a reagent to a substrate, in this case a steroid, in a specific manner. Crown ethers rely essentially upon ionic or dative covalent bonding. However, weaker chemical forces can also be significant; cyclodextrins use predominantly hydrophobic binding and micellar systems mix ionic and non-polar interactions with hydrogen bonds. In other words, all the forces available to enzymes are used in some form by biomimetic chemistry.

As I have said, biomimetic synthesis is a relatively young field and many of the examples to be described are pioneering. From the point of view of laboratory-scale synthesis, conventional chemical techniques and enzymic procedures will often have the advantage but if simple and effective selective biomimetic systems can be devised, they will be very significant on an industrial scale and will compete with immobilized enzymes and genetically engineered preparations. It may seem that the target for selectivity in biomimetic systems is unattainably high at first sight. However, as was first

Table 4.4 A substance S is transformed into a product P and unwanted by products:

k_2/k_1	Yield without catalysis, %	Catalysis required for 95% yield	Typical reaction
1	50	9	Aromatic substitution
9	10	81	
99	1	891	Aliphatic radical subsitution
199	0.5	1791	
0.25	80	2	
0.11	90	11 (for 99% yield)	

pointed out by Guthrie (1976), even a modest increase in yield may be important commercially. Consider the data in Table 4.4 which illustrates the simple case of a substance S being transformed into a product P and unwanted by-products. Taking an arbitrary 90% yield as acceptable, Table 4.4 clearly shows that improvements in selectivity of as little as 10- to 1000-fold may well be useful. Since the catalytic effects of enzymes are commonly estimated as being between 10^5 and 10^9, to harness but a small fraction of their selectivity in a biomimetic system could be valuable. In addition, reactions unknown in nature can be considered. With this encouragement, we can proceed to discuss synthetic biomimetic chemistry.

4.6.1   Covalent control of selectivity

The use of ring systems to control stereoselectivity in synthesis has been a major strategy (Norman, 1982). Six-membered rings, whose conformations lead to well-understood stereoelectronic effects have been especially important as have the constraints geometrically imposed by the opening of small rings. However, little attention had been paid to the possibility of using large rings in the control of regioselectivity in synthesis until biomimetic studies of steroid functionalization were begun by Breslow. It had been known for many years that micro-organisms were capable of carrying out specific oxidations at chemically non-activated positions in the steroid-ring system. Clearly control of selectivity is provided by specific binding of the substrate to the enzyme's active site and Breslow's target was to construct a simple chemical system that would locate the reagent specifically with respect to the rigid steroid-ring system (Fig. 4.29). The most susceptible saturated carbon atoms for attack are the tertiary centres at C-9, C-14 and C-17. The hydrogen atoms at these centres are all α and axial and can therefore be abstracted by a radical approaching from the α-face. The problem then resolves itself into positioning the reagent appropriately. Experiments in which substituted benzophenones were attached via ester linkages to a 3-hydroxyl function and hydrogen abstraction was initiated photochemically showing that the strategy was feasible. The technique was then developed into a synthetic chlorination process by means of substituted iodobenzenes or diphenylsuphides which place a chlorine atom close to either of the target hydrogen atoms. The reaction sequence is initiated through hydrogen abstraction by the strategically placed chlorine atom which removes the nearby hydrogen atom. Pairing of the tertiary radical with another chlorine atom then leads to the selectively substituted products. Provided that dilute solutions were used to minimize the chance of intermolecular reactions, high selectivity was obtained. Indeed if the pendant chain bearing the chlorine atom was longer than the steroid rings, no reaction occurred. The chlorinated steroids could then be further

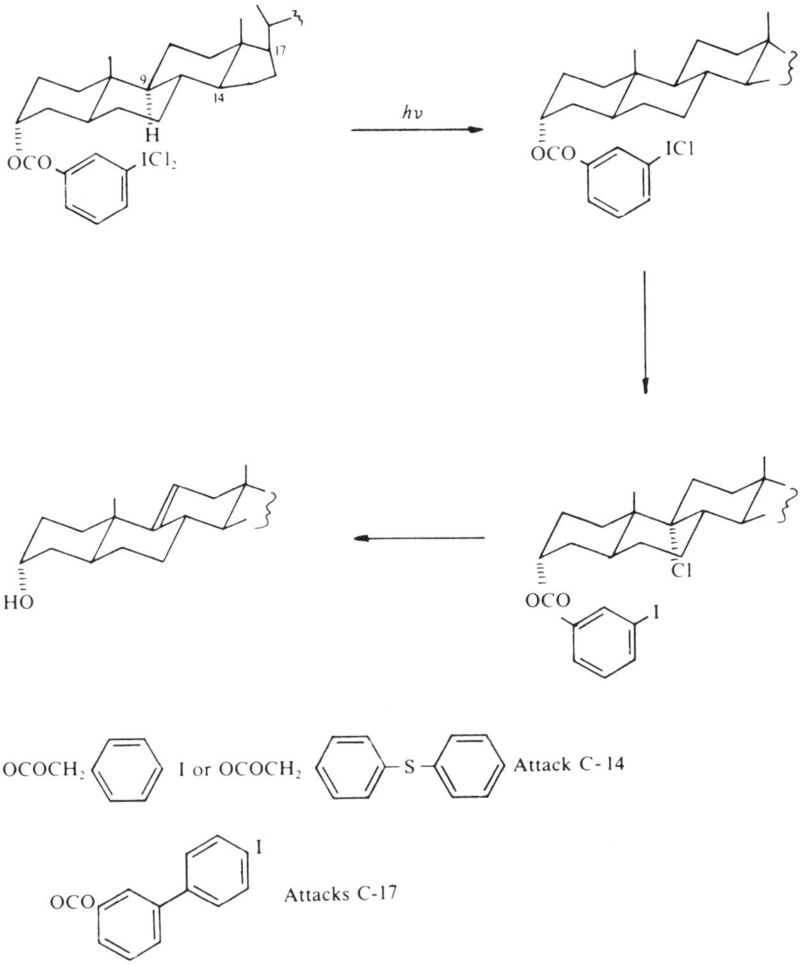
9
17
14
H
OCO
ICl2
hν
OCO
ICl
Cl
OCO
I
HO
OCOCH2
I or OCOCH2
S
Attack C-14
OCO
I
Attacks C-17

Fig. 4.29 Breslow's directed chlorination in steroid synthesis.

transformed to alkenes and, in the case of 17-chloro compound, it was possible to degrade the side chain leading to valuable precursors for the synthesis of steroid hormones. Further developments of these techniques have resulted in a catalytic multiple chlorination of steroids (Breslow and Heyer, 1982) and in an improved experimental procedure using pyridine ester derivatives in place of the iodo compounds (Breslow *et al.*, 1987).

To make a direct comparison with enzyme-catalysed reactions, it would be nice to be able to insert oxygen atoms as well as chlorine. Although hydroxylation has not yet been realized, largely because of the difficulty of

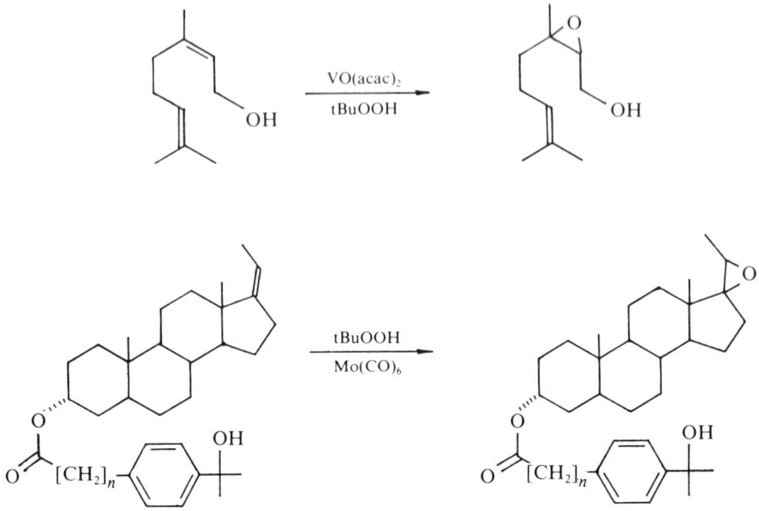

Fig. 4.30 Template-directed epoxidations: $n = 1$ epoxidation occurred, $n = 2$ no reaction.

designing a suitable reagent, epoxidation can be controlled selectively over remarkably long distances. Sharpless and Verhoeven (1979) in an early exploitation of selective epoxidation that preceded the discovery of the celebrated Sharpless oxidation discussed earlier showed that *t*-butyl hydroperoxide was a very useful epoxidizing agent in the presence of transition-metal complexes of vanadium and molybdenum. In these reactions, the metal acted as a template and co-ordinated with both the hydroperoxide and a hydroxyl group in the substrate. For instance, the double bond of farnesol or geraniol closest to the alcohol could be specifically epoxidized (Fig. 4.30). Breslow adapted his template to include a tertiary alcohol to which a metal could bond. It was then possible to direct epoxidation to a remote position in the steroid substrate.

Steroids, with their rigid ring structure, are ideal substrates for this kind of biomimetic strategy. The predictability of the behaviour of the template system is also an advantage. However, the application of similar techniques to more flexible molecules has invariably resulted in the production of mixtures. The problem of selective functionalization of simple alkanes, a problem of industrial significance in the production of primary alcohols for detergents and plasticizers, has still to be solved by biomimetic chemistry, although selective terminal hydroxylation of alkanes can be accomplished by micro-organisms. As has already been mentioned, hydroxylation reactions of steroids have been used in industrial synthesis of corticosteroids for many years, but despite considerable interest, the microbiological hydroxylation of simple aliphatic compounds has not been successful so far on a commercial scale.

4.6.2 Cyclodextrins

The major area of application of the binding properties of cyclodextrins has been the development of systems aimed at mimicking the high rate enhancements of enzymes. In this respect, research has been spectacularly successful (Breslow, 1980; Tabuishi, 1982; Komiyama and Bender, 1984); however, synthetic studies have been less extensive. The cavity in the cyclodextrin torus is a suitable size to bind mono- and bicyclic aromatic rings. For example, anisole and other benzene derivatives with one small substituent, bind so that the substituent is buried in the cavity. The *para* position is usually the most exposed in the cyclodextrin complex and can be attacked by a suitable reagent (Breslow, 1972, 1980). Thus when the anisole complex was reacted with a solution of hypochlorous acid, the peripheral hydroxyl groups of the cyclodextrin exchanged with the hypochlorite and formed a chlorinating agent ideally placed to cause *para* substitution with high selectivity. Interestingly, the most closely related enzyme-catalysed reaction to this chlorination occurs with inferior selectivity close to the ratio obtained with hypochlorous acid in the absence of the cyclodextrin. This chemistry can be extended to construct a simple reactor for selective chlorination using a cyclodextrin crosslinked with epichlorohydrin. Disubstituted aryl ethers can also be chlorinated successfully; however in the case of 4-methylanisole, the methyl substituent preferentially resides in the cavity and chlorination *ortho* to the methoxy group occurs.

A rare case of selective substitution by means of cyclodextrin complexation was described by Tabushi *et al.* (1979); the nearest enzyme-catalysed relative to the reaction studied is the phenylation of phenolic natural products, but the enzymes that catalyse this reaction do not appear to have been isolated. Tabushi was interested in synthesis of vitamin K and he required to catalyse and control the alkylation of a phenol with respect of O- or C-alkylation. His solution was to alkylate the dinaphthol (Fig. 4.31) complexed with a cyclodextrin; concomitant oxidation led to the required intermediate naphthoquinone. A kinetic analysis showed that alkylation was catalysed by the cyclodextrin and that complexation also prevented further oxidation of the product. Selectivity in the alkylation of naphthols can also be controlled by micellar systems as will be described below.

Stereoselectivity is also possible using cyclodextrin derivates although the origin of the selectivity is unclear. Breslow found that pyridoxal attached to β-cyclodextrin was effective in catalysing transamination of aromatic α-keto acids to the corresponding α-amino acid (Breslow *et al.*, 1980). Preferential

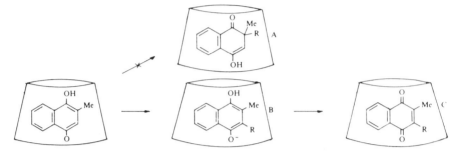

Fig. 4.31 Multiple catalytic effects of a cyclodextrin on naphthalene diol oxidation. R = CH$_2$CH = CH$_2$, CH$_2$, CH = CHCH$_3$ or CH$_2$CH = CMe$_2$. A, Alkylation to form this complex is too hindered. B, Cyclodextrin increases anion concentration and catalyses alkylation. C, Oxidation of diol occurs by O$_2$ affording H$_2$O$_2$: further oxidation of quinone by H$_2$O$_2$ is prevented by cyclodextrin.

reactivity towards keto acids with aromatic groups that can bind in the cyclodextrin cavity was observed and in such cases, the enantiomeric selectivity was also high. A potentially very significant synthetic application of cyclodextrin chemistry concerns the Diels–Alder reaction (Rideout and Breslow, 1980). In reactions where both reactants can complex together within the cyclodextrin's cavity, substantial catalysis occurred and the possibility clearly exists of controlling regioselectivity in such complexed cycloadditions. Similar rate enhancements were also observed in solutions of ionic detergents, an interesting observation in the context of micellar systems described below.

These few examples prove the synthetic utility of cyclodextrins. It might seem that the cyclodextrin system is somewhat limited in the range of compounds that it can accept as guests. Since there are three ring sizes of cyclodextrin produced by micro-organisms, the diameter of the cavity can be chosen within reasonable limits, although it cannot yet be easily varied. However, synthetic methods have been developed to extend or decrease the depth of the cavity and hence to modify the substrate specificity at least as far as ester hydrolysis reactions are concerned. The application of modified cyclodextrins to synthesis is a subject of active research in several laboratories.

4.6.3 Crown ethers

The ability of the first crown ethers to bind inorganic ions was recognized as long ago as 1967 and it was soon shown that the strongest binding was obtained for those cations whose ionic radii most closely matched the available space between the donor oxygen atoms. Their discoverer,

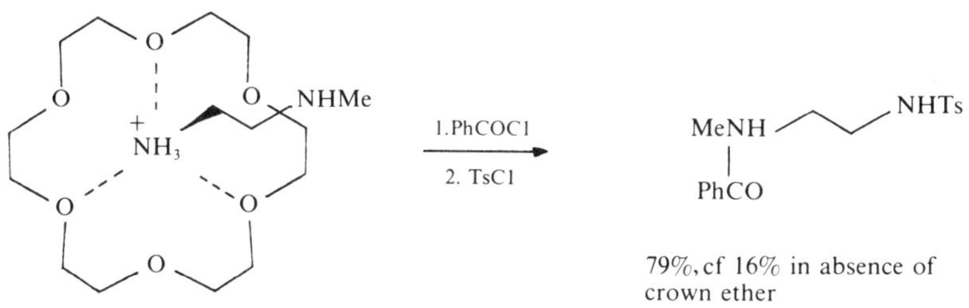

Fig. 4.32 Crown ether-controlled acylation of a diamine.

Pedersen, and their principal exploiters, Cram and Lehn, were jointly awarded the Nobel Prize in 1987 (Pedersen, 1988; Cram, 1988; Lehn, 1988). Potassium bound especially strongly to ethers containing six donor oxygen atoms spaced equally between dimethylene bridges. Since the ammonium cation has an ionic radius very similar to potassium, it was no surprise to find that stable ammonium complexes could be prepared. It was then a small step to show that primary alkyl ammonium cations would also bind, and from this result, the organic chemistry of crown ethers has grown (Stoddart, 1984). A direct synthetic application of the ability of crown ethers to bind primary ammonium cations was demonstrated by Barrett et al. (1980). Although a primary ammonium cation binds, largely with the aid of hydrogen bonding of the three hydrogen atoms, with three of the donor oxygen atoms of the ligand, secondary ammonium cations are both too bulky and too ineffective in hydrogen bonding to complex with the ligand. It was therefore possible to protect a primary ammonium cation selectively in the presence of a secondary cation with a crown ether. Unsymmetrical acylation was then carried out in vastly improved yield over a comparable uncontrolled reaction (Fig. 4.32). A similar binding strategy this time using H_3NBH_3 as the guest enabled Stoddart to effect enantioselective reduction (Allwood et al., 1984).

A major area of interest in crown ether chemistry has been the introduction of chirality into the macrocyclic ether. Diols from carbohydrate derivatives can be converted into crown ethers, Cram's use of chiral binaphthyl diols (Fig. 4.33; Cram, 1976) has been especially effective. The chirality in these compounds, like biphenyls, is due to restricted rotation around the central C–C bond. The first implication of these ligands was in the formation of enantioselective complexes with derivatives of amino acids. By means of a U-tube transport apparatus, it was possible to effect a resolution of the racemic amino acid analogous to the use of the specific esterase enzymes described earlier. The potential of these chiral crown ethers has been greately extended by the demonstration of asymmetric Michael addition reactions to methyl vinyl ketone (Cram and Sogah, 1981).

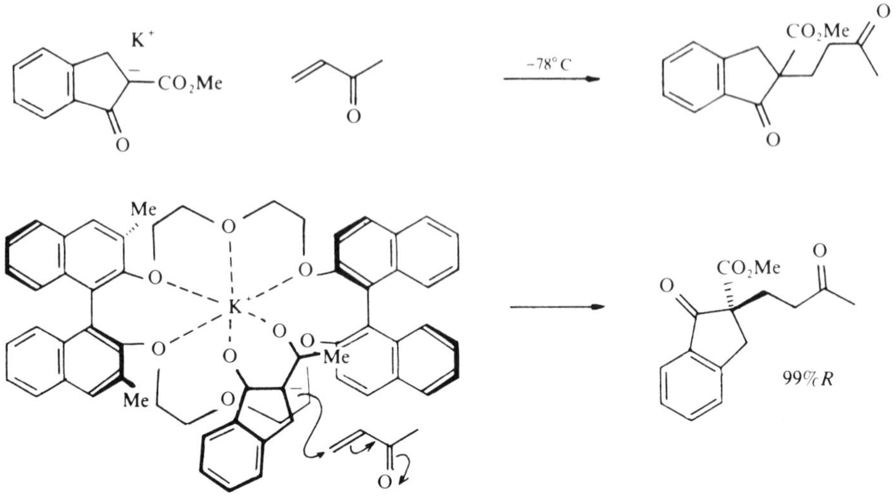

From front

Fig. 4.33   Enantioselective Michael addition mediated by a crown ether.

Although similar reactions have been carried out before, using chiral alkaloids to provide the asymmetric induction, the best examples of crown ether-mediated reactions proceed with almost complete stereoselectivity. It is essential to use the potassium enolate as the nucleophile so that a host–K^+-substrate⁻ complex can be formed. Since potassium ions bind strongly to the crown ether, the enolate can be maintained in a chiral environment and will add to methyl vinyl ketone to give a product in a high state of optical purity. As with the resolution experiments, the asymmetric induction relies upon the attainment of the crown ether complex in which steric interactions between bulky groups on the host and guest molecules are minimized. It was also extremely important to operate at low temperature and to choose reaction conditions carefully so that the differences in energy between the possible diastereoisomeric complexes could have most effect. Thus in the reaction illustrated in Fig. 4.33, at −78°C, the product was obtained in 99% enantiomeric excess whereas at 25°C, the optical purity was reduced to 67%.

Crown ethers, unlike cyclodextrins, have the advantage of being man-made molecules whose structures can be designed to suit the problem at hand (Stoddart, 1984). Although not directly relevant to synthesis, crown ethers capable of mimicking enzymic enantioselective hydrolysis and reduction reactions have been synthesized. Some ingenious molecules have been designed incorporating external control of complexing ability; photo-sensitive groups permit the switching on and off of complexation in some compounds and, in others, a second metal ion can control the conformation of the crown ether again controlling complexations. The latter example can be compared to allosteric control of the activity of an enzyme. In addition to the crown ethers that bind main-group cations, derivatives containing β-diketones have been prepared to interact with transition metal ions. Further variants bind phosphates through macrocyclic amidines. Such compounds mimic a range of biological receptors and can be constructed to accommodate bifunctional compounds with structures similar to neuro-transmitters (Sutherland, 1982). Molecules that are capable of binding two functional groups might clearly bind two substrates. In this way, a class of artificial enzymes might be designed. Recent work has shown that it is possible to synthesize functional analogues of ionophores, the components of cell membranes, responsible for the transport of inorganic ions such as potassium and chloride. Thus the potential of this family of compounds is much wider than the synthetic field and applications to switching devices and to selective artificial transport systems are conceivable.

4.6.4 Micellar and related systems

In contrast to the biomimetic systems that have been considered so far, micelles are structurally ill-defined. It has, of course, been known for many years that surfactants with greater than ten carbon atoms in a straight aliphatic chain will readily agglomerate in aqueous solution to form micelles. The gross physical chemical properties of micelles have been well established by such techniques as light-scattering measurements. In this and in other ways it has been possible to demonstrate that typically 50–100 molecules make up a micelle and that the surfactant molecules enter and leave the micelle at rates greater than 10 s^{-1}. The electrical properties of micelles have also been well studied; micellar catalysis of organic reactions has to a large part been attributed to the ability of micelles to develop unusually high concentrations of counterions in the region of the head group. For example, a hydrolysis reaction mediated by hydroxide anion would be catalysed by a micelle bearing positively charged head groups but inhibited by one with negatively charged heads (Bunton, 1976). Despite careful studies with skilfully selected probes, the detailed structure of micelles is uncertain, which is a major disadvantage to the design of synthetic systems incorporating micelles as the binding group. One of the most useful representations is due to Menger (1979) who suggested on the basis of spectroscopic and model-building studies that micelles are irregular spheroids in which the component surfactant molecules enter and leave through grooves in the surface. Water penetrates up to seven carbon atoms into these grooves, and from the synthetic point of view, it is attractive to imagine the grooves as convenient binding loci for substrates.

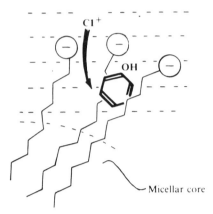

Fig. 4.34 Orientation of phenol in micellar solution and its effect upon chlorination.

With these difficulties in mind, it is not surprising that studies of the application of micellar systems to biomimetic synthesis have proceeded empirically (Suckling, 1981). Micellar systems have the virtue that they are simple to construct but the price for this facility is the lack of a clear structure, like a cyclodextrin or crown ether, on which to design a binding and an active site. The most encouraging suggestions that micelles might be useful in synthesis came from several laboratories where it was shown that in aliphatic chemistry, spectacular control of reductive mercuration of terpenoid molecules could be achieved (Link *et al.*, 1980), and orientation in electrophilic and radical aromatic substitution can be influenced by micelles. We were able to demonstrate by means of high-field nuclear magnetic resonance (NMR) spectroscopy that micelles cause small polar aromatic substrates to be oriented as shown in Fig. 4.34. If the reagent approaches the substrate from the polar aqueous phase, then it is easy to see how an increase in *ortho* substitution can result. In a number of cases, the chemical selectivity corresponded well to the orientation of the substrate suggested by the NMR spectra. These observations begin to provide the basis for the design of more effective selective substitution systems.

There are two obvious approaches both of which add a degree of control to the semiordered micellar system. Firstly, several molecules of surfactant can be covalently bonded together to form a molecule bearing tentacles. Menger *et al.* (1981) was the first to describe such a molecule (Fig. 4.35); his six-armed 'hexapus' was capable of solubilizing a molecule as large as cholesterol in water. Our own tentacle molecules contained three arms and can be shown to bind small aromatic molecules like phenol both by inhibiting chlorination reactions and by substantial changes in the NMR

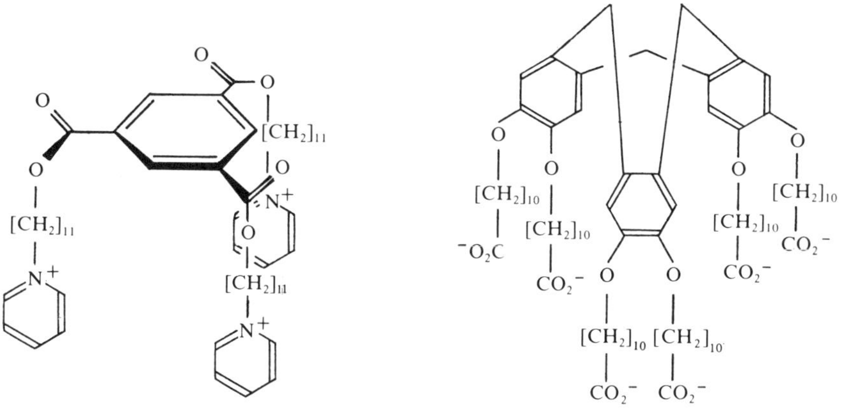

Fig. 4.35 Tentacle molecules that could become the basis of biomimetic systems in synthetic or other applications.

spectrum analogous to those found in micellar solution (Suckling, 1982; Fig. 4.35). Both of these molecules are, however, prototypes for selective functionalization systems in that they possess binding sites but have still to be equipped with active sites.

The second way to improve the selectivity of micellar systems is to localize the reagent at a defined position within the micelle. The clearest indication of the importance of this comes from our work on the chlorination of phenol in micellar solution. As Fig. 4.35 illustrated, we were able to show that

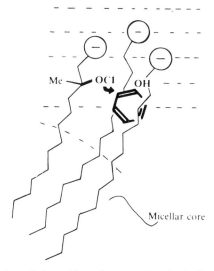

Fig. 4.36 Chlorination of phenol by a functionalized micelle.

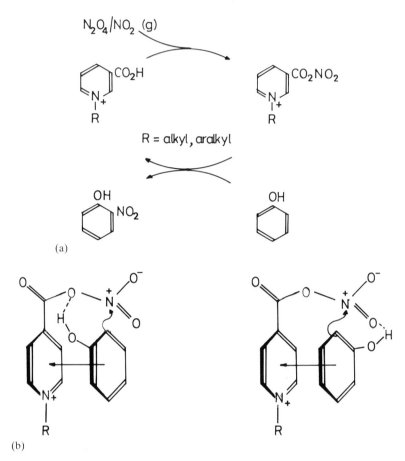

(a)

(b)

Fig. 4.37 Selective nitration of phenol a, carrier recycling; b, possible origin of selectivity.

phenol is solubilized on average close to the head group such that the polar hydroxyl group points into the aqueous phase. Clearly to obtain selective *ortho* substitution, the reagent must be inserted at about C-3 of the micelle. This was achieved using a functionalized stearic acid bearing a tertiary alcohol (Fig. 4.36; Onyiriuka and Suckling, 1982, 1986); the hypochlorite from this alcohol caused chlorination of phenol to occur with close to complete selectivity for *ortho* substitution. The most important conclusion from this work is that it is possible to obtain high selectivity even with a relatively poorly defined binding site provided that the reagent is properly positioned with respect to the substrate. When compared with conventional reagents, these micellar systems have the major potential of being able to carry out a selective transformation in principle in one step.

One of the observations made in our study of tentacle molecules was when an especially strong association between phenols and pyridinium head groups took place. We wondered whether this interaction would be sufficient to induce selective substitution if a suitably located reagent was included. The reaction selected for study was a non-natural one that has great industrial importance but also has technical and environmental problems, namely aromatic nitration. Firstly single chain amphiphilic molecules with pyridinium carboxylate head groups were examined and selective *ortho* nitration of phenol was demonstrated using nitronium tetrafluoroborate as nitrating agent (Pervez *et al.*, 1988). We quickly showed that the long alkyl chains were superfluous and that the expensive nitronium tetrafluoroborate could be replaced by the cheap nitrogen dioxide. Spectroscopic studies suggested that a combination of charge-transfer interactions and hydrogen bonding was responsible for the selectivity, as in related reactions (Lemaine *et al.*, 1987). The highest level to which this system was developed was a polymer-supported pyridone as nitro group carrier. This material was firstly suspended in solvent charged with NO_2/N_2O_4 gas, washed and then treated with phenol; the product isolated after simple filtration and evaporation of solvent was crystalline 2-nitrophenol, greater than 98% pure (Fig. 4.37). The term 'biomimetic' can justifiably be used for this reaction, not simply because of its scientific origin, but because it takes advantage of a specific orientation of reactants controlled by hydrogen bonding, a force of great significance in substrate binding to enzymes.

Although this biomimetic nitration has departed from micellar chemistry, other groups have shown that the controlled environment provided by micelles promotes selectivity in a number of other reactions including ether synthesis and bromination of phenylpolyoxaalkanes (Jursic, 1988).

4.6.5   Biomimetic oxidation

Although studies of biomimetic epoxidation eventually led to important advances in synthetic chemistry as we have seen, biomimetic hydroxylation of non-reactive carbon atoms is a major goal yet to be achieved in a preparatively practical way. Such systems as have been devised are usually models of cytochromes P-450, a class of enzymes with a wide range of hydroxylating and oxidizing capabilities (see section 6.2). Groves has made a major contribution through his studies of metallotetraphenylporphyrin derivatives that transfer oxygen typcially from iodosobenzene to a substrate. Epoxidation is probably the most efficient example: porphyrin ruthenium complexes are capable of activating dioxygen catalytically through a complex cycle (Groves and Ahn, 1987). If a superstructure is erected

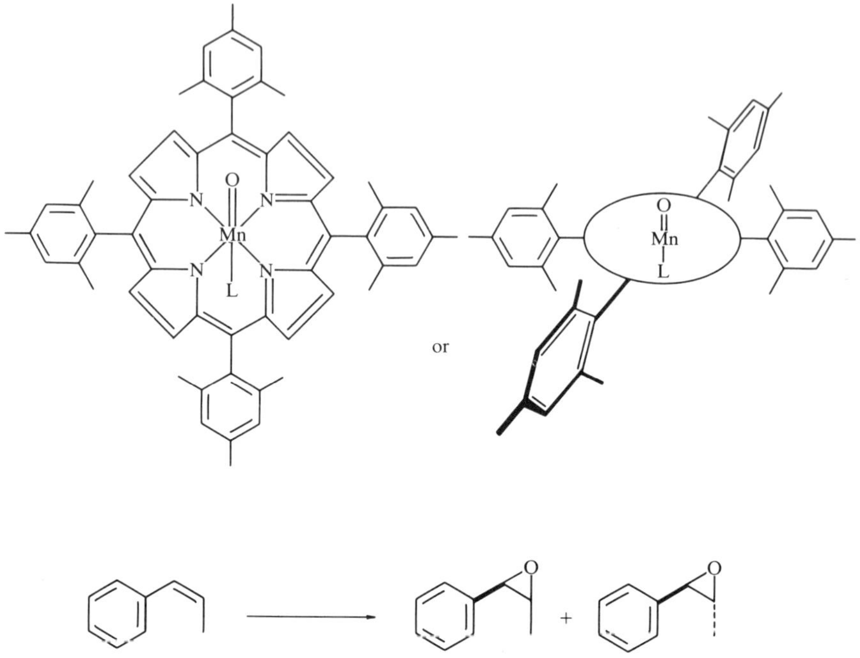

Fig. 4.38 Biomimetic oxidation using porphyrins. Best performance: 32:1 *cis/trans* ratio of epoxides.

(Fig. 4.38), stereoselective epoxidation can be achieved (Groves and Stern, 1988). Some modifications of Grove's system lead to impressive perform-ance both in terms of yield and selectivity. Mansuy has shown that with manganese porphyrins, hydrogen peroxide can serve as a suitable oxygen source for both alkane and alkene oxidation (Battioni *et al.*, 1985; Groves and Stern, 1988). Cytochromes P-450 themselves, because of their difficulty in isolation and instability, are not useful as synthetic catalysts. Nevertheless other hydroxylases can be used effectively in synthesis as some studies on the production of tryosine from phenylalanine show (Klibanov *et al.*, 1981).

Digressing briefly on the subject of haem-containing enzymes it is worth noting the use of the enzyme chloroperoxidase in selective synthesis. In the presence of hydrogen peroxide and chloride, this enzyme catalyses the insertion of chlorine as an electrophile into an organic substrate, usually an alkene or an electron-rich aromatic compound. A telling application concerns the large-scale synthesis of ethylene and propylene oxides. Conventional heterogeneous catalysed oxidation of the alkenes always leads to some losses due to over-oxidation to carbon dioxide. Chloroperoxidase, in contrast, will convert these substrates into the corresponding epoxide or halohydrin without over-oxidation. When the epoxide is the product, chloride is regenerated and thus is also catalytic (Neidlemann *et al.*, 1981). The fuel for the reaction is hydrogen peroxide and, during development work it was felt that conventional methods of hydrogen peroxide production might be bettered by oxidases that generate hydrogen peroxide from oxygen with concomitant oxidation of an organic substrate. Research has since shown that it is possible to perform this important epoxidation using two enzymes, chloroperoxidase and an oxidase. It is significant that immobilized enzymes were again favoured and also that considerable biochemical and microbiological input was required to produce suitable strains of organisms as enzyme sources.

4.6.6   Other biomimetic possibilities

In addition to the three main groups of binding species described above, other molecules have been found to have biomimetic binding properties which may in time find a use in synthesis. Some experiments with small peptides have shown that it is possible to mimic the stereospecific enzyme-catalysed addition of cyanide to benzaldehyde using a cyclic phenylalanyl-histidine peptide. (Oku and Inoue, 1981). Many chemists have attempted to achieve stereoselective epoxidation using biomimetic chiral catalysts or simpler chiral-phase transfer catalysts. Usually base catalysed reaction of αβ-unsaturated ketones have been studied, and in many cases, high chemical yields and almost 100% enantiomeric excesses were obtained (Julia *et al.*, 1980; Wynberg and Marsmann, 1980). These reactions are related to Cram's asymmetric Michael additions mediated by crown ethers.

Recently, some interesting hybrid systems have been described. Kaiser (Kaiser, 1988; Slama *et al.*, 1981) has incorporated a foreign coenzyme, a flavin, into an enzyme, papain, with a well-defined binding site. This is an attractive pairing because papain is known to contain an active-site thiol group-, and many natural flavoenzymes use a thioether to bind the flavocoenzyme to the active site. Although the work is in its infancy, he has been able to demonstrate high catalytic activities of the flavin in oxidation reactions in its unfamiliar environment. The work has come closest to synthesis in the conversion of haemoglobin to a hydroxylase (Kokubo *et al.*, 1987). It was known from X-ray crystallography that the β-chain of haemoglobin possesses a cysteine close to the haem (Fig. 4.39) and, following the pattern established with flavopapain, a flavin was introduced at this site to provide a route for electron transfer to the haem. In this way a protein capable of catalysing the conversion of aniline into 4-aminophenol was prepared. Kinetic properties typical of an enzyme were demonstrated. A further scientific signpost was the demonstration that the serine protease subtilisin, in which the serine hydroxyl group was converted into a cysteine, was an efficient catalyst for peptide synthesis (Nakatsuka, *et al.*, 1987).

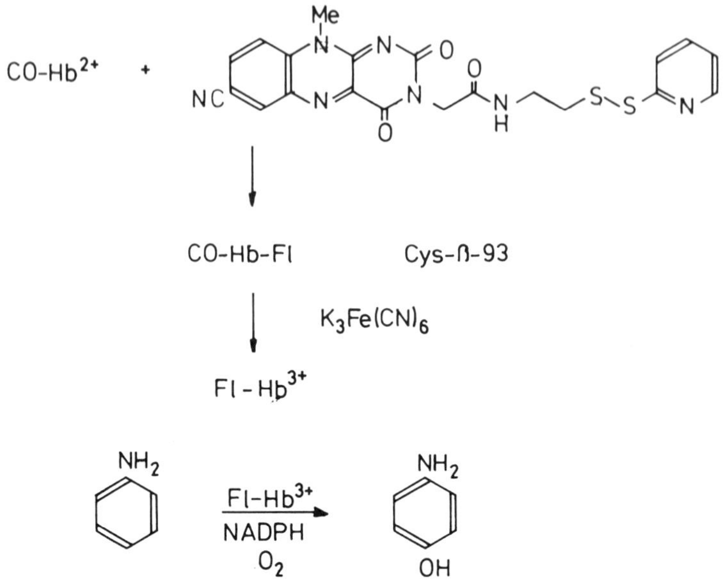

Fig. 4.39 Hydroxylation using flavohaemoglobin, a chemically modified protein.

Chemically modified proteins may thus have uses through modifications of catalytic activity or of substrate specificity in addition to improvements in stability through well established immobilization techniques (see section 4.2.2).

Alcohol dehydrogenase is, as we have seen, extremely well studied from the point of view of substrate specificity. Curiously, it also contains the charge relay system of the serine proteinases and it is therefore intriguing to wonder whether it might be persuaded to act as an esterase with suitable substrates. Also the zinc at the active site is believed to promote reduction of aldehydes and ketones by acting as a Lewis acid. Might it be capable of inducing other reactions (catalysed by Lewis acids) such as Beckmann rearrangements? Zeolites are silicate minerals that possess channels or pores with a defined diameter. Also zeolites can be manufactured to contain acidic or basic surface catalytic sites. They have been widely studied for their potential in controlling large-scale rearrangement and substitution reactions of small hydrocarbons. The analogy between the pores of the zeolite and the binding site of an enzyme is obvious. The use of a zeolite to catalyse the hydrogenation of fructose to mannitol has been combined with immobilized glucose isomerase to produce a one-pot conversion of glucose, which is readily available from corn syrup, into mannitol, which currently commands high prices for pharmaceutical and medical applications (Ruddleston and Stewart, 1981).

4.7 ENZYMES IN ORGANIC SOLVENTS

One of the main criticisms of the synthetic potential of enzymes was that it is very difficult to induce typical organic compounds to react with them because of the low water solubility of the potential substrate and the high water solubility of the enzyme. However lipases, enzymes with proved synthetic potential, have long been known to act at aqueous–organic interfaces. There was no reason, apart from received opinion, not to reduce the water content still further and to investigate virtually dry systems. 'Anhydrous' in this field has been defined as a system in which water cannot be detected by Karl Fischer titration but the optimum water concentration may lie between this limit and 10% v/v. A further stimulus was the realization that to use hydrolytic enzymes effectively in the reverse direction, i.e., to synthesize esters and amides, low water media were necessary to force the equilibrium to favour synthesis. The effect of non-aqueous environments upon enzyme stability has already been described (section 4.2.4). What is striking about the developments in this field is not that a conventional view has been overturned, but that an enormous expansion of inventiveness and creativity has been stimulated. The following examples have been selected to illustrate both practical and conceptual innovation. They refer to peptide and ester synthesis, the reactions most extensively studied in organic media.

The use of peptidases to synthesize peptides was discussed as early as 1898 and has been established now for many years (Jakubke *et al.*, 1985); the method was promoted essentially to avoid racemization commonly encountered when non-enzymic coupling reagents were used. Enzyme-catalysed reactions can be used to couple pairs of amino acids to yield dipeptides or peptides to prepare larger oligopeptides. Most of the commonly available peptidases have been applied to synthesis, including trypsin, chymotrypsin, subtilisin, pepsin, and papain, with varying degrees of success, success that with hindsight can often be related to the reaction conditions used.

An example of a simple peptide-forming reaction that can be improved by use of an organic solvent is the chymotrypsin catalysed coupling of Z-Trp (1 mM) with GlyNH$_2$ (100 mM). At pH 6.7 the equilibrium constant for peptide formation is 0.45 M^{-1} in water, 2.21 M^{-1} in 60% v/v glycerol and 38 M^{-1} in 85% 1,4-butanediol (Jakubke *et al.*, 1985). Not only do organic solvents widen the range of accessibility of substrates through solubility, and alter the position of equilibrium, they also change the balance of interactions in substrate binding. In aqueous solution, peptide binding is essentially a hydrogen bond exchange between water and the binding cleft of the enzyme; the net binding energy due to hydrogen bonding is thus small because almost as many hydrogen bonds are made as are broken. In contrast, if a peptide is transferred from a non-polar phase to a binding site, all the hydrogen bonds formed are energetic gains. Similarly the hydrophobic interactions of non-polar amino acid side chains will be weaker in a non-polar medium. One way of expressing the outcome of these arguments is that the substrate requirements of the enzyme are relaxed in non-polar media. Two examples illustrate this point.

Klibanov (Zaks and Klibanov, 1986) has determined the reactivity of phenylalanine derivatives, typical hydrophobic chymotrypsin substrates, in comparison with that of serine derivatives. In aqueous solution, hydrolysis at phenylalanine in peptides is 46 000 times more efficient than that for serine sites. In contrast the esterification of acyl serine is 3.5 times faster than that of phenylalanine in organic solvent (octane).

Wong has achieved the ultimate heterodox reaction in peptidase chemistry. The stereospecificity of peptidases for the $S(L)$ series of α-amino acids is always emphasized but Wong has shown that it is possible to prepare dipeptide esters containing D-amino acids using chymotrypsin in up to 80% yield (Blair and Wong, 1986). For example, Z-TryOMe acylates D-MetOMe in aqueous carbonate/dimethyl formamide solution. There is, however, a clear rationale for the success of this reaction (Fig. 4.40). The non-polar acylating amino acid binds in its usual pocket but the binding requirements of the site marked n, the nucleophilic amino acid, are sufficiently relaxed under these conditions that many groups can be accommodated including D-amino acid esters. This fact had been realized earlier but it was left to Wong to exploit it in synthesis. The most practical version of this reaction used an immobilized preparation (on a polyacrylamide gel) of a chemically modified chymotrypsin: a methionine (Met-192), which is close to the active site, was oxidized to the sulphoxide thus improving greatly the stability of the enzyme.

One of the main ways of extending the usefulness of an enzyme in synthesis has been to investigate transformations of non-natural substrates. In peptide synthesis this device has taken the form of using lipases, whose normal substrates are fatty acid esters, to prepare peptides in organic solvents (West and Wong 1987; Margolin and Klibanov, 1987). Naturally, these reactions work best if hydrophobic substrates are used

$$ZPhe\text{-}OMe + Ala\text{-}OsBu \rightarrow ZPheAlaOsBu$$

$$AcPheOCH_2CH_2Cl + LeuNH_2 \rightarrow AcPheLeuNH_2$$

Yields of greater than 80% are obtained but reaction times are long ($\sim$24 h) and large quantities of enzymes are required. This is typical of enzyme-

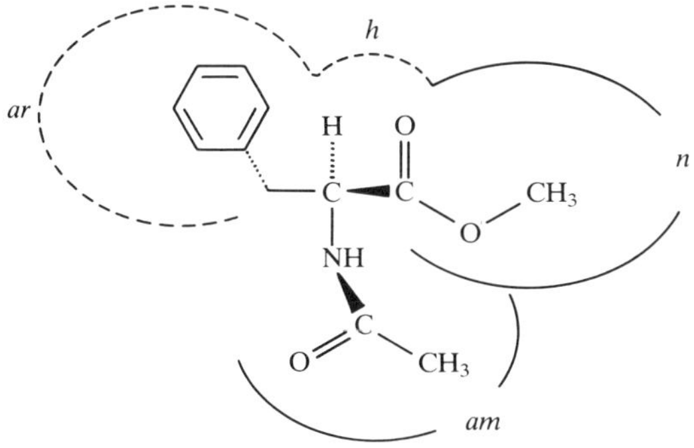

h accommodates small groups H Cl OH F, not Me

N is active site

am amide site – also takes, without binding energy gain
   H bonding   Me, Cl, $CO_3CO_2.EtOCOCH_2$ etc.

ar big ar pocket

n and *am* are open ended and will take up to tripeptides.

Fig. 4.40   A schematic model of the binding sites of chymotrypsin.

catalysed reactions in organic media: the enzyme is usually present as a crude powder in suspension in the solvent.

Lipases can also be induced to catalyse acylation of carbohydrates, another unexpected reaction. In this case, pyridine has been found to be a suitable solvent (Therisod and Klibanov, 1987) and the least hindered alcohol of mono- or disaccharides, typically a primary alcohol, is selectively acylated. An expecially neat way to carry out such selective acylations is to use isopropenyl esters and protease N in anhydrous DMF (Wong *et al.*, 1988; Kim *et al.*, 1988). The leaving group from the acylating agent is acetone; hence the reaction is essentially irreversible and, more importantly, because acetone is not recognized by the enzyme, product inhibition of acylation is obviated.

Alcohol dehydrogenase has also been shown to be active in organic solvents retaining its chirality (Grunwald *et al.*, 1986). In this case the enzyme was cosupported on glass beads with NAD or NADH by lyophilization. With both enzyme and coenzyme restricted to the glass beads high enantioselectivity of oxidation and reduction was obtained. The efficiency of cofactor recycling was exceptionally high and one molecule reacted with up to 10^6 substrate molecules. However reactions were slow and required large samples of enzyme.

The results obtained from the studies described above have stimulated much interest as well as argument. There is a lack of basic mechanistic and physical chemical understanding of the influence of non-natural environments upon enzymes, especially their chemoselectivity. For example, why does the esterase activity of peptidases remain high at pH 9 whereas the amidase activity is greatly reduced (Barbas and Wong, 1987) and why does dioxan inhibit peptidase activity but leave esterase activity? Progress will be facilitated if answers to these questions can be provided.

4.8   ADVANCES IN PROTEIN CHEMISTRY AND MOLECULAR BIOLOGY

In several of the previous sections I have discussed what can be achieved when chemical modifications are made to enzymes and proteins or when non-natural reaction conditions are employed. The development of enzyme chemistry applied to synthesis has, however, been greatly enriched by the contributions made by biological sciences (see Chapter 9). Perhaps the use of genetic engineering to redesign an enzyme comes most quickly to mind but, as will be seen below, so far its success in terms of generating practical catalysts has not matched that of chemically modified enzymes. Of greater significance, especially for industry, has been the search for new enzymes to tackle specific tasks. When coupled with an appreciation of reaction mechanism, this approach can lead to major extensions in the synthetic capability of an enzyme as in the case of penicillin biosynthesis. The most recent addition to the biological developments concerns catalytically active antibodies. In this work, an antibody is raised to an analogue of the transition state of the reaction in question; since enzymes are believed to catalyse reactions through stabilizing the transition state by binding at the active site, a similar event at the binding site of an antibody should also lead to catalysis. This has been demonstrated in several cases. Of course each of the approaches above is in competition with conventional chemistry but as will be seen, there are many examples where the biological catalyst has a substantial advantage. In these new ventures, chemistry and biology are not in competition, they are in a productive partnership.

4.8.1   Genetic engineering

The biological strategies for genetic engineering are described in Chapter 9 and in order to approach the modification of an enzyme for synthetic purposes, several criteria must be met. From the chemical point of view, there must be an accurate 3-dimensional structural analysis of the subject enzyme from X-ray crystallography and the reactivity of many substrates must be interpretable in the light of this model. From this information, the shortcomings of the native enzyme and the improvements required can be identified down to the detail of the amino acid substitutions to be made. To achieve the biological transformation, it is necessary to know the sequence of the gene specifying the enzyme and to have available a method for modifying it as required. Modification is usually achieved by excising part of the gene with specific enzymes (restriction enzymes). A short strand of chemically synthesized gene bearing the required information is then inserted by enzymic ligation. Finally the modified gene must be incorporated into a suitable host and expressed as protein.

So far, few enzymes of synthetic interest meet all these criteria although many impressive scientific feats have been achieved using genetic engineering to probe the mechanism of action of enzymes (see Chapters 2 and 9). The first enzyme to be modified with the aim of enhancing its synthetic utility was yeast alcohol dehydrogenase (YADH) (Murali and Creaser, 1986). YADH differs in its synthetic profile from HLADH in that it accepts a very narrow range of substrates (typcially, small primary alcohols), but has a thousand-fold higher catalytic activity. The aim of the modification was to expand the substrate acceptability of YADH to take advantage of its high catalytic activity.

Although an X-ray structure was not available for YADH, the gene sequence for it and for HLADH was. From the homology of the two sequences and the crystal structure of HLADH, it was suggested that the substrate restrictions were due to the presence of Trp and Thr in YADH in place of Phe and Ser in HLADH (Fig. 4.41). The required modifications

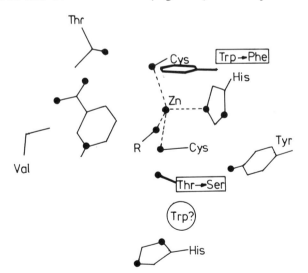

Fig. 4.41   The active site region of yeast alcohol dehydrogenase and some potentially significant mutations. Heteroatoms are identified by ●.

were made and the genetically engineered enzyme isolated. Only small differences in substrate reactivity were observed. For exmple v_{max} for doubly modified enzyme acting upon ethanol was unchanged but the v_{max} with octanol as substrate was six times larger. There are many possible reasons why the effect was disappointingly small. Perhaps the test substrate was an inappropriate choice in that it failed to take advantage of the new structure by filling the space made available; selection of a better substrate could be possible using molecular graphics. Perhaps the chosen modifications were irrelevant to the required enzyme function; there is a large Trp residue in the region of the HLADH substrate binding pocket that might be a candidate for replacement by a smaller residue. Whatever the reason, much has still to be learned about protein structure and activity before improvements to enzymes, especially with regard to synthesis, can become routine.

4.8.2   New enzymes

The difficulties encountered in genetically improving enzymes for synthesis heighten the scientific challenge but for immediate practical purposes, especially in industrial applications, the most effective strategy is likely to be a search for a micro-organism that contains the required enzyme. Having isolated such an organism, the enzyme can then be isolated, the gene cloned, and all the powerful techniques of protein chemistry brought to bear. There is an increasingly large number of examples of new enzymes. The synthetic targets vary widely from monomers for speciality polymer manufacture to pharmaceutical and agrochemical intermediates. The production of such materials by biotransformations is becoming an integral part of the fine chemicals industry.

The first example of an industrially important new enzyme was driven by a requirement to prepare 5,6-dihydroxycyclohexa-1,3-diene, trivially known as benzene *cis* glycol (BCG) in kilogram quantities, principally for conversion into high value polymers such as polyphenylene. An inefficient three-stage process was available to this material and an additional disadvantage was that no aromatic compounds other than benzene could be used. Scientists at ICI therefore sought an enzymic route (Ballard *et al.*, 1983). An organism that grows on benzene as carbon source was discovered. This organism *Pseudomonas putida*, contains an enzyme, benzene dioxygenase, that converts benzene into BCG using oxygen and NADH. In intact cells, NADH is regenerated by other enzymes. The normal strain of *Ps. putida* oxidizes BCG further (Fig. 4.42) but mutants were obtained by standard microbiological procedures of strain selection that accumulate BCG. Fermentation of these strains thus provides an efficient process to BCG.

The conversion of BCG into polyphenylene was accomplished via its

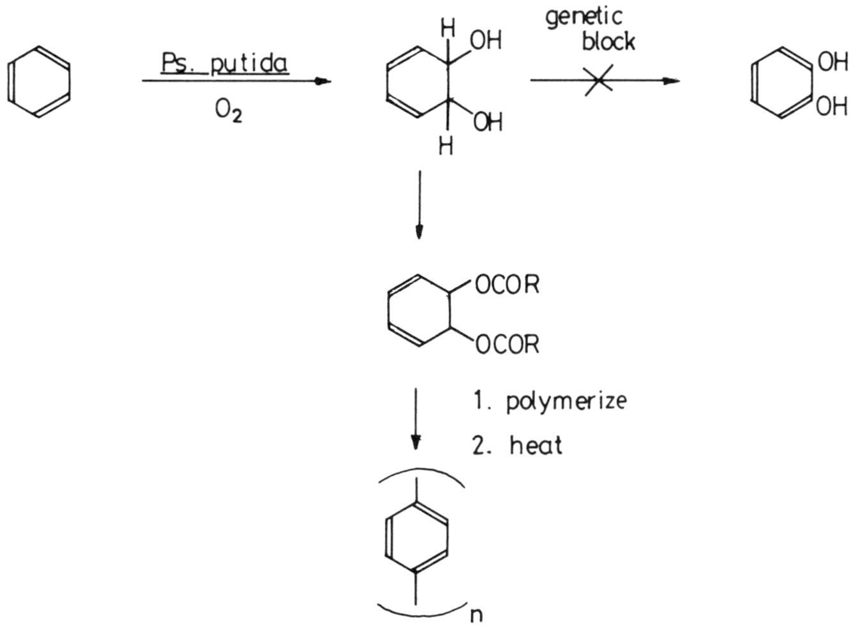

Fig. 4.42   Benzene oxygenase and polyphenylene production.

diacetate; this compound is a conjugated diene and hence amenable to polmerization. The highly polymerized diester contains the carbon skeleton of polyphenylene and by thermal elimination of acetic acid, polyphenylene itself was obtained. There are further uses for BCG, however. The *cis*-addition provides a stereo-chemically defined polyfunctional molecule ideal as a building block for stereoselective synthesis as has been exploited by Ley in a synthesis of (±)-pinitol (Ley *et al.*, 1987). Moreover, benzene dioxygenase will accept other aromatic compounds including monosubstituted benzene derivatives like toluene and bromo-benzene. In these cases, chiral diols with high functionality are obtained which are excellent materials for the foundation of a stereocontrolled synthesis (Taylor *et al.*, 1987).

It is common in such developments for a major effort into the biology of the chosen organism to be required. Additional enzymology is also likely to be necessary before extensive chemical applications can begin. For example, penicillins were one of the first important drugs to be produced at least in part by a fermentation process. However the products available were limited by the natural specificity of the fungal enzymes. Strain selection led to a very efficient production. As the clinical demand for more efficacious β-lactam antibiotics has grown with the development of resistant organisms,

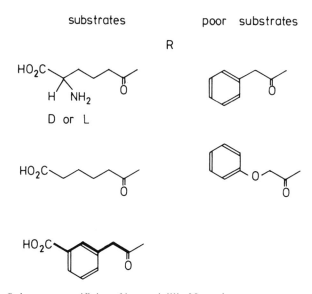

Fig. 4.43 Substrate specificity of isopenicillin N synthase.

so the demands on the manufacturer have increased. Modern chemistry and biology is equipped to meet this challenge and it has been led by Abraham and Baldwin at Oxford (Baldwin and Abraham, 1988).

A long-standing investigation of the mechanism of penicillin ring formation including stereochemical studies (see section 4.4.1) had led Baldwin to believe that radical intermediates played a key role. In order to study this at the enzymic level it was necessary to have supplies of the purified enzymes. Abraham was able to make available sufficient quantities of the required enzyme which was called isopenicillin N synthase (IPNS) after the normal product of its action (Fig. 4.43). By synthesizing many dozens of potential substrates, Baldwin's team was able to show that IPNS would catalyse the synthesis of a very wide range of β-lactams. Modifications

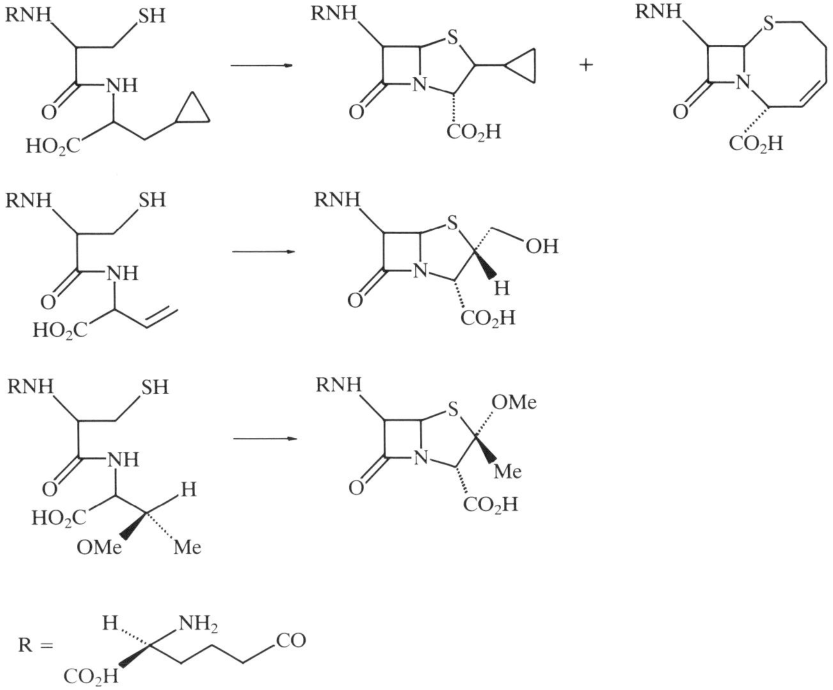

Fig. 4.44   New penicillins obtained using isopenicillin N synthase.

of the natural substrate were introduced in the valine or aminoadipoyl residues (Fig. 4.43). IPNS showed a strong preference for a six-carbon chain with a terminal carboxylic acid as the acyl group of cysteine; this led to more than 100 penicillins with unusual side chains. With regard to the radical mechanism, when substrates containing unsaturated groups or cyclopropanes in the valine residue were used, products characteristic of radical delocalization or rearrangement were obtained (Fig. 4.44(a)). In one case, a compound with significant antibiotic activity was prepared. Assuming that a radical mechanism operates, a radical stabilizing substituent was introduced; *O*-methylthreonine was substituted for valine (Fig. 4.44(b)). This compound was converted efficiently into a penicillin with the normal ring sizes, an elegant confirmation of the scientific power of chemical mechanistic logic allied to enzymology. The work has further extended into related studies of the conversion of penicillins into the related important antibiotics, the cephalosporins (Baldwin and Abraham, 1988). Doubtless, this outstanding scientific endeavour will be the basis of successful commercial projects.

4.8.3 Catalytic antibodies or abzymes

Fischer expressed the desire to be able to construct catalysts related to enzymes as long ago as 1894 (see Chapter 1) although he expressed his views in terms of small molecules. Biomimetic chemistry (section 4.6) has to some extent fulfilled his prophecy and will no doubt improve. The closest approach to the spirit of his views, however, has been the discovery of catalytically active antibodies. Antibodies are proteins that recognize and bind molecules that invade cells; they are characterized by high affinity and specificity for the molecule or part of a molecule, the antigen, with which they react. Since enzymes are believed to catalyse reactions by binding most tightly to the transition state for the reaction, it was suggested that an

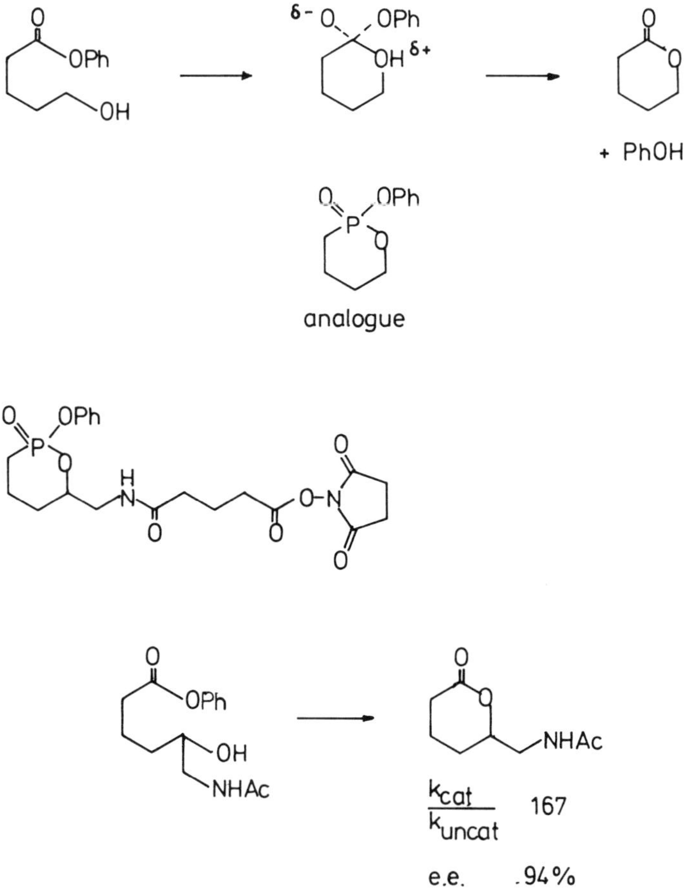

Fig. 4.45 Stereospecific lactone synthesis mediated by catalytic antibodies.

antibody to a transition state analogue might have catalytic properties (Massey, 1987). Transition state analogues are familiar to medicinal chemists and enzymologists in the design of drugs based upon enzyme inhibitors (Silverman, 1988). The skill required of the chemist is to design and synthesize a transition state analogue in a form that will induce a small rodent to generate antibodies. This usually entails attaching the transition state analogue through a linking arm to a large protein molecule. A further discussion on the biological aspects of this procedure is found in Chapter 9.

The potential for synthetic applications of catalytic antibodies has quickly been extended from the exploratory hydrolyses of active esters to reactions with synthetic relevance. The first signpost study concerned the lacto-nization of 5-hydroxypentanoic esters (Napper *et al.*, 1987). As in most acyl transfer reactions, a tetrahedral intermediate is on the reaction path and a transition state leading to this intermediate can be represented as in Fig. 4.45. The conventional analogue of this intermediate replaces the tetra-hedral carbon with tetrahedral phosphorus(V). The conjugate required to elicit antibodies was prepared from the phosphonate *X* bearing a linking arm activated at its terminus. Notice that this linking arm introduces a chiral centre on the six-membered ring. Antibodies that possessed catalytic

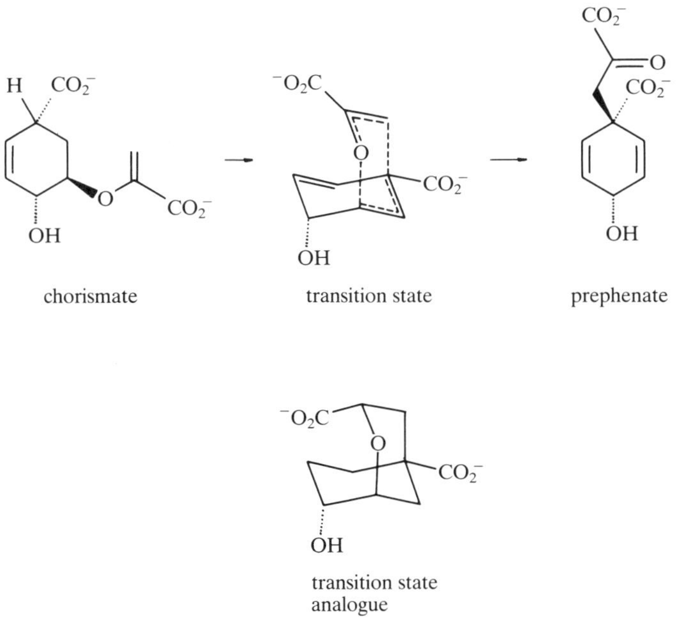

chorismate   transition state   prephenate

transition state
analogue

Fig. 4.46 A biosynthetically important Claisen rearrangement catalysed by anti-bodies raised to transition state analogue.

activity were isolated. Although the rate enhancement due to antibody catalysis observed was modest (167-fold) the most significant observation was that the product was optically active (94% e.e.). The transition state analogue used was racemic and each enantiomer would elicit a separate complementary antibody. One such antibody was isolated and characterized leading to the exciting discovery of a new asymmetric catalyst. This work has now been extended to bimolecular reactions (Benkovic *et al.*, 1988).

The low rate enhancement is not typical of what can be achieved using catalytic antibodies. The conversion of chorismic acid into prephenic acid catalysed by chorismate mutase is academically conspicuous as the only authenticated example of a biological electrocyclic reaction. It has been shown by mechanistic studies to proceed through a chair conformation transition state and this allowed the design of a suitable transition state analogue (Hilvert *et al.*, 1988) (Fig. 4.46). Once again the strategy was successful; a catalyst with a rate enhancement of 10^4 was obtained and this value compares well with the 10^6 rate enhancement achieved by chorismate mutase. These results are extremely encouraging for the development of the field and they suggest that catalytic antibodies may be especially useful for promoting reactions that are entropically unfavourable. Clearly, the full range of modification and derivatization strategies that have been applied to enzymes can be applied to catalytic antibodies to improve their practicality.

4.9 CONCLUSIONS

Writers on synthesis in recent years have stressed such qualities as convenience, predictability, and reliability in their evaluation of reagents. I have tried to show how far enzymes and other biochemical systems match up to these requirements. In the case of alcohol dehydrogenase, esterases, and lipases, a positive response can certainly be given to each criterion. Although few other enzymes have been so well assessed, the type of selective transformation that enzymes are good at is usually crucial to the success of a synthesis and accordingly it must be worthwhile to consider seriously whether a biochemical method might be suitable. Increasingly, examples of syntheses are appearing in the literature in which the basic stereochemical template was prepared by an enzyme-catalysed reaction, so-called 'chemo-enzymatic' syntheses. There is no reason today why enzymes should not be thought of on equal terms to chemical reagents for selective transformations. Indeed if a good analogy can be found, a microbiological method might be valuable too, especially in the industrial context. What about biomimetic methods? At present synthetically useful examples are to be found in only a few cases. It may be that their best field of operation will be in industrial chemistry where it is often necessary to carry out relatively few reactions supremely well. New biology has created real opportunities for imaginative work. In the first edition of this book I suggested that a suitable protein could provide a well-defined chiral environment within which a non-natural synthetic reaction could take place. This has now been achieved with catalytic antibodies. Several promising growth areas can be identified, in particular, those in which new chemistry and new biology are partners as described in the previous section. In addition, hybrid systems in which the proteins provide the selectivity and reagents the reactivity will probably emerge. No doubt the increasing challenge from such science and technology will provoke the chemist concerned with synthetic methodology to redouble his efforts to achieve selectivity and efficiency. Such competition is healthy for scientific development. It is up to the academic chemists to create the opportunities from this challenge; their industrial colleagues can then seek the most cost-effective way to manufacture their increasingly sophisticated and increasingly pure products.

REFERENCES

Allwood, B. L., Shahriari-Zavareh, H., Stoddart, J. F., and Williams, D. J. (1984) *J. Chem. Soc., Chem. Commun.*, 1461.

Baldwin, J. E., Loliger, J., Rastetter, W., Neuss, N., Huckset, L. L., and de la Higuera, N. (1973) *J. Am. Chem. Soc.*, **95**, 3796.

Baldwin, J. E., and Abraham, E. (1988) *Natural Product Reports*, 129.

Ballard, D. H. G., Courtis, A., Shirley, A. M., and Taylor, S. C. (1983) *J. Chem. Soc., Chem. Commun.*, 954.

Barbas, C. F. and Wong, C.-H. (1987) *J. Chem. Soc., Chem. Commun.*, 533.

Barrett, A. G. M., Lana, J. C. A. and Tograie, S. (1980) *J. Chem. Soc., Chem. Commun.*, 300.

Bashiardes, G. and Davies, H. G. (1987) *Tetrahedron Lett.*, **28**, 5563.

Battersby, A. R. and Staunton, J. (1974) *Tetrahedron*, **30**, 1707.

Battersby, A. R., Staunton, J. and Wiltshire, H. R. (1975) *J. Chem. Soc., Perkin Trans 1*, 1156.

Battioni, P., Renaud, J.-P., Bartol, J.-F., and Mansuy, D. (1985) *J. Chem. Soc., Chem. Commun.*, 341, 888.

Baxter, A. D., and Roberts, S. M. (1986) *Chemistry and Industry*, 510.

Becker, W., Freund, H. and Pfeil, E. (1965) *Angew. Chem.*, **77**, 1139.

Bednarski, M. D., Chenault, H. K., Simon, E. S., and Whitesides, G. M. (1987) *J. Am. Chem. Soc.*, **109**, 1283.

Benkovic, S. J., Napper, A. D., and Lerner, R. A. (1988) *Proc. Natl. Acad. Sci. USA*, **85**, 5355.

Blair, J. B., and Wong, C.-H. (1986) *J. Chem. Soc., Chem. Commun.*, 417.

Breslow, R. (1972) *Chem. Soc. Rev.*, **1**, 553.

Breslow, R. (1980) *Acc. Chem. Res.*, **13**, 170.

Breslow, R. and Heyer, D. (1982) *J. Am. Chem. Soc.*, **104**, 2045.

Breslow, R., Hammond, M. and Lauer, M. (1980) *J. Am. Chem. Soc.*, **102**, 421.

Breslow, R., Brandl, M., Hunger, J., and Adams, A. D. (1987) *J. Am. Chem. Soc.*, **109**, 3799.

Breitgoff, D., Laumen, K., and Schneider, M. P. (1986) *J. Chem. Soc., Chem. Commun.*, 1523.

Brown, J. M. (1989) *Chem. in Britain*, **25**, 268.

Brown, S. L., Davies, H. G., Foster,. D. F., Seeman, J. I., and Warner, P. (1986) *Tetrahedron Lett.*, **27**, 623.

Bunton, C. A. (1976) in *Applications of Biochemical Systems in Organic Chemistry, Techniques of Organic Chemistry Series* (eds J. B. Jones, C. J. Sih and D. Perlman), Wiley Interscience, New York, Vol. 2, p. 731.

Chenault, H. K., Simon, E. S. and Whitesides, G. M. (1988) *Biotech. Genetic Eng. Rev.*, 1.

Chibata, I. (ed.) (1978) *Immobilised Enzymes*, Halsted Press for Kodanasha, New York.

Christen, M. and Crout, D. H. G. (1988) *J. Chem. Soc., Chem. Commun.*, 264.

Corey, E. J., Becker, K. B. and Varma, R. K. (1972) *J. Am. Chem. Soc.*, **94**, 8616.

Corey, E. J., Bakshi, R. K., Shibata, S., Chen, C.-P., and Singh, V. K. (1987) *J. Am. Chem. Soc.*, **109**, 7925.

Cram, D. J. (1976) in *Applications of Biochemical Systems in Organic Chemistry Series* (eds, J. B. Jones, C. J. Sih and D. Perlman), Wiley Interscience, New York, Vol. 2, p. 815.

Cram, D. J. and Sogah, G. D. Y. (1981) *J. Chem. Soc., Chem. Commun.*, 625.

Cram, D. J. (1988) *Angew. Chem. Int. Edn. Engl.*, **27**, 1009.

Davies, S. G., (1989) *Chem. in Britain*, **25**, 268.

Davies, S. G., Dordor-Hedgecock, I. M., Walker, J., and Warner, P. (1984) *Tetrahedron Lett.*, **25**, 2709.

Davies, S. G., Dordor-Hedgecock, I. M., and Warner, P. (1985) *Tetradedron Lett.*, **26**, 2125.

Deardoff, D. R., Mathews, A. J., Scott McMeekin, D., and Craney, C. L. (1986) *Tetrahedron Lett.*, **27**, 1255.

DiCosimo, R., Wong, C.-H., Daniels, L. and Whitesides, G. M. (1981) *J. Org. Chem.*, **46**, 4622.

Durrwachter, J. R., and Wong, C.-H. (1988) *J. Org. Chem.*, **53**, 4175.

Estell, D. A., Light, D. R., Rasteter, W. H., Lazarus, R. A., and Miller, J. V. (1984) *Eur. Pat. Appl.* EP 132, 308, *Chem. Abs.*, **102**, P144140t.

Ferraboschi, P., Grisenti, P., Casati, R., Fiecchi, A., and Santaniello, E., (1987) *J. Chem. Soc., Perkin Trans. 1*, 1743, 1749.

Fleet, G. W. J. (1989) *Chem. in Britain*, **25**, 287.

Fleming, I. (1973) *Selected Organic Syntheses*, Wiley, London, p. 36.

Fronza, G., Fuganti, C. and Graselli, P. (1980) *J. Chem. Soc., Chem. Commun.*, 442.

Fronza, G., Fuganti, C. Graselli, P., and Servi, S. (1985) *Tetrahedron Lett.*, **26**, 4961.

Fuganti, C. and Graselli, P. (1979) *J. Chem. Soc., Chem. Commun.*, 995.

Fuganti, C. and Graselli, P. (1982) *J. Chem. Soc., Chem. Commun.*, 205.

Fujisawa, T., Watanake, M., and Sato, T. (1984) *Chem. Lett.*, 2055.

Gao, Y., Hanson, R. M., Klunder, J. M., Ko, S. Y., Masamune, H., and Sharpless, K. B. (1987) *J. Am. Chem. Soc.*, **109**, 5765.

Groves, J. T. and Ahn, K. K. (1987) *Inorganic Chem.*, **26**, 3831.

Groves, J. T. and Stern, M. K. (1988) *J. Am. Chem. Soc.*, **110**, 8628.

Grunwald, J., Wirz, B., Scollar, M. P., and Klibanov, A. M. (1986) *J. Am. Chem. Soc.*, **108**, 6732.

Guthrie, J. P. (1976) in *Applications of Biochemical Systems in Organic Chemistry,*

Techniques of Organic Chemistry Series (eds, J. B. Jones, C. J. Sih and D. Perlman), Wiley Interscience, New York, Vol. 2, p. 627.

Gutsche, C. D., and Alam, I. (1988) *Tetrahedron*, **44**, 4689.

Han, C.-Q., DiTullio, D., Wang, Y.-F., and Sih, C. J. (1986) *J. Org. Chem.*, **51**, 1253.

Hilvert, D., Carpenter, S. H., Nared, K. D., and Anditor, N.-T. (1988) *Proc. Natl. Acad. Sci. USA.*, **85**, 4953; Hilvert, D. and Nared, K. D. (1988) *J. Am. Chem. Soc.*, **110**, 5593.

Iguchi, S., Nakai, H., Hayashi, M. and Yamamoto, H. (1979) *J. Org. Chem.*, **44**, 1363.

Itoh, T., Yonekawa, Y., Sato, T., and Fujisawa, T. (1986) *Tetrahedron Lett.*, **27**, 5405.

Jakubke, H.-D., Kuhl, P., and Konnecke, A. (1985), *Angew. Chem. Int. Edn. Engl.*, **24**, 85.

Jones, J. B. (1980) in *Enzymic and Nonenzymic Catalysis* (eds. P. Dunnil, A. Wideman and N. Blakebrough), Ellis and Horwood, Chichester, p. 54.

Jones, J. B. and Beck J. F. (1976) in *Applications of Biochemical Systems in Organic Chemistry, Techniques of Organic Chemistry Series* (eds, J. B. Jones, C. J. Sih and D. Perlman), Wiley Interscience, New York, p. 107.

Jones, J. B. and Davies, J. (1979) *J. Am. Chem. Soc.*, **101**, 5405.

Jones, J. B. and Jakovac, I. J. (1982) *Can. J. Chem.*, **60**, 19.

Jones, J. B. and Lok, K. P. (1979) *Can. J. Chem.*, **57**, 2533.

Jones, J. B. and Schwartz, H. M. (1981) *Can. J. Chem.*, **59**, 1574.

Jones, J. B. and Taylor, K. E. (1976) *Can. J. Chem.*, **54**, 2969; *J. Am. Chem. Soc.*, **98**, 5689.

Jones, J. B., Finch, M. A. W. and Jakovac, I. J. (1982) *Can. J. Chem.*, **60**, 2007.

Jones, J. B., Sih, C. J. and Perlman, D. (eds) (1976) *Applications of Biochemical Systems in Organic Chemistry, Techniques of Organic Chemistry Series*, Wiley Interscience, New York.

Jones, J. B. (1986) *Tetrahedron*, **42**, 3351.

Julia, S., Masana, J. and Vega, J. C. (1980) *Angew. Chem. Int. Edn. Engl.*, **19**, 929.

Jursic, B. (1988) *Tetrahedron*, **44**, 1553.

Kaiser, E. T. (1988) *Angew. Chem. Int. Edn. Engl.*, **27**, 913.

Kawanow, H., Ishii, Y., Saburi, M. and Uchida, Y. (1988) *J. Chem. Soc., Chem. Commun.*, 87.

Keinan, E., Harfeli, E. K., Seth, K. K., and Lamed, R. (1986) *J. Am. Chem. Soc.*, **108**, 162.

Kim, M.-J. and Whitesides, G. M. (1988) *J. Am. Chem. Soc.*, **110**, 2959.

Kim, M.-J., Hennen, W. J., Sweers, H. M. and Wong, C.-H. (1988) *J. Am. Chem. Soc.*, **110**, 6481.

Klibanov, A. M., Berman, Z., and Alberti, B. N. (1981) *J. Am. Chem. Soc.*, **103**, 6263.

Kluender, H., Huang, F.-C., Fritzberg, A., Schnoes, H., Sih, C. J., Fawcett, P. and Abraham, E. P. (1973) *J. Am. Chem. Soc.*, **95**, 6149.

Klunder, J. M., Ko, S. Y. and Sharpless, K. B. (1986) *J. Org. Chem.*, **51**, 3710.

Kokubo, T., Sassa, S., and Kaiser, E. T. (1987) *J. Am. Chem. Soc.*, **109**, 606.

Kolb, M., Van Hijfte, L. and Ireland, R. E. (1988) *Tetrahedron Lett.*, **29**, 6769.

Komiyama, M. and Bender, M. L. (1984) in *The Chemistry of Enzyme Action*, (ed. M. I. Page), Elsevier, Amsterdam, p. 405.

Kori, M., Itoh, K. and Sugihara, H. (1987) *Tetrahedron Lett.*, **29**, 2319.

Krishnamurthy, S. (1974) *Aldrichim. Acta.*, **7**, 55.

Krishnamurthy, S. (1981) *J. Org. Chem.*, **46**, 4629.
Kurihara, M., Kouiyama, K., Kobayashi, S. and Ohno, M. (1985) *Tetrahedron Lett.*, **26**, 5831.
Laumen, K. and Schneider, M. (1984) *Tetrahedron Lett.*, **25**, 5875.
Lehn, J. M. (1988) *Angew. Chem. Int. Edn. Engl.*, **27**, 99.
Lemaine, M., Guy, A., Roussel, J., and Guette, J.-P. (1987) *Tetrahedron*, **43**, 835.
Ley, S. V., Sternfeld, F., and Taylor, S. C. (1987) *Tetrahedron Lett.*, **28**, 225.
Lewis, J. M., Haynie, S. L. and Whitesides, G. M. (1979) *J. Org. Chem.*, **44**, 864.
Link, C. M., Jansen, D. K. and Sukenik, C. N. (1980) *J. Am. Chem. Soc.*, **102**, 7798.
Margolin, A. L. and Klibanov, A. M. (1987) *J. Am. Chem. Soc.*, **109**, 3802.
Massey, R. J. (1987) *Nature*, **328**, 457.
Menger, F. M. (1979) *Acc. Chem. Res.*, **12**, 111.
Menger, F. M., Takeshita, M. and Chow, J. F. (1981) *J. Am. Chem. Soc.*, **103**, 5938.
Midland, M. M., Tramontano, A., Zederic, S. A. and Greer, S. (1979) *J. Am. Chem. Soc.*, **101**, 2352.
Mohr, P., Rosslein, L. and Tamm, C. (1987) *Helv. Chim. Acta.*, **70**, 142.
Mori, K. and Otsuka, T. (1985) *Tetrahedron*, **41**, 547.
Mosbach, K. and Larsson, P. M. (1970) *Biotechnol. Bioeng.*, **12**, 19.
Murali, C. and Creaser, E. H. (1986) *Protein Engineering*, **1**, 55.
Nakatsuka, T., Sasaki, T. and Kaiser, E. T. (1987) *J. Am. Chem. Soc.*, **109**, 3808.
Napper, A. D., Benkovic, S. J., Trammontano, A., and Lerner, R. A. (1987) *Science*, **237**, 1041.
Neidlemann, S. L., Amon, W. F., Jr. and Geigert (1981) *US Patent 4 247 641*, *Chem. Abstr.*, **94**, 190 337.
Norman, R. O. C. (1982) in *The Chemical Industry* (eds, D. Sharp and T. F. West), Ellis and Horwood, Chichester, p. 347.
Noyori, R., Tomino., I. and Tanimoto, Y. (1979) *J. Am. Chem. Soc.*, **101**, 3129, 5843.
Oku, J. and Inoue, S. (1981) *J. Chem. Soc., Chem. Commun.*, 229.
Onyiriuka, S. O. and Suckling, C. J. (1982) *J. Chem. Soc., Chem. Commun.*, 833.
Onyiriuka, S. O. and Suckling, C. J. (1986) *J. Org. Chem.*, **51**, 1900.
Oppolzer, W., (1988) *Pure Appl. Chem.*, **60**, 39.
Pedersen, C. J. (1988) *Angew. Chem. Int. Edn. Engl.*, **27**, 1021.
Perlman, D. (1976) in *Applications of Biological Systems to Organic Chemistry, Techniques of Organic Chemistry Series* (eds J. B. Jones, C. J. Sih and D. Perlman), Wiley Interscience, New York, Vol. 1, p. 47.
Pervez, H., Onyiriuka, S. O., Rees, L., Rooney, J. and Suckling, C. J. (1988) *Tetrahedron*, **44**, 4555.
Pollak, A., Baughn, R. L., Adalstensson, O. and Whitesides, G. M. (1978) *J. Am. Chem. Soc.*, **100**, 304.
Pollak, A., Baughn, R. L. and Whitesides, G. M. (1977) *J. Am. Chem. Soc.*, **99**, 2366.
Pratt, A. J. (1989) *Chem. in Britain*, **25**, 282.
Prelog, V. (1964) *Pure Appl. Chem.*, **9**, 126.
Rebek, J. Jr., Askew, B., Killoran, M., Nemeth, D. and Lin, F.-T. (1987) *J. Am. Chem. Soc.*, **109**, 2426; Rebek, J. Jr., Askew, B., Nemeth, D. and Parris K, *ibid.*, 2432.
Rees, L., Valente, E., Suckling, C. J. and Wood, H. C. S. (1986) *Tetrahedron*, **42**, 117.
Rees, L., Suckling, C. J. and Wood, H. C. S. (1987) *J. Chem. Soc., Chem. Commun.*, 470.
Rideout, D. C. and Breslow, R. (1980) *J. Am. Chem. Soc.*, **102**, 7816.

Ruddleston, J. F. and Stewart, A. (1981) *J. Chem. Res. (S)*, 378.

Sabbioni, G. and J. B. Jones (1987) *J. Org. Chem.*, **52**, 4565.

Scott, A. I. (1976) in *Applications of Biochemical Systems in Organic Chemistry, Techniques of Organic Chemistry Series* (eds J. B. Jones, C. J. Sih and D. Perlman), Wiley Interscience, New York, Vol. 2, p. 555.

Seebach, D. and Henradon, B. (1987) *Tetrahedron Lett.*, **28**, 3791.

Sharpless, K. B. and Verhoeven, T. R. (1979) *Aldrichim. Acta*, **13**, 13.

Short, R. P. and Masamune, S. (1987) *Tetrahedron Lett.*, **28**, 2481.

Sicsic, S., Ikbal, M. and Legoffic, F. (1987) *Tetrahedron Lett.*, **28**, 1887.

Sih, C. J. and Rosazza, J. P. (1976) in *Application of Biochemical Systems to Organic Chemistry, Techniques in Organic Chemistry Series* (eds J. B. Jones, C. J. Sih and D. Perlman), Wiley Interscience, New York, Vol. 1, p. 69.

Sih, C. J., Heather, H. B., Sood, R., Price, P., Peruzzotti, B., Hsu-Lee, L. F. and Lee, S. S. (1975) *J. Am. Chem. Soc.*, **97**, 865.

Silverman, R. B. (1988) in *Topics in Medicinal Chemistry* (ed. P. R. Leeming), Royal Society of Chemistry, London, p. 73.

Simon, H., Bader, J., Gunther, H., Neumann, S. and Tharos, J., (1985) *Angew. Chem. Int. Edn. Engl.*, 539.

Slama, J. T., Oruganti, S. R. and Kaiser, E. T. (1981) *J. Am. Chem. Soc.*, **103**, 6211.

Stoddart, J. F. (1984) in *The Chemistry of Enzyme Action*, (ed. M. I. Page), Elsevier, Amsterdam, p. 529.

Suckling, C. J. (1977) *Chem. Soc. Rev.*, **7**, 215.

Suckling, C. J. (1981) *Ind. Eng. Chem. Prod. Res. Dev.*, **20**, 434.

Suckling, C. J. (1982) *J. Chem. Soc., Chem. Commun.*, 661.

Suckling, C. J. and Suckling, K. E. (1974) *Chem. Soc. Rev.*, **4**, 387.

Suckling, C. J., Suckling, C. W. and Suckling, K. E. (1978) *Chemistry through Models*, Cambridge University Press, Cambridge, pp. 203–210.

Sutherland, I. O. (1982) in *The Chemical Industry* (eds D. Sharp and T. F. West), Ellis and Horwood, Chichester, p. 421.

Szajewski, R. P. and Whitesides, G. M. (1980) *J. Am. Chem. Soc.*, **102**, 2011.

Tabushi, I., Kimura, Y., and Yamamura, K. (1981) *J. Am. Chem. Soc.*, **103**, 6486.

Tabushi, I. (1982) *Acc. Chem. Res.*, **15**, 66.

Tabushi, I., Yamamura, K., Fujita, K. and Kawakubo, H. (1979) *J. Am. Chem. Soc.*, **101**, 1019.

Taylor, S. J. C., Ribbons, D. W., Slavin, A. M. Z., Widdowson, D. A., and Williams, D. J. (1987) *Tetrahedron Lett.*, **28**, 6391.

Thanos, I. and Simon, H. (1986) *Angew. Chem. Int. Edn. Engl.*, **25**, 462.

Therisod, M. and Klibanov, A. M. (1987) *J. Am. Chem. Soc.*, **109**, 3977.

Tomuto, G. M. R., Schar, H. P., Busquets, X. F., and Ghisalba, O. (1986) *Tetrahedron Lett.*, **27**, 5707.

Trost, B. M. and Klun, T. P. (1979) *J. Am. Chem. Soc.*, **101**, 6756.

Trost, B. M. and Klun, T. P. (1981) *J. Am. Chem. Soc.*, **103**, 1864.

West, J. B. and Wong, C.-H. (1987) *Tetrahedron Lett.*, **28**, 1629.

Wandrey, C., Wichmann, R., Berke, W., Morr, M., and Kuler, M. R. (1984) *3rd Eur. Congr. Biotechnol*, **1**, 239, Verlag Chemie, Weinheim.

Wandrey, C. and Wichmann, R. (1985) *Biotechnol. Ser.*, **5**, 177.

Wang, Y.-F., Lalonde, J. J., Momongan, M., Bergbreiter, D. E., Wong, C.-H., (1988) *J. Am. Chem. Soc.*, **110**, 7200.

Whitlock, H. W., Jr. (1976) in *Applications of Biochemical Systems in Organic Chemistry, Techniques of Organic Chemistry Series* (eds, J. B. Jones, C. J. Sih and D. Perlman), Wiley Interscience, New York, Vol. 2, p. 1045.

Whitesides, G. M. and Wong, C.-H. (1985) *Angew. Chem. Int. Edn. Engl.*, **27**, 622.

Wong, C.-H. and Whitesides, G. M. (1981) *J. Am. Chem. Soc.*, **103**, 4890.
Wynberg, H. and Marsmann, B. (1980) *J. Org. Chem.*, **45**, 158.
Yamada, H. and Schimizu, S. (1988) *Angew. Chem. Int. Edn. Engl.*, **27**, 622.
Zaks, A. and Klibanov, A. M. (1984) *Science*, **224**, 1249.
Zaks, A. and Klibanov, A. M. (1985) *Proc. Natl. Acad. Sci. USA.*, **82**, 3192.
Zaks, A. and Klibanov, A. M. (1986) *J. Am. Chem. Soc.*, **108**, 2767.

5 | Enzymes as targets for drug design

Philip D. Edwards, Barrie Hesp,
D. Amy Trainor and Alvin K. Willard

5.1 INTRODUCTION

Alfred Burger (1980) dates the modern era of medicinal chemistry from the discovery of the antibacterial properties of sulphanilamide in 1936. The subsequent observation that sulphanilamide antagonized the incorporation of *p*-aminobenzoic acid into folic acid marked the beginning of a shift in the medicinal chemist's attention from endogenous metabolites *per se* to the biochemical processes which control the flux of these metabolites in the organism. Many of these important biochemical processes are the enzyme-catalysed reactions of biosynthesis and catabolism. Most others are receptor-mediated processes in which interaction of an endogenous ligand with a receptor in a cell membrane initiates a complex sequence of biochemical events. The research topics in our own organization's portfolio of projects appear to be based almost equally on receptor and enzyme mechanisms. This balance is unlikely to be greatly different elsewhere.

In this chapter we shall be concerned only with enzymes as targets for drug design. For the organic chemist, raised on a diet of reaction mechanisms, enzymes offer some advantages over receptors since our understanding of receptor organization and mechanism lags far behind the detailed information available for many enzyme-catalysed reactions. Although the techniques of molecular biology have revealed a good deal of structural information for several receptors (see Chapter 9), there is as yet no detailed X-ray structural data available and we can only speculate as to their geometry on the basis of our knowledge of the three-dimensional structures of molecules which bind to them. Enzymes have often been purified and characterized, and fine details of structure and mechanism are sometimes available from X-ray studies. Enzyme mechanisms conform to the basic rules of organic chemistry, and some knowledge of the mechanism of the uncatalyzed reaction is often a starting point for the design of inhibitors, as was discussed in Chapter 2.

Selection of the target enzyme is of critical importance. There are numerous enzymes involved in life processes but many, indeed perhaps the majority, are unsuitable targets for drug intervention. As a result of the economics involved, most of today's clinically useful drugs have resulted from research conducted in pharmaceutical companies. Once a new drug has been identified, the development phase is of the order of 7–10 years and the cost $100–125 million in the United States alone.

Consider for example the enzymes of the catecholamine biosynthetic pathway (Fig. 5.1). Inhibitors of tyrosine hydroxylase and dopamine (3,4-dihydroxyphenethylamine, 5.4) β-hydroxylase lower blood pressure primarily by reducing synthesis of the neurotransmitter, norepinephrine (noradrenaline, 5.5). However, such substances are unacceptable as drugs for antihypertensive therapy because of their intrinsic lack of selectivity in control of adrenergic function. They interfere with norepinephrine synthesis in blanket fashion, leading to unacceptable side effects, for example orthostatic hypotension, resulting from blockade of cardiovascular reflexes. Inhibitors of tyrosine hydroxylase are unlikely to receive serious attention in the present-day pharmaceuticals industry for this reason. There are alternative, selective ways of controlling blood pressure *via* enzyme inhibition, as will be described in section (5.2.5).

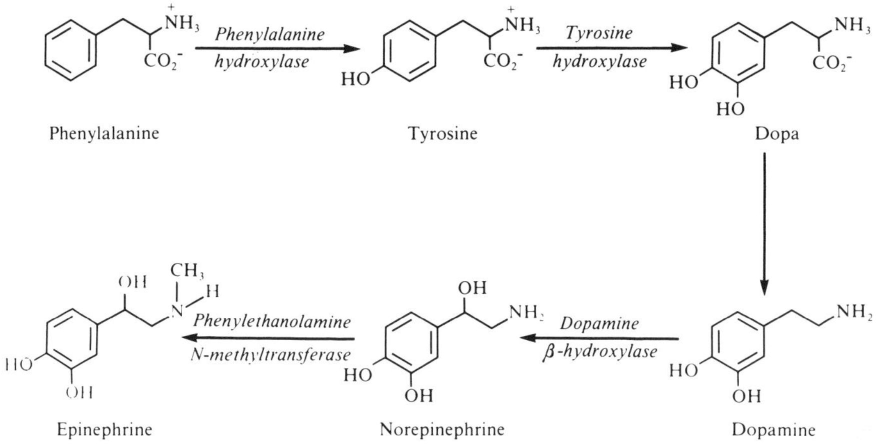

Fig. 5.1   Catecholamine biosynthesis.

Our intent in this chapter is to use five case studies to exemplify the evolution of the medicinal chemist's appreciation of enzymes as targets for drug design. Each concerns drugs of established importance or research topics of great current interest. They span the range from drugs developed prior to any knowledge of their target enzymes, to today's attempts at *ab initio* design. Each is a rich source of information for the aspiring medicinal chemist, though this will be neither a comprehensive review of enzyme inhibitors nor shall we address but a few disease areas. Finally, we discuss recent developments in tools and techniques, such as X-ray structure determination, protein engineering, computational methods, and molecular modelling which are providing new perspectives on protein structure, molecular recognition, and indeed, on rational drug design.

5.2   CASE STUDIES IN DRUG DISCOVERY

5.2.1   Inhibitors of dihydrofolate reductase

Several clinically important drugs produce their therapeutic effects through inhibition of the enzyme dihydrofolate reductase (DHFR). DHFR catalyses the reduction of dihydrofolate (5.7) to tetrahydrofolate (5.8) via transfer of hydride from NADPH (Fig. 5.2). A number of mammalian and microbial cellular processes utilize tetrahydrofolate derivatives as cofactors in one-carbon transfer reactions. Transfer of the one-carbon moiety from 5,10-methylenetetrahydrofolate (5.9) to deoxyuridylate occurs concomitantly with oxidation of the cofactor to the dihydrofolate level. Thymidylate synthetase is cricially dependent upon continual regeneration of tetrahydro-folate (5.8) from dihydrofolate (5.7) by DHFR (Hitchings and Roth, 1980). Since cellular DNA synthesis cannot be maintained in the absence of functioning thymidylate synthetase, inhibitors of DHFR effectively prevent cell division throughout the microbial and animal world.

Before considering the mechanism of action of DHFR inhibitors in depth, we shall briefly discuss their origins. The search for antimalarial agents proceeded apace throughout World War II. Curd and colleagues at ICI based their synthetic eforts on pyrimidine analogues because of the key role of pyrimidines in cell metabolism and the known antimalarial activity of sulphonamides, a class which included pyrimidine derivatives. Their work was based on the premise that biochemical differences between humans and parasites could serve as a basis for the rational design of chemotherapeutic agents. Synthetic exploration of pyrimidines quickly led to a new series of antimalarial agents, typified by the diaminopyrimidine derivative (5.11), which was conceived as incorporating the aniline moiety of the sulpho-namides, the basic side chain of earlier alkaloid antimalarials, and the pyrimidine nucleus. Further speculative modification of such structures was

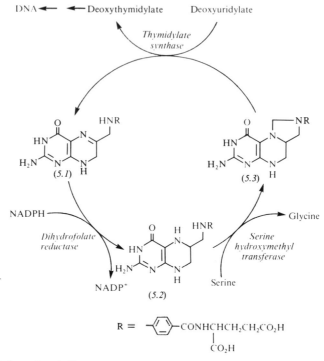

Fig. 5.2   The role of dihydrofolate reductase in the biosynthesis of deoxyribonucleic acids (DNA).

aimed at the synthesis of compounds resembling the various tautomeric forms available to these heterocyclic amines. The presence of a ring system was not considered to be obligatory. A compound which subsequently became a successful drug, proguanil (*5.12*), was an acyclic biguanide. It was later found however that proguanil was acting as a prodrug, since it was active only through metabolism to cycloguanil (*5.13a*), the first member of the dihydrotriazine series (*5.13*).

In the United States further chemical endeavours centred on diamino-pyrimidines (*5.14*) by the Burroughs Wellcome Groups led initially to pyrimethamine (*5.14a*) and then trimethoprim (*5.14b*). Also in the 1940s, a mechanistic approach to new antibacterial agents conducted by scientists at American Cyanamid, aimed at the design of antagonists of the known bacterial growth factor, folic acid, led directly to another diamino hetero-cycle methotrexate (*5.10*).

Though not fully appreciated at the time, all of the active compounds described above work by inhibiting DHFR. Considering the knowledge

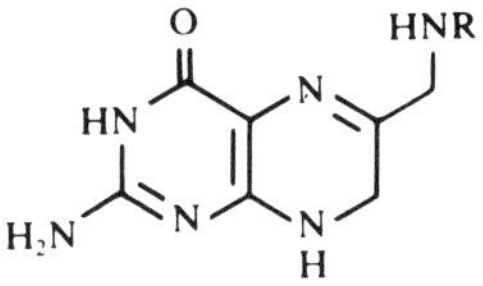

Dihydrofolate (5.7)

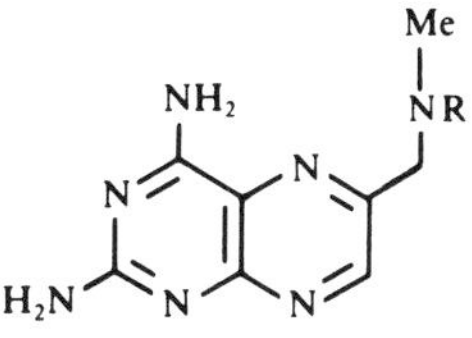

Methotrexate (5.10)

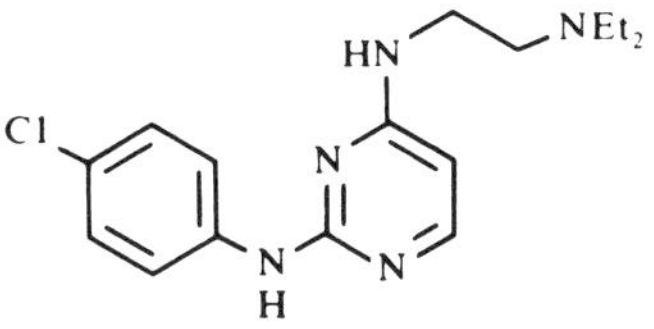

R = ⬡—CONHCHCH$_2$CH$_2$CO$_2$H
            |
            CO$_2$H

(5.11)

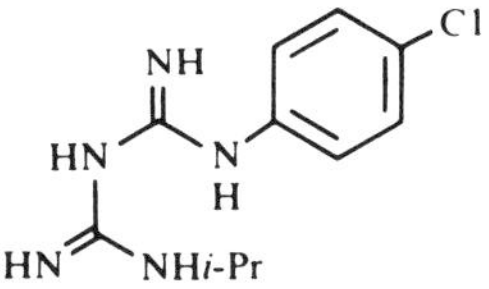

Proguanil (5.12)

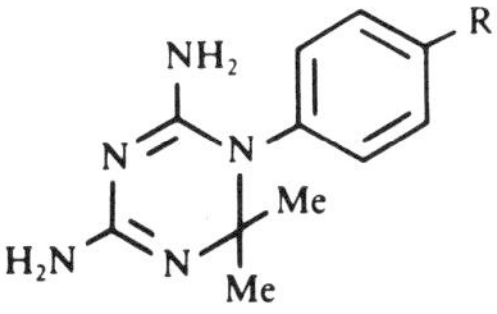

Dihydrotriazines (5.13)

(5.14)

a R = Cl
b R = OMe

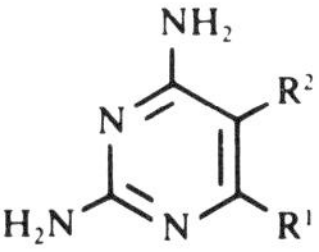

a R^1 = Et, R^2 = *p*-ClC$_6$H$_4$
b R^1 = H, R^2 = (MeO)$_3$C$_6$H$_2$CH$_2$
c R^1 = H, R^2 = Adamantyl
d R^1 = Et, R^2 = Adamantyl

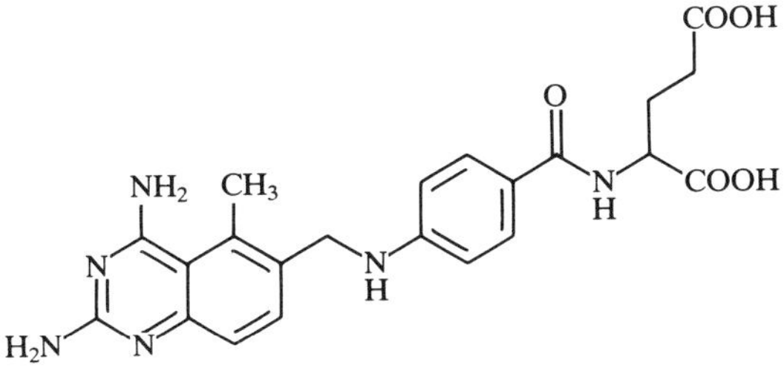

(5.15)

base and constraints of the day, the early examples of DHFR inhibitors are truly excellent examples of rational drug design.

Because of the ubiquitous distribution of the enzyme, the therapeutic value of DHFR inhibitors would be limited but for the selectivity differences observed for the various chemical subtypes. Some members of the triazine and pyrimidine classes of inhibitors display remarkable specificities for the enzymes from different species. Pyrimethamine (5.14a) is extremely effective as an inhibitor of the enzyme from the malaria parasite and is 2000–5000 times less potent against the bacterial and mammalian enzymes: hence its clinical utility in malaria. Trimethoprim (5.14b) is most potent against bacterial enzymes, being some 14 times less effective against enzyme from the malaria parasite and essentially inactive against mammalian enzyme. This selectivity for the bacterial enzyme underlies the efficacy and widespread antibacterial usage of this drug. Finally, several members of the dihydrotriazine (5.13) and quinazoline (e.g. 5.15) class of antifolates are potent and selective inhibitors of mammalian DHFR (Cody, 1985; Calvert, 1980); indeed, some of the former are 1000 times more potent against mammalian than bacterial enzymes.

Methotrexate (5.10) is non-selective in vitro, inhibiting enzymes from vertebrates and bacteria almost equally. Its structure differs from that of folic acid only in the replacement of a 4-oxo group by an amino group. It may be that species differentiation is more difficult to achieve in enzyme inhibitors which are close structural analogues of the substrate than in inhibitors from more distant chemical classes (Hitchings and Roth, 1980; Hitchings and Smith; 1980). Methotrexate's primary clinical use is to kill cancerous cells, the rationale being that such cells are rapidly proliferating, greatly dependent on DNA synthesis, and somewhat more susceptible to DNA synthesis inhibitors than are normal cells.

Modern biochemical, physical organic, spectroscopic and X-ray crystallographic techniques have been used to elucidate the fine details of enzyme mechanisms, including DHFR. X-ray structure determinations of complexes of each of the three major classes of inhibitors with DHFR have been particularly useful in advancing our understanding of the interaction of inhibitors with the enzyme. These studies include the ternary complexes, viz. bacterial enzyme (L. casei)-NADPH-methotrexate (Matthews et al., 1978), the avian enzyme-NADPH-dihydrotriazine (Volz et al., 1982), the mouse tumour cell line L1210 enzyme-NADPH-methotrexate or -trimethoprim complexes (Stammers et al., 1987) and the binary complex, bacterial (E. coli) enzyme-trimethoprim (Baker et al., 1981). The X-ray crystal structure of the human enzyme is not available. A model of the human enzyme for use in drug design has been constructed using the L1210 enzyme structure and the sequence of human DHFR (Stammers et al., 1987).

Methotrexate (5.10) binds in an open conformation in a deep cavity which

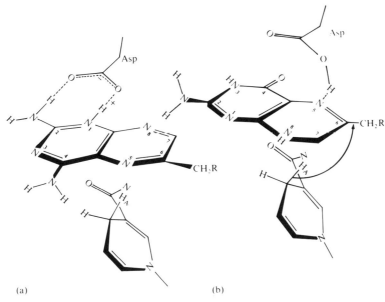

Fig. 5.3 Orientation of methotrexate (a) and dihydrofolate (b) in the active site of dihydrofolate reductase.

cuts across one face of the enzyme. The pteridine ring is almost perpendicular to the aromatic ring of the *p*-aminobenzoyl group. There are seventeen interactions between methotrexate and the enzyme, ten of which involve the pteridine ring. The charge interaction between Asp-26 (*L. casei*) and N_1 of the pteridine appears to be of key importance in the binding of the pteridine fragment (Volz *et al.*, 1982). The geometric relationship between NADPH and the pteridine ring of methotrexate is consistent with the experimental evidence that transfer of hydride is from the A face of the reduced nicotinamide ring (Fig. 5.3(a)).

The various inhibitors are unprotonated at neutral pH (unbound methotrexate N_1 pKa $\geq$ 5.73) but are protonated at N_1, with pKa > 10, in the bound form (Volz *et al.*, 1982; Cocco *et al.*, 1981; Ozaki *et al.*, 1981; Roberts *et al.*, 1981). For the *E. coli* enzyme, a charge interaction exists between N_1 of trimethoprim and Asp-27. The analogous interaction for the avian enzyme is between N_1 of the triazine and Glu-30. In contrast, evidence suggests that the substrate, dihydrofolate, is unprotonated, both free and bound to DHFR. The proposed productive binding mode of dihydrofolate to the enzyme is shown schematically in Fig. 5.3(b). The aspartate carboxyl is shown unionized in line with current thinking that the bound form of dihydrofolate is unprotonated. Clearly, there is formal transfer of a proton to N_5 at some point on the reaction co-ordinate and it is possible that aspartate is the source of the proton. Consistent with Baker's earlier suggestion, Cocco *et al.*, (1981) have reasoned that the greater than 1000-fold difference in affinities of methotrexate and dihydrofolate for the enzyme is almost entirely due to the high pKa of N_1 in bound methotrexate.

The structural similarity of methotrexate (5.10) to the natural substrate, dihydrofolate (5.7), leads one to expect that the substrate and the inhibitor would bind at the active site in very similar, if not identical, orientations. However, stereochemical evidence suggests that dihydrofolate binds to the active site in a manner very different from methotrexate. Thus, the stereochemistry at C_6 of tetrahydrofolate as determined implies that NADPH delivers the hydride to the face opposite to that found in the DHFR-methotrexate complex. An alternative binding mode was suggested in which the pteridine ring in dihydrofolate is rotated 180° to interchange N_1 and N_8 with C_4 and N_5, respectively, leaving the side chain from N_{10} essentially unchanged: such a binding mode would lead to tetrahydrofolate with the correct configuration at C_6 (Roth, 1986).

The different binding modes of such closely related analogues as methotrexate (5.10) and dihydrofolate (5.7) exemplify the ramifications of what frequently appear to be trivial structural modifications in the interactions between small molecules and enzymes or receptors. In this particular instance the minor structural changes of 4-oxo for 4-amino both switches the binding mode and confers enhanced affinity for the active site.

Major advances have been made in elucidating the structural differences among the dihydrofolate reductases from different species to aid in the design of new species-selective inhibitors. Based on X-ray studies a model has been described which permits a detailed accounting, in structural terms, for the selectivity of trimethoprim for the bacterial enzyme (Matthews, 1987). Of the five vertebrate reductases of known sequence, the chicken enzyme has been studied the most extensively. The geometry of trimethoprim binding to chicken enzyme is significantly different from bacterial enzyme (Matthews *et al.*, 1985). There are two potential binding sites for the side chains of the heterocyclic antifolates, the upper and the lower clefts. In *E. coli* the benzyl side chain of trimethoprim points into the lower cleft (Fig. 5.4), whereas in the chicken enzyme the benzyl group of trimethoprim fits more tightly into the upper cleft. In the latter binding mode the extra bridging methylene group of the side chain forces the pyrimidine ring into a non-planar conformation and results in the disruption of key hydrogen-bonding interactions between the enzyme and the pyrimidine ring. In contrast, this same ring with a phenyl or adamantyl (5.14c) group at the 5 position appears to fit easily into the upper cleft without disrupting the key intermolecular hydrogen bonds (Matthews *et al.*, 1985, 1987). Indeed, a lipophilic 2,4-diaminopyrimidine, substituted with 5-adamantyl and 6-ethyl

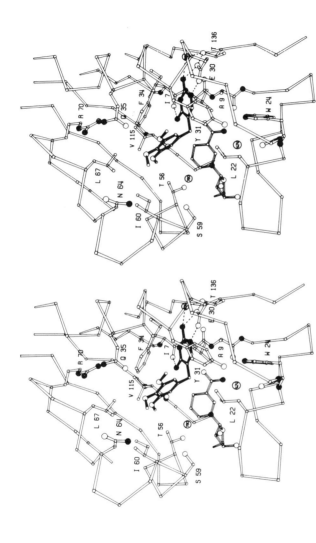

Fig. 5.4 Binding of trimethoprim to *E. coli* dihydrofolate reductase. Trimethoprim is indicated by solid bonds, and protein by open bonds. Carbon atoms are represented by smaller open circles, oxygen atoms by larger open circles, and nitrogen atoms by blackened circles. Large numbered circles represent fixed solvent molecules. Hydrogen bonds are indicated by dashed lines. (From Matthews *et al.*, 1985.)

groups (5.14d), is one of the most potent inhibitors of vertebrate DHFR known (Cody et al., 1984).

The phenyldihydrotriazines (5.13) are selective inhibitors of the vertebrate enzyme and appear to bind in a similar manner to the 5-phenyl- and 5-adamantylpyrimidines. The upper cleft in the E. coli is smaller in width by 1.5-2.0 Å and cannot accommodate the phenyl ring of the phenyldihydrotriazines (5.13) without altering the position of the dihydrotriazine ring.

It is evident that much effort continues to be devoted to the design of inhibitors of DHFR. In this context the detailed structural and mechanistic information which has emerged in recent years will be of great assistance (for example, Haddow et al., 1987). Recent emphasis has been on cancer and on two targets in particular: inhibitors which work against methotrexate-resistant cells, and inhibitors which are selective for cancer cells. There are four mechanisms by which malignant cells become resistant to DHFR inhibitors: impaired uptake of the inhibitor, an altered target enzyme, an elevated level of DHFR, or decreased polyglutamylation of methotrexate (Bertino et al., 1986). Methotrexate is transported into tumour cells by an active transport system that normally functions to transport reduced folates. Cells resistant to methotrexate due to a malfunction in their transport mechanism can be treated with certain methotrexate analogs, such as those containing albumin or polylysine sidechains, or with a 'non-classical' inhibitor such as trimetrexate (5.15). These agents appear to enter the cell by passive diffusion.

The recent recognition that polyglutamate derivatives of methotrexate rapidly accumulate in mammalian cells and are even more potent inhibitors than the parent compounds brings a new dimension to cancer chemotherapy (Fry et al., 1983). Methotrexate polyglutamates have important pharmacological consequences due to their ability to accumulate in tumour cells rapidly and selectively, their potent inhibition of DHFR, and their extended retention in cells. Scientists are beginning to recognize the potential for therapy inherent in this phenomenon.

5.2.2 β-Lactam antibiotics

The discovery of penicillin by Fleming, and the pioneering efforts of Florey and colleagues which led to its introduction into medical practice in the early years of the second world war, is a story familiar to most science graduates. The β-lactams now constitute by far the most important class of antibacterial agents and continue to be a focus of attention for the medicinal chemist. (See Chapter 7 for a discussion of the biosynthesis of β-lactams.) No other class of drug has been the subject of such intensive investigations in the history of the pharmaceutical industry. Scarcely a year passes without a major new discovery which shatters earlier views on the structural prerequisites for antibacterial activity – compare, for example, the structure of penicillin G (5.16) with those of two newer antibiotics, thienamycin (5.18) and the monocylic β-lactam, known as a monobactam, SQ 26455 (5.19).

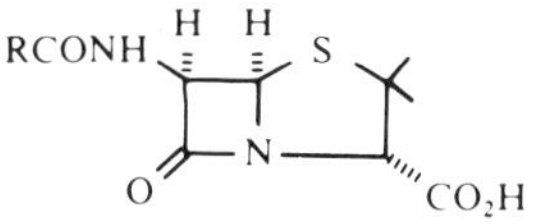

(5.16) R = PhCH₂
(5.17) R = PhOCH₂

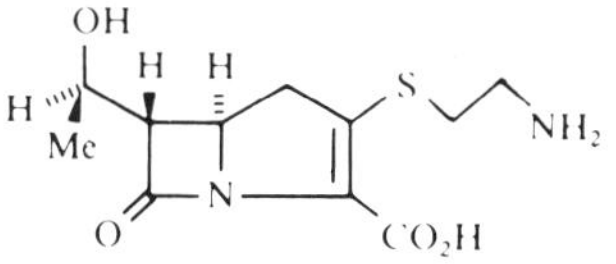

(5.18)

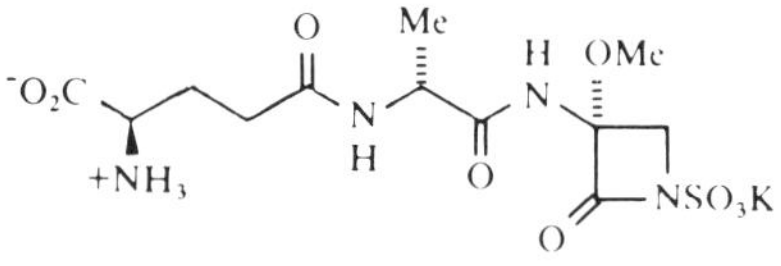

(5.19)

The key target enzymes of the β-lactams are the bacterial transpeptidases. The β-lactams kill bacteria by disrupting cell wall synthesis specifically by inhibiting the transpeptidases which cross-link peptidoglycan chains in the final steps of the biosynthetic sequence (Fig. 5.5). The enzyme formally cleaves a D-alanyl-D-alanine bond at the terminus of one peptidoglycan chain with formation of a new amide linkage between the remaining D-alanine and the terminal glycine of a pentaglycine sequence of a second peptidoglycan chain (Tipper, 1979; Tomasz, 1979).

Tipper and Strominger first proposed that the penicillins inhibit the transpeptidases by mimicking the structure of the D-alanyl-D-alanine moiety (Fig. 5.6) cleaved by the transpeptidase: more recently the suggestion that penicillin inhibits the transpeptidases by mimicking the transition-state of the substrate (see Section 2.8) has recieved a good deal of attention. The lack of effective conjugation in the amide bond of the β-lactam, which results in increased reactivity, is thought to be parallelled in the substrate at a point along the reaction coordinate.

Although the early penicillins were exciting therapeutic agents, there were a number of serious shortcomings associated with their use: they were poorly active when given orally; their spectrum of antibacterial activity was very limited, particularly *versus* Gram-negative bacteria; their *in-vivo* half-lives were short owing to rapid renal excretion; and they were very susceptible to inactivation by bacterial β-lactamases, enzymes which open the β-lactam ring of the antibiotic. These deficiencies have been overcome

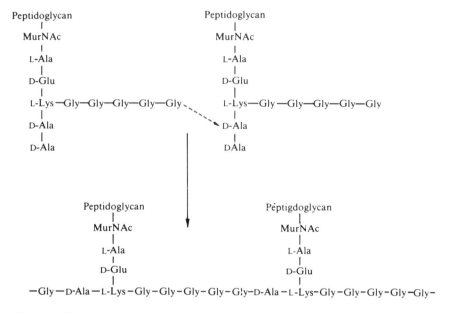

Fig. 5.5 Transpeptidation to two linear peptidoglycan chains in bacterial cell wall synthesis (Strominger, 1970; copyright 1970, Academic Press; used with permission).

for the most part in the newer agents and this, together with their excellent safety margin, underlies the commercial success of this class of antibiotics. Nevertheless, the search for newer, improved agents continues unabated with emphasis being given to broadening the profile of activity yet further and improving the stability of the antibiotics to the action of β-lactamases. The latter remains a serious concern in view of the heterogeneity of the β-lactamases and the propensity for rapid transmission of plasmid-borne genes through the bacterial world.

Perhaps the most intriguing and challenging aspect of the resistance problem is the similarity, real or apparent, between the enzymes which inactivate the antibiotics and those which are themselves the targets for the inhibitory action of the drugs. Studies with radiolabelled β-lactams have shown that the antibiotics bind to one or more penicillin-binding proteins (PBPs) in bacterial cytoplasmic membranes. Seven PBPs have been identified and numbered Ia, Ib, II, III, IV, V and VI in order of decreasing molecular weight. The PBPs are a mixture of transpeptidases and carboxypeptidases, with the latter predominating. The carboxypeptidases differ from the transpeptidases in the nucleophile which serves as acceptor for the penultimate D-alanyl moiety in the peptidoglycan; for the former the

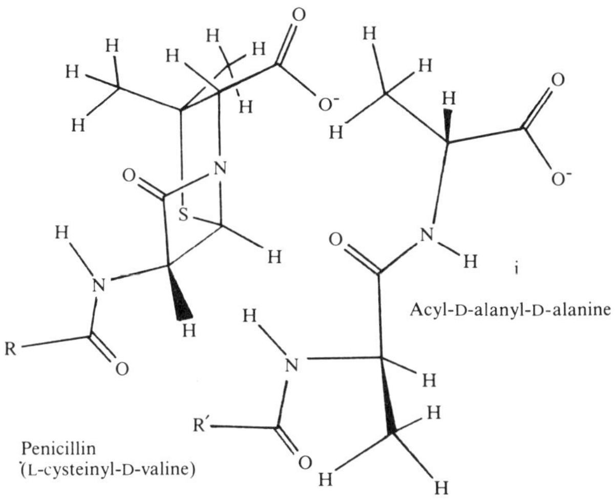

Fig. 5.6 Structural resemblance of penicillin to the acyl-D-alanyl-D-alanine fragment of bacterial cell walls (modified from Stromiger, 1970, copyright 1970, Academic Press; used with permission).

acceptor is water, whereas for the latter it is the glycine terminus of a second peptidoglycan. PBPs Ia, Ib, II and III are transpeptidases, some or all of which are essential for cell survival. PBPs IV, V and VI are carboxypeptidases and are not critical for bacterial cell growth.

The β-lactams are believed to inhibit all of the PBPs, both the transpeptidases and the carboxypeptidases, by very similar mechanisms. Most of our information concerning the inhibitory mechanisms of the penicillins has been obtained from studies with carboxypeptidases because of the great technical difficulties in obtaining the transpeptidases. The available evidence also implies that penicillin interacts with lactamases and the PBPs in a similar fashion (Fig. 5.7).

However, it is the subtle differences in the mechanisms by which each class of enzyme interacts with the penicillins that are responsible for each inhibitor's biological activity. Following initial formation of the Michaelis complex, a second complex (*5.21*) forms in which the penicillin has become bound covalently to the enzyme through acylation of a serine residue in the active site (Yocum *et al.*, 1982; Knowles, 1982). In the case of lactamase, enzyme is regenerated by transfer of the penicilloyl moiety to water with formation of a penicilloic acid (*5.22*) (Nu = OH). With the penicillin-binding proteins, hydrolytic regeneration of enzyme usually occurs concomitant with cleavage of the 5,6-bond of the penicillin to yield an acylated glycine (*5.23*) and N-formylpenicillamine (*5.24*); though penicilloic acid

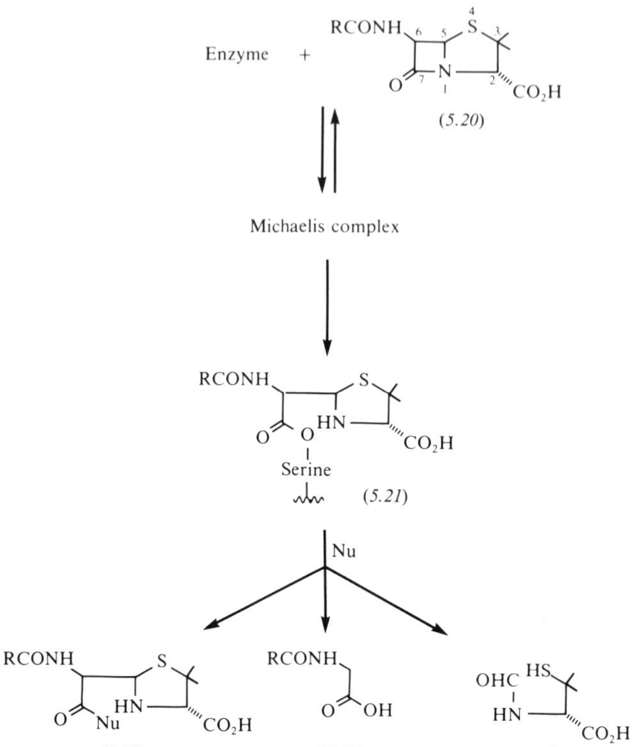

Fig. 5.7 The interaction of penicillins with bacterial β-lactamases and carboxy-peptidases: decomposition pathways available to the acyl-enxyme complex. Nu = water, amino acid, alcohol.

derivatives (5.22) are obtained when the acceptor is an amine or an alcohol (Marquet *et al.*, 1979), no irreversible inactivation of the enzyme occurs. There is no evidence to suggest that the products from 5,6-bond fission are derived from an enzyme-bound intermediate distinct from the acyl-enzyme complex (5.21). The fragmentation reaction appears to result from the reactivation process, perhaps *via* mechanisms involved in the release of enzyme-bound D-alanine in the normal substrate reaction. The rate of formation and stability of the acylated complex determines lactamase resistance in the context of one class of enzyme and antibacterial activity in the context of the other class. The rate of breakdown of the penicilloylated enzyme can be extremely slow for the penicillin-binding proteins; for example the half-life of the benzylpenicillin–carboxypeptidase complex from *Streptomycete* R39 under physiological conditions is 4250 min (Frere *et al.*, 1975). The newer, lactamase-resistant antibiotics are resistant because they are poor substrates for lactamases, yet retain their inhibitor properties for the penicillin-sensitive enzymes. Thus, hydrolysis of cefoxitin (5.25) by *E. coli* lactamase is more than 150 times slower than hydrolysis of penicillin V (5.16), one of the older penicillins (Fisher *et al.*, 1980).

We are far from a complete understanding of the structure–activity relationships in the β-lactam antibiotics, an issue of outstanding importance. It is perhaps because of this lack of understanding that each advance, in terms of overturning earlier notions of such relationships, has come directly from the discovery of new compounds from Nature, rather than from the progressive application of our knowledge of enzyme mechanisms to drug design.

A new chapter in the β-lactam story began with the isolation of clavulanic acid (5.26) from *Streptomyces clavuligerus* by workers at Beecham Pharmaceuticals in the early 1970s (Brown, 1981). Clavulanic acid differed from all other known β-lactams in being only a weak antibacterial agent but a potent inhibitor and irreversible inactivator of β-lactamases. Good synergistic antibacterial effects result from mixtures of clavulanic acid and lactamase-sensitive penicillins. Indeed, one such combination with amoxicillin, *Augmentin*®, is now marketed worldwide. The discovery of clavulanic acid stimulated the search for other inhibitors, both from natural sources and through synthesis. As a result, a number of such compounds have emerged

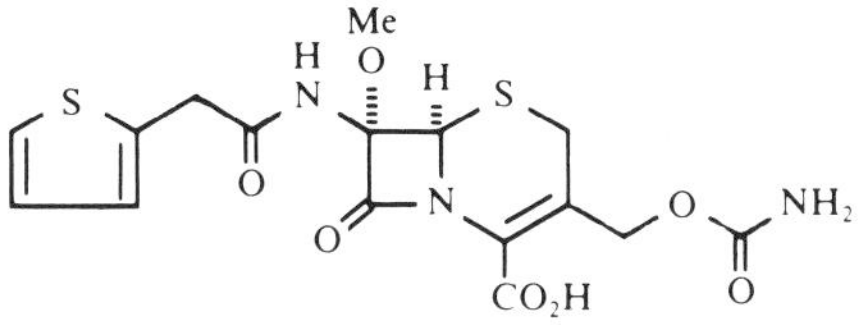

(5.25)

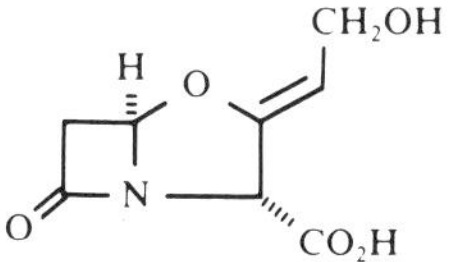

(5.26)

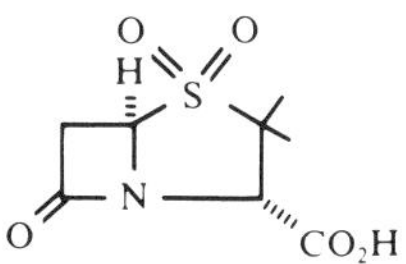

(5.27)

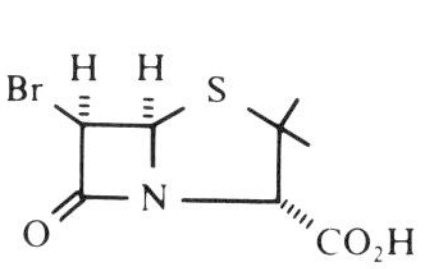

(5.28)

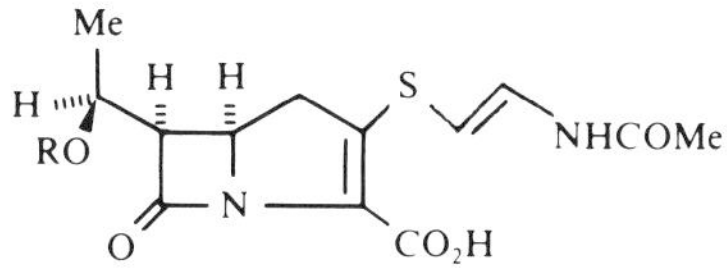

(5.29) R = SO₃H
(5.30) R = H

in recent years including the semisynthetics penicillanic acid sulphone, sulbactam (*5.27*), and 6-β-bromopenicillanic acid (*5.28*), and the naturally occurring carbapenems, for example MM 13902 (*5.29*) and thienamycin (*5.18*). The carbapenems differ from the other agents in that they are also good broad-spectrum antibacterial agents. For each of the β-lactamase inhibitors, the inhibited species is a transformation product, or products, of an initial acyl–enzyme complex, rather than the acyl–enzyme complex *per se*. This contrasts with the mechanism by which penicillins inhibit the transpeptidases and carboxypeptidases.

The complex mechanisms by which these compounds inhibit lactamase differ in specific details. Knowles (1985) has summarized the various mechanisms as shown in Fig. 5.8. Each inhibitor may be turned over by the normal pathway to give products derived from the acyl–enzyme intermediate (*5.33*). Often such products also involve cleavage of the five-membered heterocyclic ring.

The presence of an electron sink X in the sulphones and clavulanic acid ensures diversion of a fraction of the tetrahedral intermediate or acyl–enzyme complex to products (*5.34*) which may fragment further, or more interestingly, capture a nucleophilic site on the enzyme and thereby inactivate the enzyme irreversibly. For penicillanic acid sulphone approximately 7000 turnovers are required before the enzyme is inactivated.

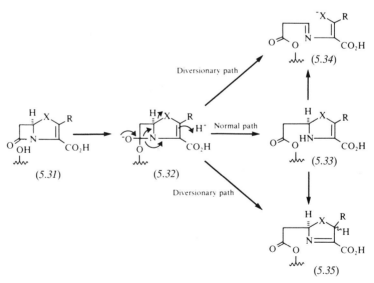

Fig. 5.8 Diversionary pathways that may lead to inhibition or inactivation of β-lactamase (Knowles, 1982; copyright 1982, Springer Verlag; used with permission).

Sulphone analogues which are poorer substrates require fewer turnovers to inactivate the enzyme.

An alternative diversionary path, which does not lead to irreversible inactivation, appears to be available for those carbapenems which can function as lactamase inhibitors (Knowles, 1985). In such cases the inhibitory species is the $\triangle^1$-pyrroline (5.35), an intermediate which is released from the enzyme much more slowly than the normal acyl–enzyme complex (5.33).

Within the same chemical series the availability of a diversionary pathway can be influenced profoundly by minor structural modifications. This is perhaps most evident in the case of MM 22382 (5.30) and its sulphate ester MM 13902 (5.29). The former is rapidly hydrolysed by lactamase with no detectable participation of the diversionary pathway; the latter is a good inhibitor.

The β-lactam series encompasses compounds which are good inhibitors of the transpeptidases but good substrates for lactamase (e.g. penicillin G), good inhibitors of the transpeptidases which are poor substrates for lactamase (e.g. cefoxitin), good inhibitors of both transpeptidases and lactamases (e.g. MM 13902), and finally, compounds which are very poor inhibitors of the transpeptidases but irreversible suicide inactivators of lactamase (e.g. clavulanic acid). What is conspicuous by its absence is a good inhibitor of the transpeptidases which also irreversibly inactivates lactamase.

The sparsity of X-ray data on the penicillin-sensitive enzymes, their substrates and inhibitors, contrasts with the situation for dihydrofolate reductase. However, there are encouraging signs that this situation is changing. The X-ray crystal structure has been solved for the β-lactam-sensitive bifunctional carboxypeptidase–transpeptidase (CPase-TPase) from *Streptomyces* R61 bound to a variety of β-lactams, including a penicillin, cephalosporin and monobactam (Kelly *et al.*, 1985). These studies revealed that the β-lactam inhibitors were all bound to a common binding site (Fig. 5.9). While the crystal structure has not been resolved in sufficient detail for full characterization of the β-lactam binding site, certain features are evident.

Both the orientation and position of the bound cephalosporin are consistent with acylation of a serine residue within the active site by the inhibitor. This serine hydroxyl was shown to be the same serine necessary for catalytic activity. As will be discussed in section 5.2.5, another class of enzymes, the elastases, also utilize an active site serine for catalytic hydrolysis of proteins. In this well studied class of enzymes, the nucleophilicity of the serine hydroxyl is increased through interaction with an active-site histidine and aspartate. A histidine does not appear to be present in the β-lactam binding site of CPase-TPase. Thus the question of whether

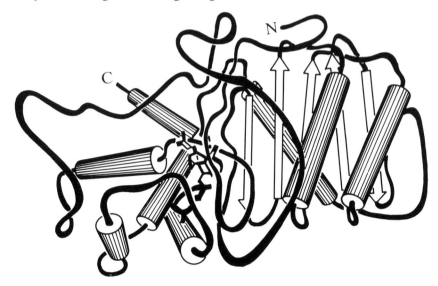

Fig. 5.9 Schematic drawing of the polypeptide chain of *S*. R61 CPase-TPase. Amino- and carboxyl-termini are indicated. Alpha helices and beta strands are represented by cylinders and arrows, respectively. Cephalosporin C is positioned in the β-lactam binding site. (Reprinted with permission from the special publication, The Royal Society of Chemistry, 1985, **52**, 319. Copyright 1985, The Royal Society of Chemistry; used with permission.)

the penicillin-sensitive enzymes contain a serine–histidine–aspartate 'catalytic triad', as usually found in the serine proteases, remains equivocal. However, the finding that β-lactams can be transformed into inhibitors of other serine proteases (section 5.2.5) reinforces the mechanistic similarities between these two classes of enzymes.

Earlier studies (Kelly *et al.*, 1981) demonstrated that substrate analogues of D-alanyl-D-alanine bind to the β-lactam binding site, thereby lending strong support to the Strominger hypothesis. Additional support has come from an analysis of the X-ray structures of several β-lactamases (Kelly *et al.*, 1986). While the primary amino acid sequences of these proteins are quite different from that of the CPase-TPase from *Streptomyces* R61, their overall tertiary structures, including the β-lactam binding sites, are remarkably similar. Thus, it would appear that the β-lactamases have originated through divergent evolution from the penicillin-binding proteins, as predicted by Strominger. Higher resolution X-ray data is eagerly awaited and should allow a greater understanding of the detailed mechanism of action of these enzymes.

5.2.3   Inhibitors of the renin–angiotensin system

As is evident from earlier sections in this chapter, many important drugs achieve their therapeutic effects through enzyme inhibition. The development of each of these has brought a singular nuance to the term rational drug discovery, but none more so than captopril (5.36), for it is the first clinically available drug to result from the application of an understanding of enzyme mechanisms at the molecular level to drug design.

(5.36)

In the normotensive, blood pressure is controlled by several complex mechanisms. In one such mechanism, the renin–angiotensin system (Fig. 5.10), two proteases, renin and angiotensin-converting enzyme (ACE), are of key importance (Peach, 1977; Laragh, 1978). The sequential action of these enzymes converts the circulating plasma globulin angiotensinogen (5.37), to the octapeptide angiotensin II (5.39), *via* the decapeptide angiotensin I (5.38). Angiotensin I is biologically inactive, in terms of a direct effect on blood pressure, but angiotensin II is the most potent

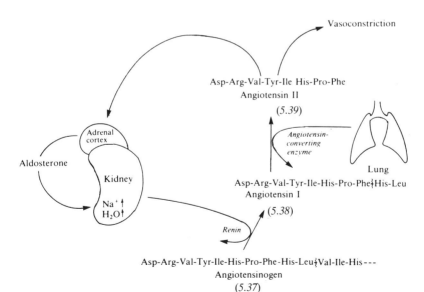

Fig. 5.10   The renin–angiotensin system.

endogenous pressor substance known. There are at least two mechanisms by which angiotensin II elevates blood pressure: first, it has a direct constricting effect on blood vessels, and second, it stimulates the adrenal cortex to release aldosterone which acts on the kidney and results in increased sodium and fluid retention.

The term hypertension refers to a single parameter, blood pressure, but it is a collection of diseases of diverse aetiologies, rather than a single entity. It appeared unlikely, therefore, that a single mechanism-based therapy would be efficacious in all patients. Scientists at the Squibb Institute for Medical Research argued that a defect in the renin–angiotensin system could underlie the disease in at least a subset of hypertensives, those with high renin levels, and that an inhibitor of ACE would lower blood pressure in such patients (Ondetti *et al.*, 1977; Cushman *et al.*, 1977). The Squibb researchers were not alone in making this deduction but most of their competitors rejected the idea, believing that the subgroup of patients most likely to benefit would be difficult to identify and too few in number to ensure commercial success. As sometimes and deservedly happens with bold research programmes, a far broader cross-section of patients respond to the therapy than was first envisaged and the success of the Squibb scientists has demonstrated that ACE is an excellent target for drug design. Indeed, there are currently three ACE inhibitors on the market and thirteen more in phase II and III clinical trials.

$$Glu — Trp — Pro — Arg — Pro — Gln — Ile — Pro — Pro$$

$$(5.40)$$

In the early phase of the program, several snake venoms were found to inhibit ACE, and one of these, the nonapeptide SQ 20881 (*5.40*), was shown to lower blood pressure in man following parenteral administration. SQ 20881 was inactive by the oral route and was not a practical drug for treatment of hypertension, but it demonstrated the validity of the approach. The problem to be faced was one which seemingly is becoming increasingly common in medicinal chemistry – that of finding small-molecule alternatives to relatively large peptides which achieve the desired pharmacological effect but do not suffer from poor absorption, rapid metabolism, or both.

Little was known about ACE except that it was a zinc-containing exodipeptidase. It was reasoned that ACE might be similar to pancreatic carboxypeptidase A, also a zinc-containing protease. In contrast to ACE, carboxypeptidase A is one of the better understood metalloproteases (Quiocho and Lipscomb, 1971). It is an exopeptidase which selectively cleaves an aromatic amino acid from the C-terminus of its substrates. The

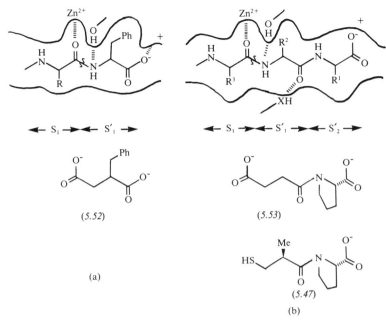

Fig. 5.11 Substrates and inhibitors in the active sites of metalloproteases: (a) carboxypeptidase A; (b) angiotensin-converting enzyme.

active site contains a cationic binding site for the carboxylate anion, a lipophilic cleft which accommodates the aromatic amino acid, a proton donor which protonates the N-terminus of the departing amino acid, and a zinc cation which activates the amide carbonyl to hydrolysis (Fig. 5.11(a)).

After a lengthy spell of non-productive random screening, the much-needed breakthrough was inspired by the report from Byers and Wolfenden (1973) that L-benzylsuccinate (5.41) is a potent competitive inhibitor of carboxypeptidase A. Like Byers and Wolfenden, the Squibb scientists reasoned that the binding of this inhibitor to the active site had much in common with the binding of the substrate and products of the reaction catalysed by carboxypeptide A. Since ACE cleaves the C-terminal dipeptide from angiotensin I (Fig. 5.11(b)), in contrast to carboxypeptidase A which cleaves a single amino acid from the C-terminus of its substrate, they further argued that a succinoyl amino acid, rather than benzylsuccinate, would be a more appropriate starting point for the design of inhibitors of ACE. Proline was chosen as the terminal amino acid in the target inhibitors because it was the C-terminal amino acid in all the naturally occurring peptide inhibitors, e.g. (5.40).

The first of these bi-product inhibitors to be synthesized, succinoylproline

Fig. 5.12 Comparison of the products of the hydrolysis of the P1-P1' bound of a substrate with succinoyl proline.

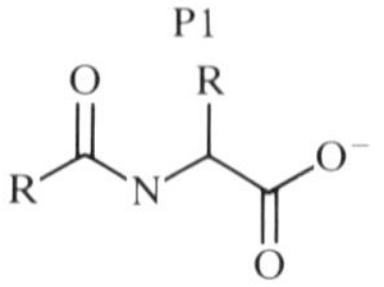

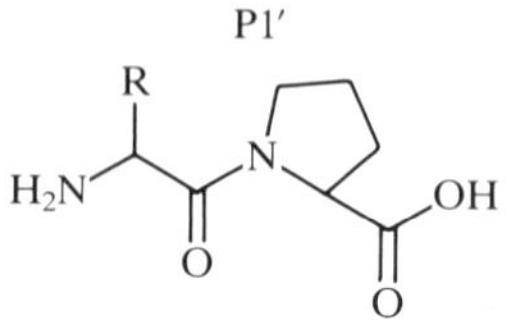

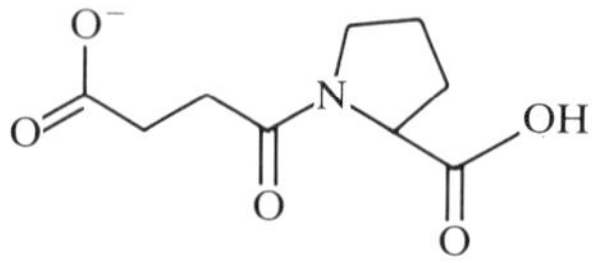

(5.42), was a weak competitive inhibitor of ACE. There followed a number of small structural changes, the most significant of which was replacement of the carboxylate anion in succinoylproline by a thiol group. The drug which quicky emerged from these studies was captopril (5.36), which with a K_i of 1.7×10^{-9}M, was not only as potent an inhibitor of ACE as the nonapeptide (5.40) first taken to the clinic, but was also orally active.

Captopril was approved by the regulatory authorities in the United States in 1981 and has captured a substantial part of the market for the treatment of hypertension in spite of early concerns over possible side effects associated with its use (blood dyscrasias, loss of taste and skin rashes). These side effects, as well as metabolic instability resulting from facile oxidation of the sulfur atom, are common to thiol containing drugs, and stimulated an intensive search for alternative inhibitors (Wyvratt and Patchett, 1985).

Scientists at Merck also opted for 'biproduct inhibition' as the basis for their design efforts. A biproduct inhibitor is one which incorporates features of both cleavage products from the enzyme-catalysed reaction. As illustrated in Fig. 5.12, succinoylproline mimics only part of the structural features found in the products of peptide hydrolysis: the amino acid side chain of the P_1 residue, as well as the basic amino group of the P_1' residue are missing. Incorporation of these features into an AlaPro backbone ultimately resulted in the production of the potent inhibitor enalaprilat (5.46) (Patchett et al., 1980). While enalaprilat is poorly absorbed following oral administration, its ethyl ester, enalapril (5.45) is well absorbed and subsequently hydrolysed in vivo to the active inhibitor enalaprilat. Captopril and enalapril are now among the most commercially successful drugs currently available.

SQ 29,852 (5.47) is a recent inhibitor from the Squibb group which is not yet available commercially. This phosphonate derivative possesses features of the transition-state of the enzyme-catalysed reaction in addition to

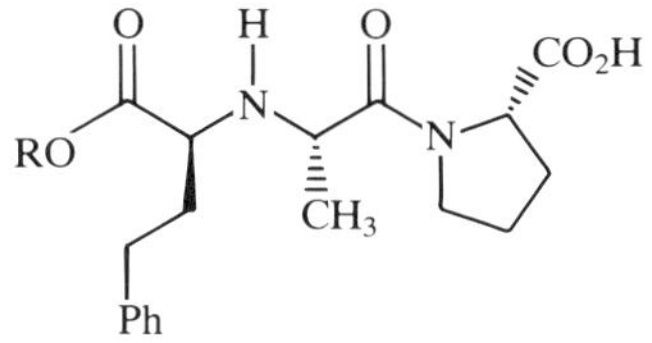

(5.45) R=CH$_2$CH$_3$
(5.46) R=H

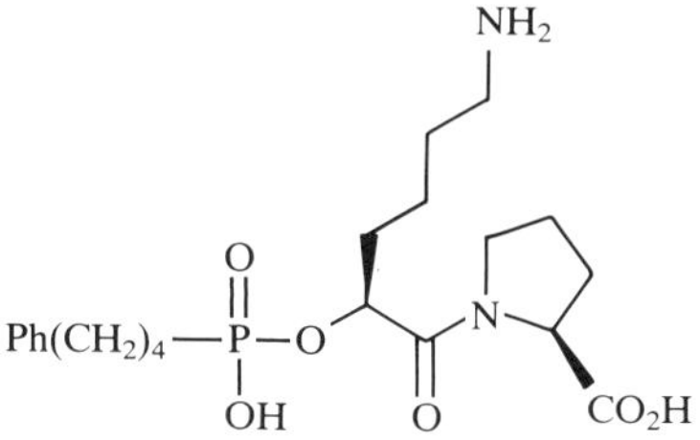

(5.47)

features of the products (Karanewsky *et al.*, 1988). As we shall see, this additional property is the foundation for development of a second class of inhibitors of the renin–angiotensin system.

Several of the side effects of captopril appear to be shared by other ACE inhibitors and, in any case, have lessened in importance at the lower doses typically used today. Nevertheless, that there are any side effects in a drug competing in such an enormous market as hypertension, has spurred the search for alternative therapies acting at the level of the renin–angiotensin system. Inhibitors of renin should be as effective as ACE inhibitors in dampening the synthesis of angiotensin II (Fig. 5.10). However, unlike ACE, for which angiotensin I is but one of several endogenous substrates, renin appears to have only a single endogenous substrate, angiotensinogen. It is argued, therefore, that renin inhibitors will have a 'cleaner' profile.

As will be seen, the challenges involved in the search for inhibitors of renin closely parallel those faced by the Squibb scientists a decade ago. Renin is a member of the acid, or aspartyl, group of proteases (Tang, 1977). Other members of this class include the mammalian gut enzyme pepsin and several enzymes of fungal origin, for example from *Pencillium janthinellum* and *Rhizopus chinesis*.

The acid proteases cleave peptides between two hydrophobic amino acids; in the case of human renin the susceptible link is a leucine and valine bond in angiotensinogen. The catalytic mechanism always involves the participation of two aspartyl carboxyl groups at the active site.

There are great practical difficulties in finding inhibitors of renin of therapeutic value. These stem from the difference in primate and rodent substrate and enzyme, as well as from poor oral absorption, rapid systemic clearance and metabolic instability of peptide inhibitors. Nonetheless, the feasibility of the approach was clearly demonstrated by Cody and colleagues (1980) with the finding that the decapeptide (5.48) inhibits renin *in vitro* and prevents the pressor response induced by infusion of human renin in the monkey. The analogy with the Squibb nonapeptide (5.40) is obvious.

As was the case with ACE, technical difficulties have hampered detailed structural investigations of renin, and attention was focused on other members of the acid proteases, particularly three fungal enzymes, penicillopepsin, endothiapepsin and rhizopuspepsin, whose X-ray structures are available. Again, the contribution made by X-ray studies has proved to be of fundamental importance. Although there is still some debate as to the fine details of the mechanism of action of the acid proteases, a consensus appears to be emerging (Suguna *et al.*, 1987).

The proposed mechanisms by which the zinc proteases and the acid proteases hydrolyse peptides are remarkably similar (Figs. 5.13 and 5.14 respectively). In the former the electrophile which activates the amide carbonyl to hydrolysis is the zinc cation, in the latter it is a proton of one of the active site carboxyls, Asp-35. For the aspartic proteases, the attacking nucleophile is an enzyme-bound water molecule which transfers one of its hydrogens to the other active site aspartate, Asp-218, which subsequently returns the hydrogen to the *N*-terminus of the departing amino acid. The same has been proposed for the zinc proteases, with Glu-270 functioning as the proton shuttle (Christianson and Lipscomb, 1986).

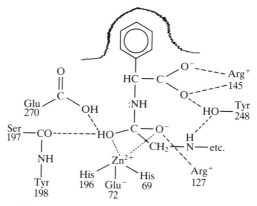

Fig. 5.13 Proposed interactions between the active site of carboxypeptidase A and the tetrahedral intermediate of the substrate. (Christianson and Lipscomb, 1986; copyright 1986, American Chemical Society; reprinted with permission.)

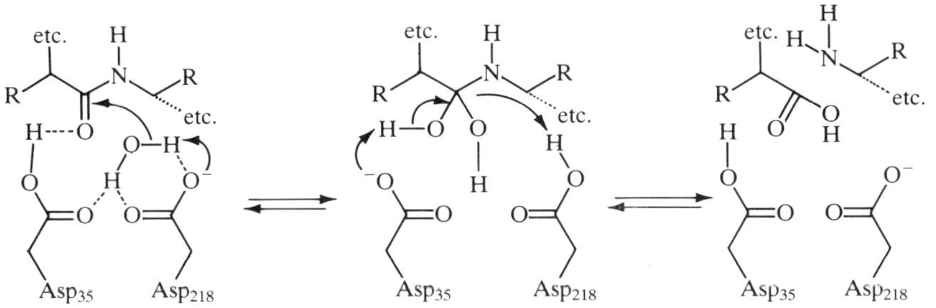

Fig. 5.14 Proposed catalytic mechanism of the acid protease rhizopuspepsin (adopted from Suguna *et al.*, 1987.)

With the demonstration that the decapeptide (*5.48*) inhibits renin *in vivo*, and the availability of detailed structural information for an acid protease, the analogy with ACE is complete but for an example of a mechanism-based inhibitor equivalent to benzylsuccinate. Pepstatin and analogues provide such examples.

Pepstatin (*5.49*) is a microbial metabolite which inhibits all known acid proteases. The unusual hydroxy amino acid statine (*5.50*) occurs twice in the pepstatin molecule. Enzymes accelerate the rate of chemical reactions by lowering the energy of the transition-state for the process. This is accomplished by maximizing favorable binding interactions between the enzyme and the transition-state intermediate. The proposed transition-state for the hydrolysis of amide bonds by aspartic proteases is the tetrahedral hydrated carbonyl. The extremely low inhibition constant observed for pepstatin ($K_i = 4.6 \times 10^{-11}$M, for pepsin) probably derives from the structural resemblance of statine to the hypothesized transition-state (*5.51*) for the enzyme-catalyzed reaction (Marciniszyn *et al.*, 1976; Marshall, 1976).

In contrast to the ACE inhibitors, which resulted from the concept of biproduct inhibition, the search for renin inhibitors has focused on the use of transition-state analogues. Szelke designed the first synthetic transition-state inhibitors, the reduced isosteres such as H142 (*5.52*). The high level of potency of pepstatin derived analogues such as 'Statine Containing Renin Inhibitory Peptide' (SCRIP, *5.53*) has spawned the search for other novel transition-state inhibitors. A recent review of these efforts has been published (Kokubu and Hiwada, 1987).

The statine-like hydroxy isostere (*5.54*), first introduced by Szelke (Szelke *et al.*, 1985), has been modified by the addition of heteroatoms to afford very potent dipeptidyl inhibitors of renin (*5.55*) (Dellaria *et al.*, 1987). These inhibitors have been shown to possess increased metabolic stability over larger peptidic inhibitors. This concept has been extended to the dihydroxy-ethylene transition-state-mimic (*5.56*) (Luly *et al.*, 1988). This inhibitor was

Pro—His—Pro—Phe—His—Phe—Phe—Val—Tyr—Lys

(5.48)

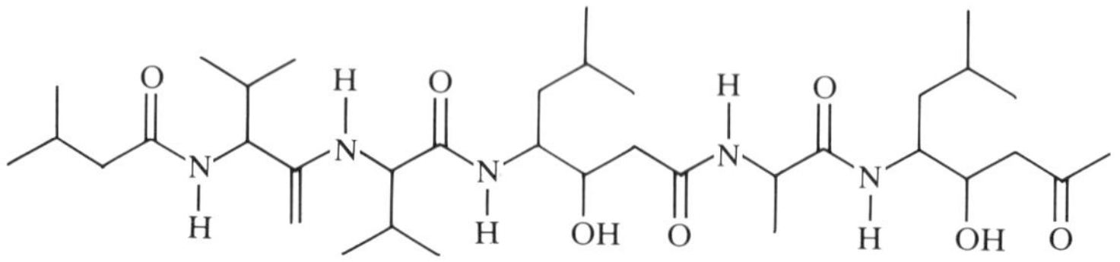

(5.49)

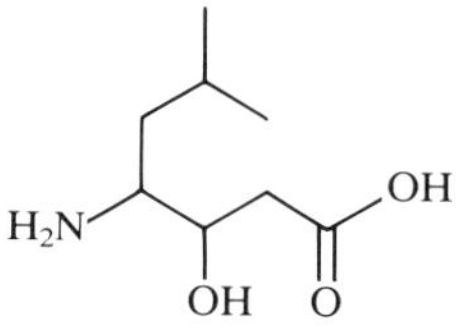

(5.50)

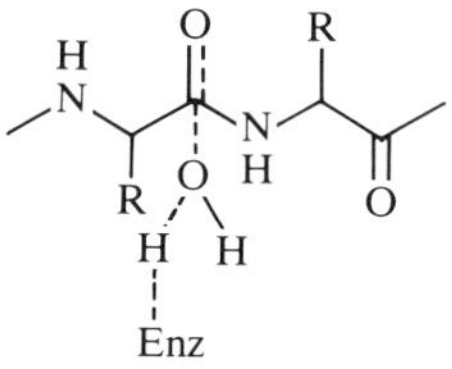

(5.51)

designed to mimic the gem diol of a hydrated aldehyde, which is believed to be the active form of aldehydic inhibitors of renin. The Abbott group determined that with the proper choice of P_1 and P_2 amino acids, introduction of the second hydroxyl afforded some of the most potent low molecular weight inhibitors of renin yet reported.

A novel approach incorporating difluorostatinone carried the concept of the tetrahedral intermediate as a transition-state mimic to its limit (Thaisrivongs et al., 1985; Gelb et al., 1985). Due to the powerful electron withdrawing ability of the fluorine atoms, these compounds (e.g. 5.57) exist predominately as the hydrated ketone in aqueous solution, and thus very closely mimic the enzyme transition-state (5.51). The difluorostatinone-containing peptides are more potent than their statinone-containing counterparts since the non-fluorinated ketones are not significantly hydrated in aqueous solution, and thus do not possess the tetrahedral geometry of the transition-state.

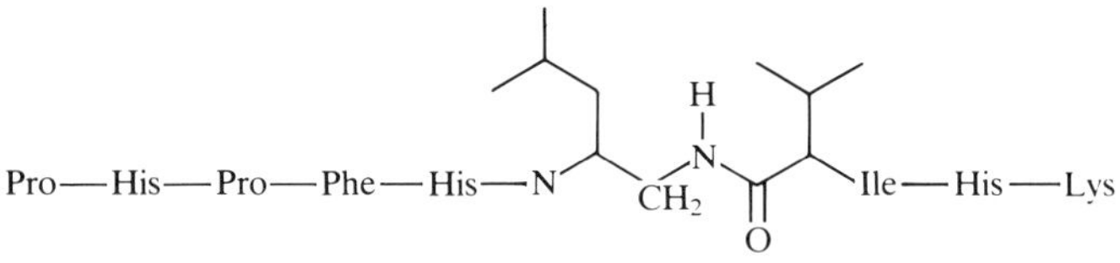

(5.52) H142

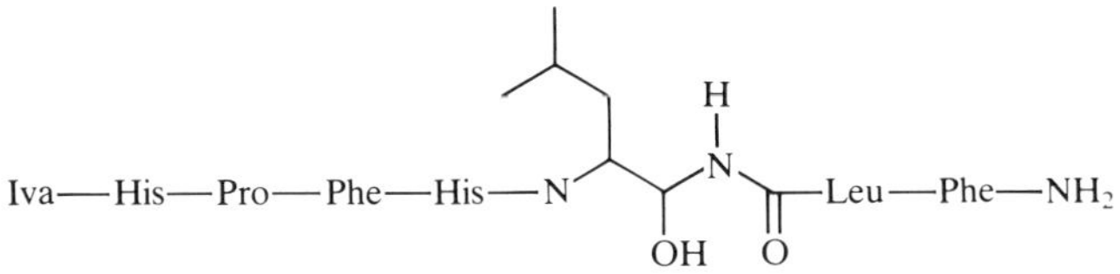

(5.53) SCRIP

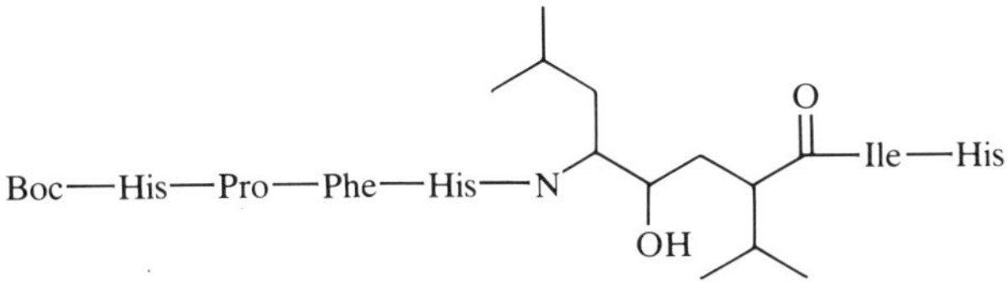

(5.54) H261

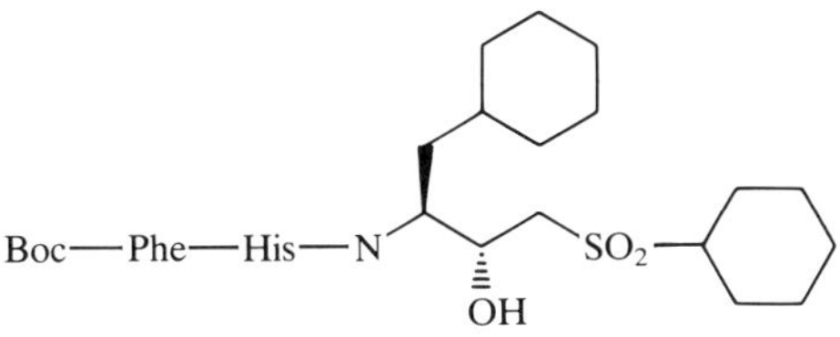

(5.55)

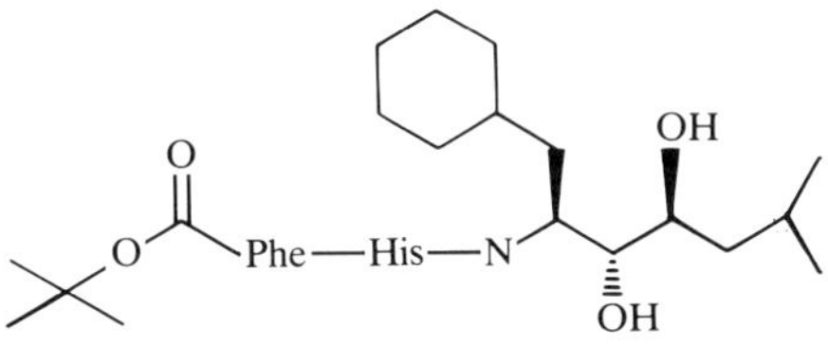

(5.56)

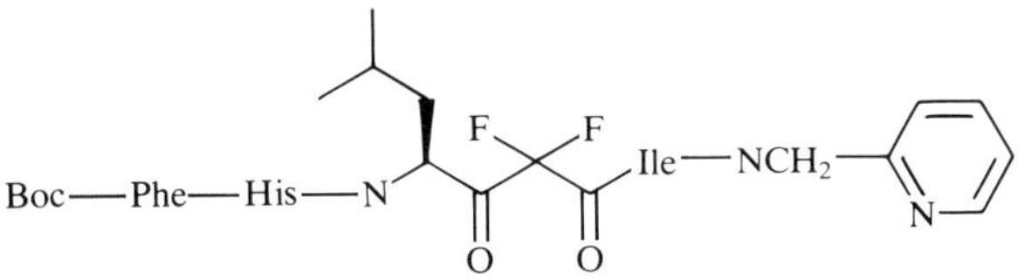

(5.57)

We have seen how protein X-ray crystallography has afforded the medicinal chemist with valuable information on the mechanism of action of enzymes. Interactive computer graphics, quantum chemical calculations and molecular modelling have also become standard tools by which medicinal chemists are able to investigate and measure enzyme–substrate interactions, and to harness molecular recognition for enzyme inhibition. Successful inhibitor design resulting from the use of such systems is becoming evermore commonplace.

One area where molecular modelling has been widely used in medicinal chemistry has been to aid the design of renin inhibitors. As was the case with ACE, no crystal structure of human renin is available to the chemist. However, X-ray crystal structures of several aspartic proteases, both unbound and bound as complexes to inhibitors, have been solved. By evaluating these structures and comparing their primary sequence with the sequence of human renin, three-dimensional models of renin and its active site have been developed (Blundell *et al.*, 1986). Several pharmaceutical companies have developed their own models of the renin active site, and used them extensively to design potent inhibitors of renin.

One of the early successes was the design of 4-amino-5-cyclohexyl-3-hydroxypentanoic acid (ACHPA, *5.58*) by the Merck group who modelled statine-containing inhibitors into a renin active site model and found that the S_1 pocket could accept a larger P_1 residue (Boger *et al.*, 1985). Using overlap algorithms, they determined that the optimal P_1 side chain was cyclohexylmethyl. ACHPA-containing inhibitors are extremely potent, and the cyclo-hexylmethyl group has found widespread use as a P_1 substitutent in other renin inhibitors.

Modelling of this type must be put in perspective, however, as the results are not always as predicted. In an attempt to take advantage of the S_1' hydrophobic pocket observed in the X-ray crystal structure of the aspartic protease (*Rhizopus chinensis*, the Merck group prepared an inhibitor containing the isobutyl statine P_1-P_1' dipeptide replacement (*5.59*). In contrast to statine (*5.50*), this dipeptide isostere has a second side chain which can bind to the S_1' subsite. The isobutyl group was predicted from their modelling to be the optimal P_1' substituent (Veber *et al.*, 1984). Although this compound turned out to be slightly more potent toward human renin than the unsubstituted statine, it was less potent against *Rhizopus chinensis*, the enzyme upon which their modelling was based. Although the experimental results are not always in agreement with the predictions, the value of modelling for generating ideas and increasing the chemist's insight cannot be over-emphasized.

One of the more elegant uses of molecular modelling resulted in the design of cyclic renin inhibitors (*5.60*) at Abbott laboratories (Sham *et al.*, 1988). Molecular modelling indicated that the P_1 and P_2 side chains of

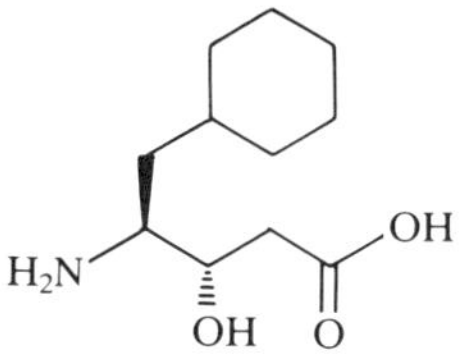

(5.58)

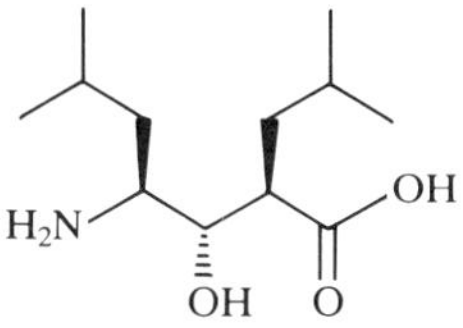

(5.59)

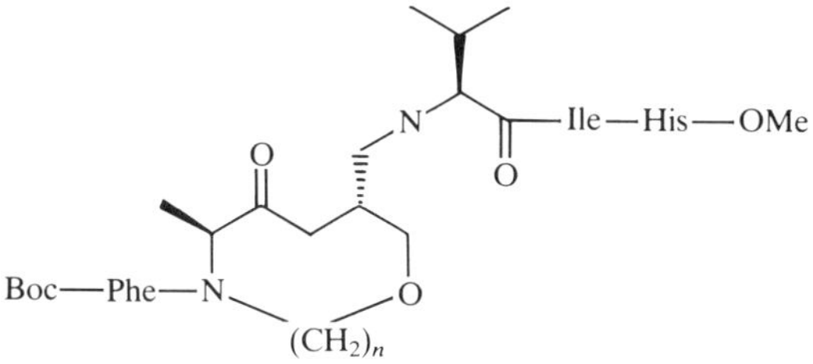

(5.60a) n=3
(5.60b) n=5
(5.60c) n=7

peptidic renin inhibitors were in close proximity and that they could be joined with an alkyl bridge. It was anticipated that this modification would increase metabolic stability of the inhibitor. Furthermore, it was predicted that inhibitors having a *trans*-conformation about the Phe–N amide bond would be more potent than those having a *cis*-conformation. Compounds 5.60b and 5.60c, which have been determined to exist predominately in the *trans*-conformation, are both potent inhibitors of renin, while compound (5.60a), which exists in the *cis*-conformation, is inactive. A recent, long sought after breakthrough will undoubtedly have a tremendous impact on future molecular modelling efforts directed towards the design of renin inhibitors. Following 1500 crystallization trials, James and colleagues (Sielecki *et al.*, 1989) have succeeded in both crystallizing recombinant human kidney renin and solving its x-ray crystal structure. As predicted from the models of human renin constructed from the structures of other mammalian and fungal aspartic proteases, the hydrophobic enzyme core and the structure of the active site, including the catalytically important aspartate residues, are very similar. However, the orientation of the loops that border the entrance to the active site differs from that of the other enzymes. This region is believed to be critical for determining substrate specificity, and knowledge of its geometry should facilitate the design of novel renin inhibitors.

Considering the amount of chemical effort which has been directed in recent years to the design of small inhibitors of human renin, it is remarkable that no compound has yet emerged as a serious candidate for commercialization. Many of the peptide-based inhibitors are very potent, but they are unsatisfactory with respect to bioavailability and plasma half-life. Only when these deficiencies have been rectified will this approach for the treatment of hypertension be fully validated in the clinical situation.

5.2.4   Inhibitors of cholesterol biosynthesis

The β-lactam antibiotics afford an excellent example of the exploitation of natural products as an important source of chemical leads – indeed, valuable drugs. In many other disease areas scientists have also looked to natural products as a rich source of structural variety in their search for drugs. Isolated enzyme assays are particularly amenable to screening of crude extracts or unfractioned fermentation broths because of the high sensitivity and high specificity of the assay. Scores of natural products in each extract or broth can be screened for activity in a single assay and the arduous tasks of isolation and structure determination can be postponed until the level of biological activity merits further effort. This approach to drug discovery has been successfully applied to the search for inhibitors of cholesterol biosynthesis.

It has been recognized for some time that elevated serum cholesterol is a primary risk factor for the development of atherosclerosis and coronary artery disease, the major cause of death in western countries. This has led to a major effort by the pharmaceuticals industry during the last decade to discover agents which lower serum cholesterol levels.

Several approaches to this goal have been pursued. A variety of compounds, exemplified by clofibrate (5.61) (Illingworth, 1987), appear to be capable of causing modest reductions in serum cholesterol although the mechanism of this effect is still not understood. Unabsorbed cationic resins, such as cholestyramine (5.62) (Grundy, 1988), bind bile acids which are derived biosynthetically from cholesterol. These bile acid sequestrants cause 15 to 30% reductions of serum cholesterol by enhancing the excretion of bile acids.

Since at least 50% of total body cholesterol in humans is derived from *de novo* synthesis, inhibition of cholesterol biosynthesis has also gained

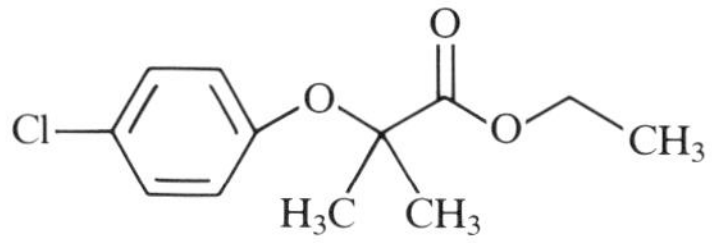

(5.61)

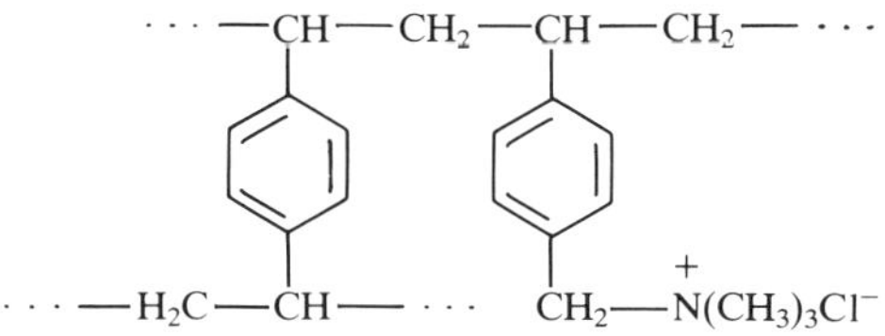

(5.62)

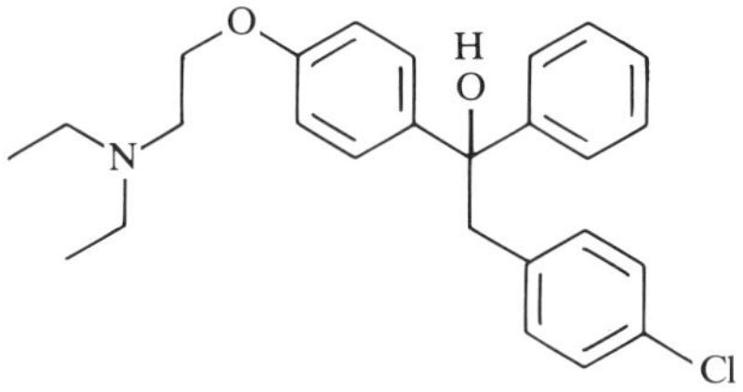

(5.63)

attention as a possible approach. The biosynthetic pathway for cholesterol involves more than 25 separate enzymes, offering many targets for the medicinal chemist (Fig. 5.15). However, the wisdom of attempting to disrupt such a fundamental metabolic pathway at any point could be questioned. Cholesterol (5.71), the ubiquinones (5.73), dolichol (5.72) and isopentyl adenosine (a component of transfer RNA), are all critical products of a common pathway. Cholesterol itself serves as a substrate for synthesis of the steroid hormones – including oestrogens, testosterone and progesterone. These metabolites play a major role in maintenance of homeostasis in man.

Assessment of the utility and safety of cholesterol biosynthesis inhibition ultimately required the evaluation of inhibitors in the clinic. An early example which was evaluated in humans was triparanol (5.63), which inhibits at a very late stage in the pathway. This inhibitor was promptly withdrawn because of serious toxicity which may have been related to the build-up of desmosterol, a biosynthetic intermediate (Tobert, 1987). This type of toxicity problem justified concerns over the appropriate choice of

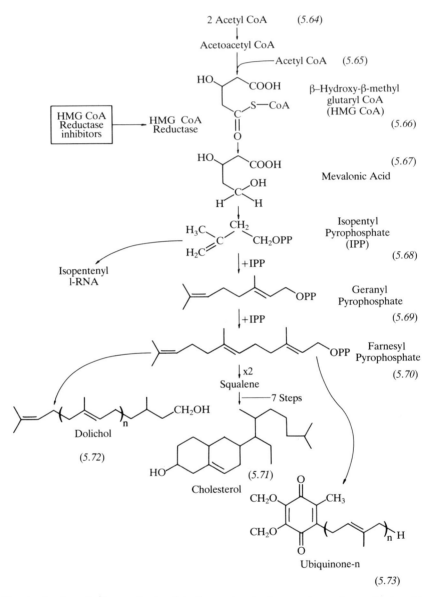

2 Acetyl CoA   (5.64)

Acetoacetyl CoA

Acetyl CoA   (5.65)

HO   COOH

S—CoA

β–Hydroxy-β-methyl
glutaryl CoA
(HMG CoA)

(5.66)

HMG CoA
Reductase
inhibitors → HMG CoA
Reductase

HO   COOH

OH

H   H

(5.67)

Mevalonic Acid

H₃C   CH₂

C   CH₂OPP

H₂C

Isopentyl
Pyrophosphate
(IPP)

(5.68)

Isopentenyl
l-RNA

+IPP

OPP   Geranyl
Pyrophosphate

(5.69)

+IPP

OPP   Farnesyl
Pyrophosphate

(5.70)

x2

Squalene

7 Steps

CH₂OH

Dolichol

(5.72)

HO

(5.71)

Cholesterol

CH₂O   O   CH₃

CH₂O   H   O   n

Ubiquinone-n

(5.73)

Fig. 5.15 Steps in the synthesis of cholesterol and other products of mevalonic acid. (Reprinted from Grundy, 1988.)

enzyme for inhibition and served to focus attention much earlier in the pathway where the intermediates, unlike desmosterol, have alternative metabolic options.

The major rate-limiting enzyme in cholesterol biosynthesis is 3-hydroxy-3-methylglutaryl-coenzyme A reductase (HMG-CoA reductase) (Brown and Rodwell, 1980) which catalyses the reduction of 3-hydroxy-3-methyl-glutaryl-coenzyme A (HMG-CoA) (5.66) to mevalonate (5.67) (Fig. 5.15). HMG-CoA reductase is a key site of feedback regulation of cholesterol synthesis in cultured human fibroblasts. Experiments in intact animals have demonstrated that under most physiological conditions it is the key regulated enzyme of cholesterolgenesis *in vivo*. This enzyme thus becomes a key target for interruption of the cholesterol biosynthesis pathway.

It was at this point that natural product screening came into play. After screening 8000 strains of micro-organisms, Endo's group at Sankyo discovered three novel inhibitors of HMG-CoA reductase, one of which was ML-236B (mevastatin) (5.74). The dihydroxy acid (i.e. ring-opened) derivative of mevastatin (5.75) is a selective, competitive inhibitor of HMG-CoA reductase with a K_i of 1 nM (Endo *et al.*, 1977; Endo, 1985). More importantly, mevastatin causes dramatic inhibition of cholesterol synthesis and decreases in serum cholesterol levels in several experimental animal species and in humans.

These observations stimulated interest in HMG-CoA reductase and paved the way for the discovery of other structurally-related inhibitors. By now lovastatin (5.76), simvastatin (5.77), and pravastatin (5.78), as well as mevastatin have been evaluated in the clinic (Tobert, 1987). The clinical and biochemical experience with these compounds has led to a more detailed understanding of their effects on serum cholesterol and, presumably, on atherogenesis.

Vertebrates normally safely transport cholesterol in the blood by esterifying the sterol with long-chain fatty acids and packaging these esters within hydrophobic cores of plasma lipoproteins (low density lipoprotein–cholesterol complexes [LDL-C] see Chapter 9). The cholesterol esters are too hydrophobic to pass through cell membranes and must be internalized by receptor-mediated endocytosis (Fig. 5.16). Once in the cell, cholesterol is liberated by cholesterol esterases and is used for the synthesis of plasma membranes, bile acids and steroid hormones. Excess cholesterol is converted into bile acids which are excreted. Normally this mechanism prevents the build-up of atherosclerotic plaque by maintaining a low concentration of blood LDL-C. An inherited defect in this process is present in familial hypercholesterolaemia, in which the normal control of cholesterol metabolism is disrupted due to defects in the LDL-C receptor gene (Grundy, 1988; Brown and Goldstein, 1986). These patients have very high levels of serum

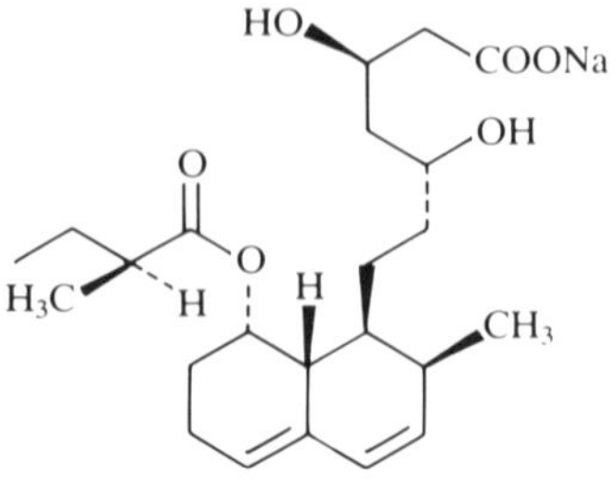

(5.75)

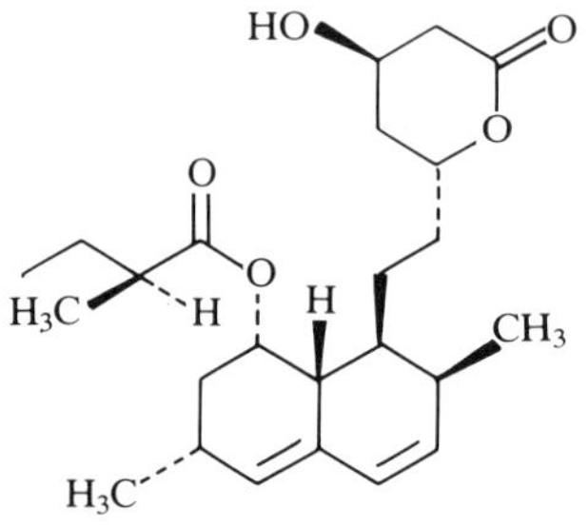

Lovastatin
(Mevinolin)   (5.76)

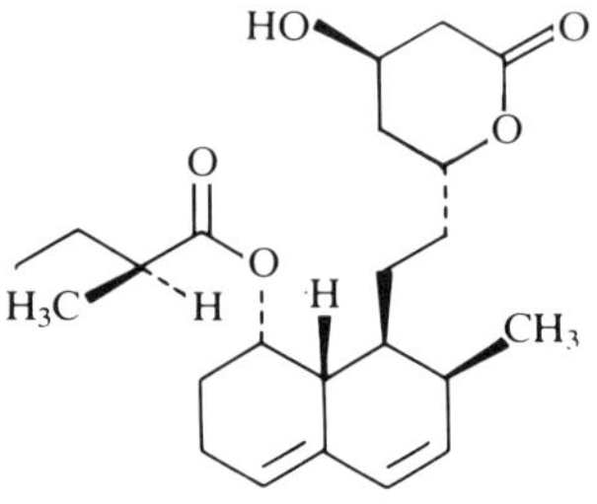

Mevastatin
(Compactin)   (5.74)

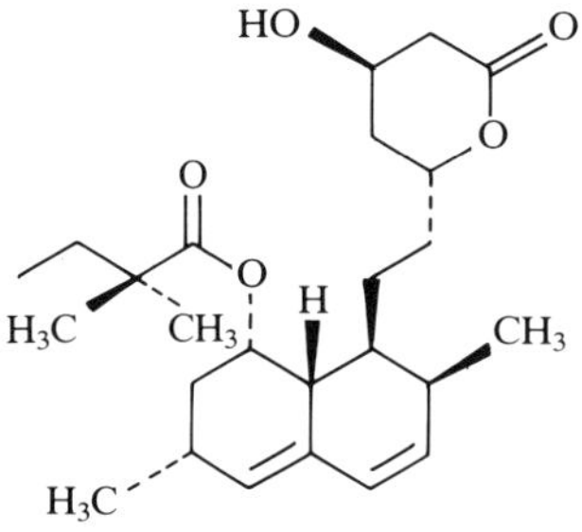

Simvastatin
(synvinolin)   (5.77)

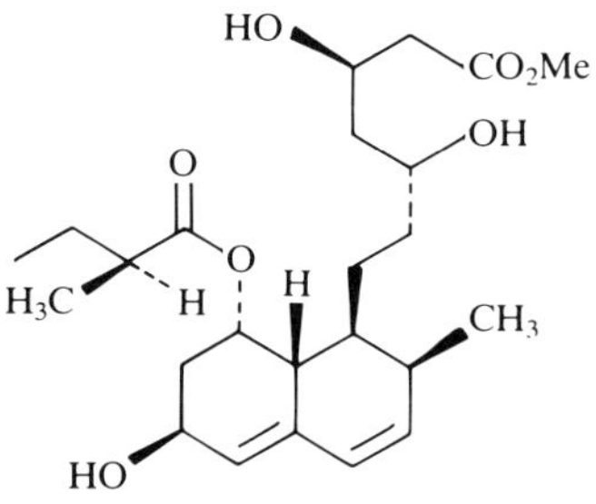

Pravastatin
(eptastatin, CS-514, SQ-31,000)
(5.78)

cholesterol and have provided important evidence for the mechanism by which HMG-CoA reductase inhibitors lower serum cholesterol levels.

Studies in various laboratories have demonstrated profound effects of HMG-CoA reductase inhibitors on serum cholesterol levels in several animal species. These inhibitors cause an initial decrease of cholesterol synthesis in the cell resulting in the stimulation of LDL-receptor activity which lowers plasma LDL-C levels. The fall in the plasma LDL-C levels is

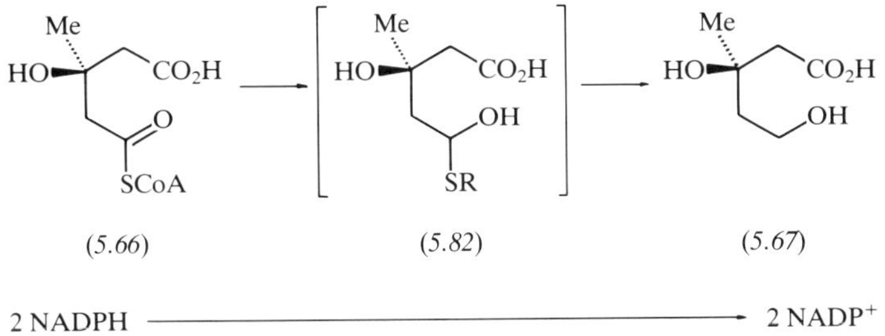

Fig. 5.16   The two-step reduction catalysed by HMG-CoA reductase.

balanced by the increase in LDL receptors thereby maintaining the amount of cholesterol entering the liver but at lower plasma concentrations. This therapy is particularly useful for patients with familial hypercholesterol-aemia in which dramatic regression of cholesterol-containing lesions have been demonstrated. These inhibitors have now been used in thousands of patients, some for as long as four years. Lovastatin (5.76) is now marketed by Merck in the United States and in other countries. Although mevastatin (5.74) was withdrawn for toxicological reasons, the fears of general toxicity due to interruption of this major pathway proved to be unfounded. Caution is still being taken in the treatment of children and pregnant women because concern for the theoretical association of the use of cholesterol biosynthesis inhibitors with impairment of endocrine function (Hoeg, 1987).

   Drug discovery in this area has not been entirely due to screening of natural products. Medicinal chemists are making their mark on HMG-CoA reductase inhibition as well. Some structural modifications of the natural products themselves have been reported (Hoffman et al., 1986b; Heathcock et al., 1987). The most important contribution to emerge from this effort so far was the discovery of simvastatin (5.77) (Hoffman et al., 1986a) through modification of the ester side chain of lovastatin (5.76). This semisynthetic inhibitor is modestly more potent in vitro and in vivo than lovastatin (5.76) and recently saw its first commercial launch in Sweden.

   Totally synthetic inhibitors have also emanated from the lead provided by the natural products. Early efforts were directed toward aromatic replacements for the hexahydronaphthalene structure in mevastatin (5.74) and its cousins. Sankyo chemists first reported the aromatic inhibitors (5.79) (Sato, 1980). Merck chemists used this approach to develop inhibitors which were more potent than lovastatin (5.76) and mevastatin (5.74), with their efforts culminating in the biphenyl derivative (5.80) (Stokker et al., 1985; Hoffman et al., 1986a, Stokker et al., 1986). Along similar lines, chemists at Sandoz have reported SRI-62320 (5.81), which is the only totally synthetic inhibitor known to be in clinical trials (Engstrom et al., 1986).

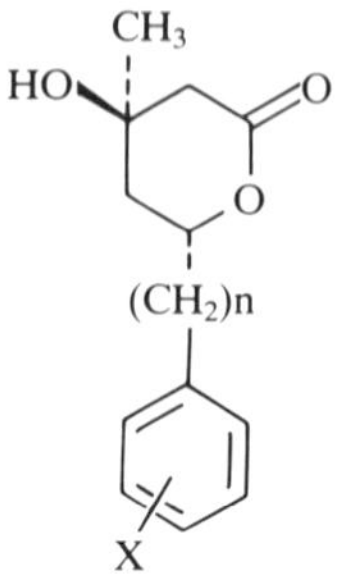

(5.79) 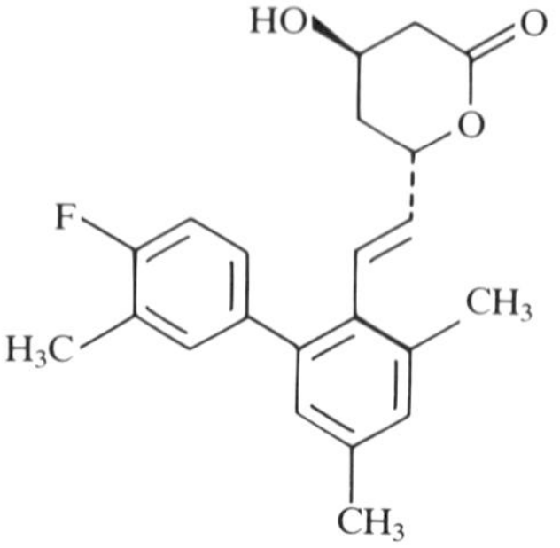   (5.80) 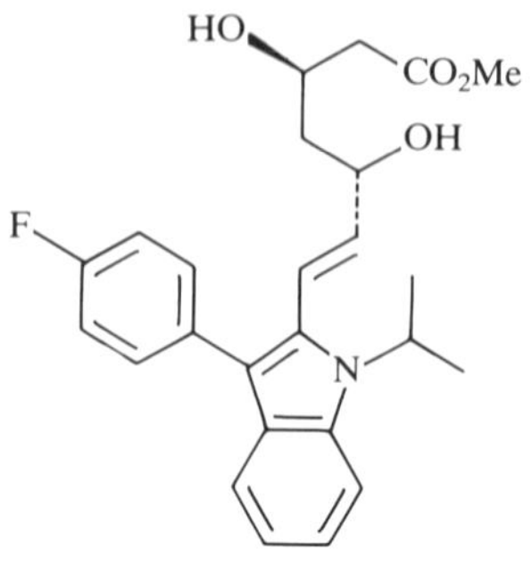   SRI-62320
                                    (5.81)

Only a few mechanistic details have been elucidated for HMG-CoA reductase. Enzymes from several sources including rat liver have been purified to homogeneity (Kleinsek, 1981). The reduction catalysed by the purified enzyme is a two-step process in which each of the two hydrides is derived from NADPH (Fig. 5.17). Following the first reduction, NADP$^+$ is believed to be released and a second mole of NADPH is bound, but the product of the first reduction step, mevaldic acid (5.82), is not released by the mammalian enzyme. Some workers believe that a cysteine residue at the active site accepts 3-hydroxy-3-methylglutaric acid from coenzyme A as the first step in the process. Using site-specific chemical probes, Dugan has reported evidence that confirms the involvement of a cysteine residue and supports the presence of an active site of histidine (Dugan and Katiyar, 1986). Surprisingly, this mechanistic information has not yet served a major role as a framework on which to base inhibitor design.

The structural similarity of the dihydroxy acid portions of these inhibitors, HMG-CoA and mevalonic acid, is believed to be the basis for recognition of the natural products by HMG-CoA reductase. It is especially intriguing to contemplate the resemblance to the enzyme-bound intermediate (5.82). Results from Abeles' laboratory suggest that the high affinity of mevastatin for HMG-CoA reductase is due to simultaneous interaction at two separate sites, the hydroxymethylglutarate domain and the hydrophobic region (Nakamura and Abeles, 1985). The dissociation constant for mevastatin is eight orders of magnitude lower than that for DL-3,5-dihydroxyvalerate demonstrating the large contribution made by the hydrophobic decalin portion of this inhibitor.

We have a case here where knowledge of the mechanism and structure of the target enzyme have contributed little or nothing to the discovery of inhibitors. The facts that these inhibitors are exquisitely potent, apparently

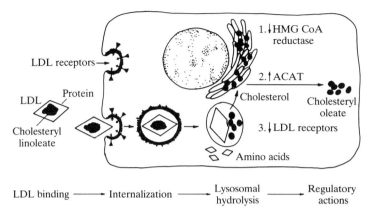

Fig. 5.17 Sequential steps in the LDL receptor pathway of mammalian cells. HMG CoA reductase, 3-hydroxy-3-methylglutaryl CoA reductase; ACAT, acyl-CoA: cholesterol acyltransferase. Vertical arrows indicate the directions of regulatory effects. (Reprinted from Brown and Goldstein, 1979.)

safe and efficacious in long term clinical trials in many patients will have blunted the drive for the medicinal chemist to exploit the few mechanistic details known in order to discover more diverse inhibitors. The current interest is in the very subtle distinctions among the inhibitors in clinical trials with regard to tissue distribution and other properties which will affect safety and efficacy in ways which can only be measured in long-term trials. It is likely that in a manner similar to the change in focus from angiotensin-converting enzyme to renin, medicinal chemists, biochemists and pharmacologists will have already begun to explore the possibility of inhibiting other early enzymes in the pathway. We can look forward to their successes.

It is accepted that HMG-CoA reductase inhibitors are highly effective hypocholesterolemic agents which will certainly be of great value in treating patients at high risk of coronary heart disease due to high levels of LDL-C in the plasma. It will be interesting to see if these inhibitors become a significant therapy for moderate hypercholesterolaemia.

5.2.5 Inhibitors of human leucocyte elastase

Many biological systems including the blood coagulation system, the complement system and the blood pressure regulation system are controlled by proteases whose activity is modulated by endogenous inhibitors. Mounting evidence suggests that several of the chronic degenerative diseases such as pulmonary emphysema, osteoarthritis and Alzheimer's disease also involve proteases, and may reflect a defect in the balance between particular proteases and their endogenous inhibitors.

Emphysema is primarily a smoker's disease. A key feature which differentiates this disease from other pulmonary afflictions is the destruction of elastin, one of the key structural proteins of the lung matrix. For this reason considerable attention has been given to the protease/antiprotease hypothesis for the aetiology of this disease. The protease involved is human leucocyte elastase (HLE), which is released from the azurophilic granules of polymorphonuclear leucocytes. Although the normal function of this enzyme has not been defined, it has a broad substrate specificity and is capable of degrading a number of structural proteins including elastin. Following release of the enzyme from leucocytes, proteolytic activity is normally held in check by an endogenous antiprotease, α_1-protease inhibitor (α_1PI). However, the delicate enzyme/inhibitor balance can be disturbed by a number of factors including, a genetic deficiency in α_1PI, an overabundance of HLE due to an inflammatory response in the lung, and oxidative destruction of α_1PI by an exogenous agent such as cigarette smoke.

Among the possible approaches for restoring the enzyme/inhibitor balance and thereby treating the disease are, direct replacement with either natural or recombinant α_1PI and enhancement of the elastase inhibitory capacity of the lung with a low molecular weight synthetic inhibitor. Substantial progress has been made in the evaluation of replacement therapy with α_1PI, but we will focus here on synthetic inhibitors of HLE which have been developed and are progressing to clinical trials (Trainor, 1987).

The design of novel synthetic inhibitors of HLE requires a detailed understanding of the catalytic mechanism responsible for elastin degradation. As a member of the serine family of proteases, HLE contains a 'catalytic triad' comprised of an aspartic acid, a histidine, and a serine residue (Fig. 5.18). The serine hydroxyl of the triad is activated for nucleophilic addition to an amide bond in the substrate (e.g. elastin). A tetrahedral adduct is formed which collapses to give an acyl enzyme and free amine. Catalytic cleavage of the acyl enzyme by water completes the cycle with release of the second product, an acid. Note that cleavage of peptide bonds by both aspartyl proteases (e.g. renin) and serine proteases (e.g. elastase) involves conversion of a carbonyl group in the substrate into a tetrahedral intermediate. The similarity stops there, however, since in the case of renin the nucleophilic agent is water, whereas in the case of elastase it is the hydroxyl group of a key serine residue of the enzyme.

The fact that elastase is a serine protease triggered quite different approaches to inhibitor design in the minds of scientists at ICI Pharmaceuticals Group and at Merck. The former group was intrigued by the possibility of intercepting the enzyme mechanism with transition-state analogues resembling the tetrahedral adduct formed in the normal enzyme reaction.

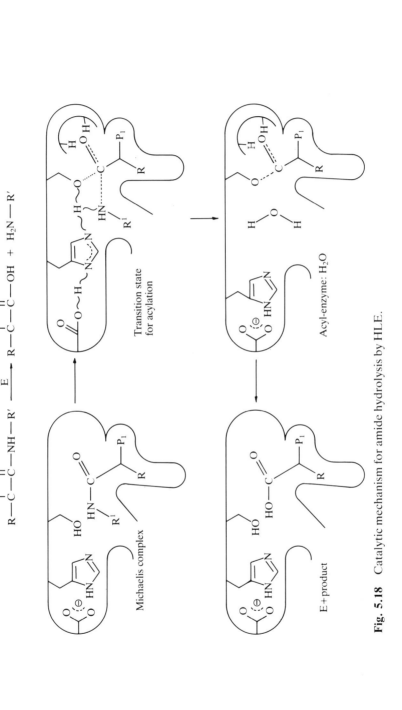

Fig. 5.18 Catalytic mechanism for amide hydrolysis by HLE.

The latter group reasoned that leads might be developed from inhibitors of other serine proteases and, in particular, from the β-lactam antibiotics. In practice both approaches led to potentially useful inhibitors.

In their search for mechanism-based inhibitors the ICI group focused on several classes of reversible inhibitors based on peptidyl electrophilic carbonyl derivatives. Electrophilic carbonyl derivatives and carbonyl mimics have been explored as transition-state analogues for several serine proteases. Recent examples targeted at HLE include aldehydes, boronic acids, and α-ketoesters. All of these inhibitors contain functional groups capable of reversibly forming tetrahedral adducts with the serine hydroxyl in the enzyme's active site and are inhibitors, rather than substrates, because they cannot be further converted into cleavage products.

The group at ICI initially designed a series of peptidyl aldehydes selective for HLE. Borrowing from the structural features of the natural substrate elastin, the natural inhibitor, α_1-PI and from the synthetic substrate work of Powers at Georgia Institute of Technology (Fig. 5.19), they were able to build the appropriate recognition features into the peptide backbone of the aldehyde inhibitors (e.g. *5.84*) to achieve the necessary selectivity and potency *versus* HLE (Yasutake and Powers, 1981). Although interesting aldehyde-based inhibitors were produced, this series was eventually abandoned due to concerns about metabolic stability and shelf life. Not infrequently, compounds with excellent pharmacological profiles are ruled out from further consideration as potential drugs due to their failure to meet pharmaceutical criteria.

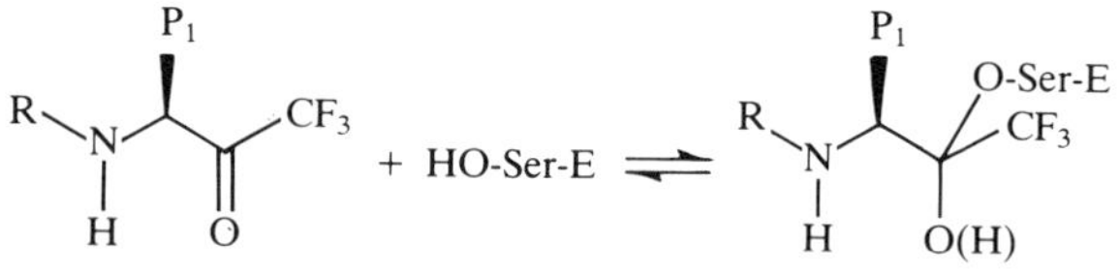

(5.85)   *(5.86)*

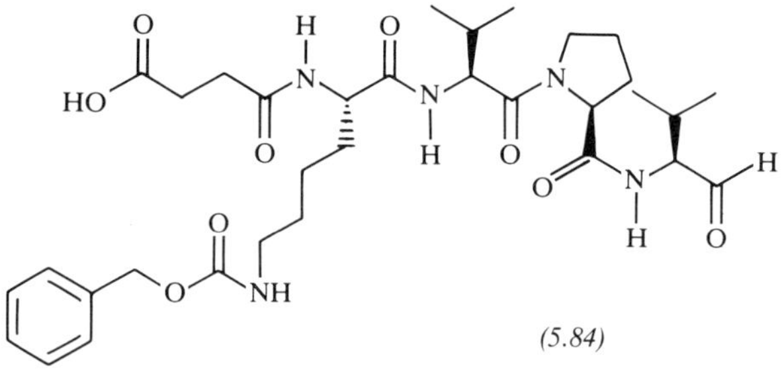

(5.84)

$K_i = 4.0 \times 10^{-9} M$

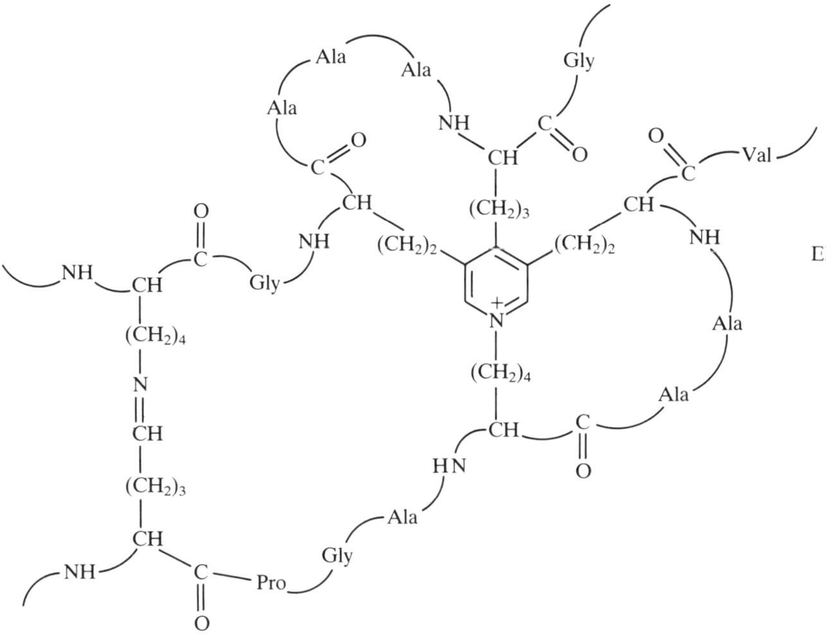

Fig. 5.19

The figure shows a structure labelled Elastin, with positions labelled below:

P_5	P_4	P_3	P_2	P_1	P_1'	P_2'	P_3'	P_4'	
Ac — Ala — Ile — Pro — Met — Ser — Ile — Pro — Pro — NH_2									α_1-Pl fragment
MeO — Suc — Lys — Ala — Pro — Val — pNA									'Powers' substrate

(Lys bears a substituent Z below it)

Concomitant with their work on aldehydes, the ICI group explored a new class of mechanism-based inhibitors, the peptidyl trifluoromethyl ketones (TFMKs) (5.85). This series contained compounds of superior potency and stability. As with the aldehydes the TFMKs were designed to inhibit HLE by the reversible formation of a covalent tetrahedral adduct (5.86) (Fig. 5.20). Simple aliphatic ketones are generally poor inhibitors of serine proteases; introduction of α-fluorine atoms increases the electrophilicity of the carbonyl, as is evident from the hydration constants of trifluoromethyl ketone and acetone (35 and 1 respectively). The trifluoromethyl group also serves to stabilize the resulting oxyanion of the tetrahedral complex such that it closely resembles the oxyanion of the actual reaction intermediate.

The TFMKs are competitive, slow-binding inhibitors of HLE. Kinetic analysis with HLE suggests that the slow-binding nature of the inhibition is a

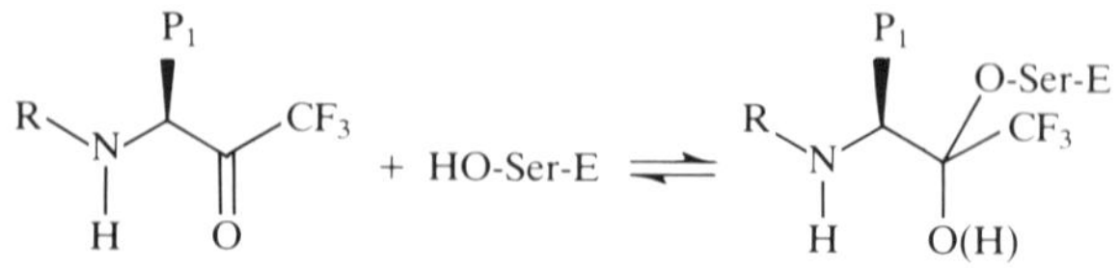

(5.85)   *(5.86)*

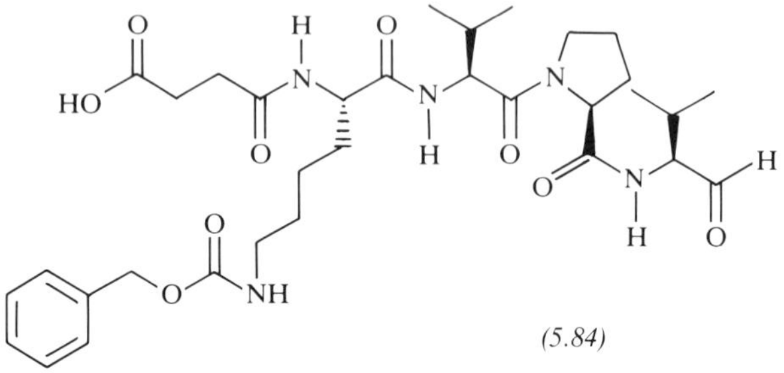

(5.84)

$K_i = 4.0 \times 10^{-9} M$

Fig. 5.20   Interaction of trifluoromethyl ketone inhibitors with the active site serine of elastase.

result of a rate-limiting conformational change of the enzyme-inhibitor (E-I) complex following the formation of the Michaelis complex. This conformational change allows for optimal interactions between enzyme and inhibitor resulting in a highly stabilized complex (Stein, R. L. *et al.*, 1987). Abeles' group at Brandeis has also investigated TFMKs as inhibitors of serine proteases including PPE, HLE and chymotrypsin. Mechanistic and NMR studies with chymotrypsin suggest that the E-I complex is present as an ionized hemiketal group with a pKa of 4.9 (Liang and Abeles, 1987).

The anticipated binding mode for the peptidyl TFMKs was confirmed by X-ray analysis of a tripeptide inhibitor (5.87) bound to the closely related enzyme, porcine pancreatic elastase (PPE). This structure unambiguously showed a tetrahedral intermediate involving covalent bonding of the inhibitor to the serine hydroxyl (Fig. 5.21) (Takahashi *et al.*, 1988).

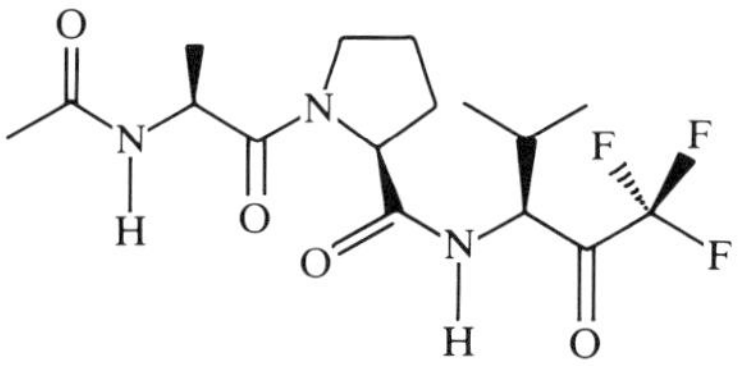

(5.87)

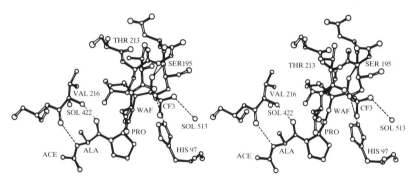

Fig. 5.21 A stereo view of the inhibitor (5.87) in the extended binding site of PPE. Hydrogen bonds are represented by broken lines. (Reprinted from Takahashi *et al.*, 1988.)

HLE has an extended binding site containing a number of remote subsites on either side of the catalytic site. Not surprisingly, the inhibitory potency of the TFMKs was found to be dependent on the length of the peptide backbone, with optimal activity (K_1 values in the nanomolar range) in the tri-and tetrapeptide analogues (Fig. 5.22). In general the most potent inhibitors *in vitro* contained very lipophilic side chains (e.g. *5.91*).

In vivo activity of inhibitors of HLE is evaluated in a hamster model of emphysema. As is often the case, the *in vitro* potency of the TFMK inhibitors did not correlate well with the *in vivo* potency (Fig. 5.23), despite

	P_5	P_4	P_3	P_2	P_1	K_i(nM)
(5.88)					Z — VAL — CF$_3$	13,000
(5.89)				Z — PRO — VAL — CF$_3$		1,800
(5.90)			Z — VAL — PRO — VAL — CF$_3$			1.6
(5.91)		Z — LYS — VAL — PRO — VAL — CF$_3$				<0.1
		$\mid$ Z				
(5.92)					MeOSuc — VAL — CF$_3$	53,000
(5.93)				MeOSuc — PRO — VAL — CF$_3$		3,200
(5.94)			MeOSuc — VAL — PRO — VAL — CF$_3$			13
(5.95)		MeOSuc — LYS — VAL — PRO — VAL — CF$_3$				<0.3
		$\mid$ Z				

Fig. 5.22 Effect of peptide length on inhibitory potency (K_i).

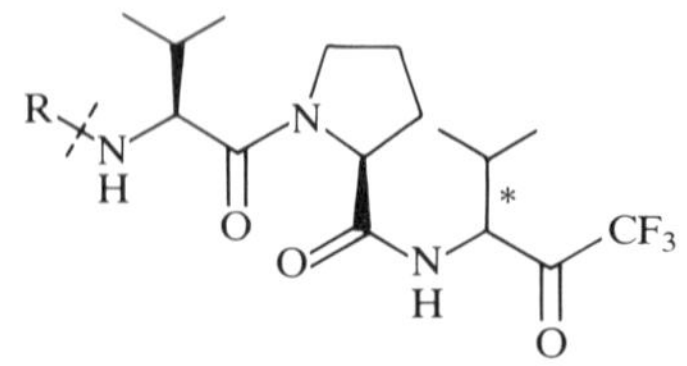

	R	* Isomer	k_i (nM)	In-vivo duration of action (h)
(5.90)	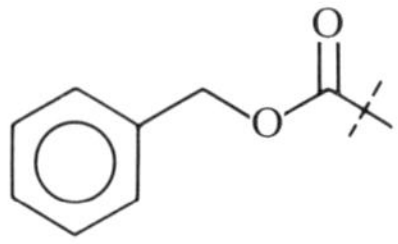	S	1.6	2
(5.96)	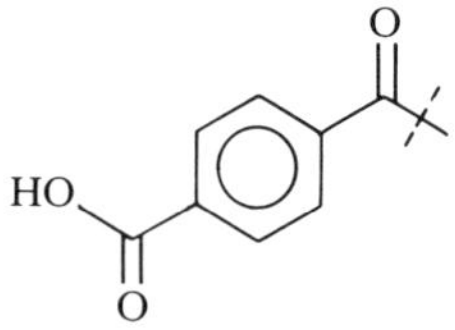	R,S	6.5	2
(5.97)	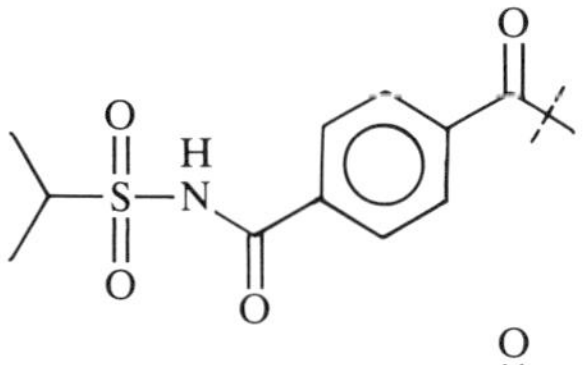	R,S	0.6	4
(5.98)	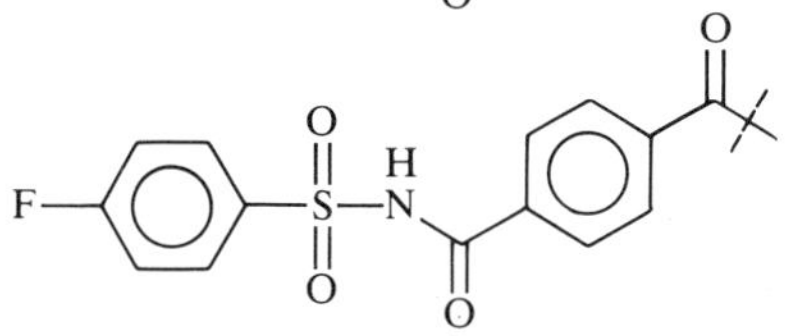	R,S	0.6	6
(5.99)	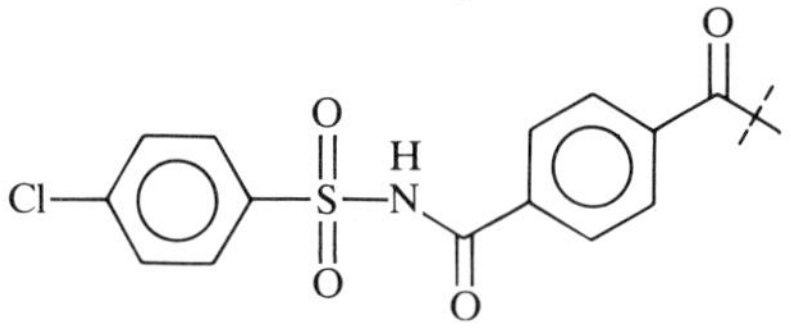	R,S	0.5	18

Fig. 5.23 Duration of action of TFMK inhibitors in a hamster model of emphysema.

the fact that the inhibitors were administered directly to the target site, the lung. Thus a separate pattern of structure–activity relationships (SAR) was established for in vivo activity and a series of acylsulphonamide derivatives emerged as the compounds of choice. Of these ICI 200,880 (5.99) proved to be very effective in vivo and had an extended duration of activity which made it particularly attractive as a potential clinical candidate. The reasons for the excellent *in vivo* profile of the acylsulphonamides remain unclear. However, an analysis of the structure–activity relationships implies that this terminal group confers the appropriate balance of acidity and water/lipid solubilities necessary for satisfactory duration of action following direct administration to the lung.

As was mentioned earlier, the β-lactam antibiotics were the starting point for a second series of elastase inhibitors. These compounds represent a radical departure from the peptide-based inhibitors previously discussed. The series has yielded potent inhibitors of HLE which, like the peptidic TFMKs, are active in the hamster model. The initial key finding at Merck was that the benzyl ester of the β-lactamase inhibitor clavulanic acid was a weak inhibitor of HLE. The scientists at Merck recognized the connection between the bacterial penicillin binding proteins (PBPs), β-lactamases, and mammalian serine proteases: the mechanism by which all three classes of enzymes hydrolyse amide bonds involves the nucleophilic addition of an active site serine hydroxyl to the carbonyl of the scissile amide bond and formation of a tetrahedral intermediate.

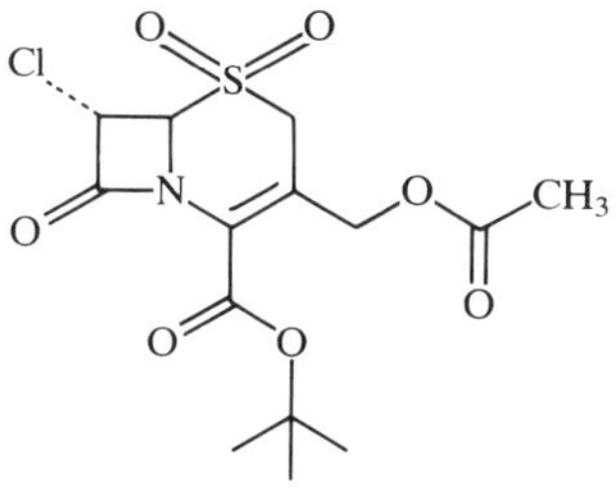

(5.100)

Following an extensive chemical program, the cephalosporin sulphone (5.100) emerged as one of the most potent β-lactam elastase inhibitors (Doherty *et al.*, 1986). Note that the stereochemistry of the cephalosporin C_7 substitutent is different in the transpeptidase and elastase inhibitors. The bacterial enzymes are inhibited by cephalosporins with the $C_{7\beta}$ configuration, whereas the mammalian enzymes require the $C_{7\alpha}$ configuration. This may be a reflection of the fact that the bacterial enzymes recognize substrates in a region containing D-amino acids while the mammalian enzymes recognize L-amino acids. As a consequence, the β-lactam antibiotics and their analogues are not effective inhibitors of elastase, nor are the β-lactam elastase inhibitors active as antibiotics or as inhibitors of β-lactamases.

Structure activity studies have confirmed that β-lactams possessing a free carboxylic acid, including clavulanic acid, are poor inhibitors of HLE. This probably stems from the fact that HLE cleaves amide bonds between internal, non-ionic amino acid residues – i.e. it is a true endopeptidase. In contrast, the bacterial enzymes are exodipeptidases – i.e. they cleave the two C-terminal residues of their substrates. Both the PBPs and β-lactamases contain a binding pocket for a carboxylic acid.

Interpretation of the mechanism by which cephalosporin A inhibits elastase has been facilitated by the availability of an X-ray crystallographic structure of the inhibitor bound to the porcine enzyme (Navia *et al.*, 1987). While the initial inhibition of elastase is reversible, there follows a time-dependent irreversible inhibition which ultimately results in covalent attachment of the dihydrothiazine ring of the inhibitor to the active site histidine (5.106) (Fig. 5.24). Peptidyl chloromethyl ketones have also been shown to inhibit serine proteases by alkylation of the active site histidine in a similar fashion.

A mechanism consistent with the facts is outlined in Fig. 5.24. The initial steps leading to the acyl enzyme complex (5.103) are as described previously for the bacterial enzymes in section 5.2.2. Again, this complex may be hydrolysed to regenerate active enzyme. Alternatively, this complex can undergo a series of eliminations to species such as (5.104) and (5.105), either of which has the potential to alkylate His-57.

The β-lactam inhibitors of HLE and PPE have added a remarkable new chapter to the story which began with Fleming's observation of the antibiotic properties of cultures of *Penicillium notatum*. Besides emphasizing the debt we owe to Nature as a source of new medicines, they exemplify the scope for applying experience gained with one target enzyme to a related enzyme involved in a quite different disease.

We have seen from these examples just how important is the choice of target enzyme. It is the keystone of the approach to a clinically useful drug. However, it is rarely an obvious decision, especially when new approaches to therapy are being considered. Uppermost in the chemist's mind are the side effects which may be associated with the mechanism of action of the new drug, the quality of response, and the percentage of responders in the patient population. The side effects associated with tyrosine hydroxylase inhibitors make them unacceptable as antihypertensive agents because alternative therapies exist, whereas quite severe side effects might be acceptable in a drug where there is no such choice. Concern that only a small proportion of hypertensive patients might benefit, not the potential side effects of drugs, caused many companies to reject the renin–angiotensin system as a target. Inhibitors of angiotensin-converting enzyme have proved to be excellent antihypertensive drugs, with minimal side effects, and are among the 'blockbusters' in terms of commercial success. Even so, there remains scope for yet further specificity, as is evident from the enormous

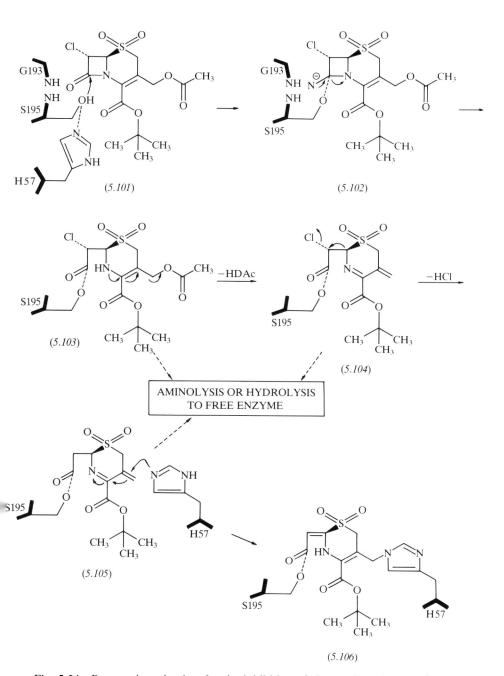

Fig. 5.24 Proposed mechanism for the inhibition of elastase by β-lactams (Navia *et al.*, 1987; copyright 1987, Macmillan Magazines Ltd.; used with permission).

research effort currently directed at finding inhibitors of renin, an alternative site for intervention in the renin–angiotensin system.

Nature has been of enormous value as a seemingly limitless source of novel structures, perhaps best exemplified by the β-lactam antibiotics. The most recent development in this remarkable story is the availability of β-lactam inhibitors of human leucocyte elastase – through a lineage which can be traced directly to Nature *via* a derivative of clavulanic acid. The first HMG-CoA reductase inhibitor to achieve commercial success, lovastatin, is also a natural product. It is remarkable how Nature has, once again, catalysed a major breakthrough in the absence of structural and mechanistic information for the target enzyme. We can expect the medicinal chemist soon to distance himself structurally from lovastatin; what is less certain is whether tomorrow's inhibitors will stem from an appreciation of enzyme structure and mechanism. It may well be that attention will turn to other steps in the biosynthetic pathway.

Emphysema is but one of several chronic degenerative diseases for which there are no drugs. Such diseases are exceptionally difficult targets for the pharmaceuticals industry, not least because of the scale and duration of the clinical programme necessary to bring a new drug to registration. The peptidic TFMK inhibitors of human leucocyte elastase are an excellent example of the chemist incorporating mechanistic and structural information in the design process. This study also serves to exemplify the problems often encountered in progressing from *in vitro* to *in vivo* activity, even when the inhibitor is administered directly to the apparent site of action.

The requirement for selectivity is always an important consideration, whether it be for a tissue or for a target enzyme. We have seen how with the various β-lactams, or the TFMKs, the chemist has been able to tailor in selectivity for a subset of serine proteases and thereby avoid the serious consequences which could result from inhibiting other members of this family, for example enzymes in the blood-clotting cascade. The dihydrofolate reductase inhibitors take this a step further in that some members of the class can distinguish between enzymes from different species. Furthermore, we are beginning to understand the reasons underlying these differences at the molecular level.

These five case studies will have served to show the diverse routes to new drug discovery. We should welcome good drugs irrespective of their origins, particularly when they meet an unfilled need, but as medicinal chemists it is natural to hope that rational drug design will play an ever-increasing role in their discovery.

5.3 RECENT DEVELOPMENTS AND THEIR APPLICATION TO INHIBITOR DESIGN

Medically and economically important drugs which exert their biological effect by modulation of enzyme activity have assured enzyme inhibition an important place in pharmaceutical research. Our understanding of the biochemistry of disease states has increased dramatically over the last decade so that the usual case, as we have seen with angiotensin-converting enzyme, renin and HMG-CoA reductase, is that the potentially important enzyme has been identified before the medicinal chemist begin to search for the drug. This has increased the importance of what we have called *ab initio* enzyme inhibitor design. Fundamental advances in our understanding of both the energetics of molecular interactions and the detailed mechanisms of enzyme-catalysed reactions will be of great importance to current and future efforts to design enzyme inhibitors.

If we ignore the more complicated issues of absorption, transport, metabolism and excretion, most problems in drug design reduce to problems of small molecule–large molecule interactions. The lock and key concept of an enzyme and its substrate is no longer a satisfactory model by which to evaluate molecular interactions. Important tools, such as nuclear magnetic resonance, protein X-ray crystallography, interactive computer graphics, molecular mechanics and quantum chemical calculations, are proving extremely valuable to the medicinal chemist in efforts to measure these interactions and to harness molecular recognition for enzyme inhibition.

The typical approach which was outlined in several of the previous examples was for the chemist to develop an idea for inhibitors from the structure of natural product inhibitors, the structures of substrates or products, or through an understanding of the enzyme's mechanism. The reduction of this idea to practice and the optimization of the lead was a process of systematic exploration of structure–activity around the lead.

The art and science of this approach have been developed to a high degree during the last two decades. Chemists have repeatedly been able to take leads represented by fairly weak inhibitors or structures which were unsatisfactory as drugs because of problems of structural complexity, lack of oral activity or physicochemical drawbacks and turn them into pharmaceutically useful drugs. Witness the numerous extremely potent inhibitors which have been developed over the past few years and the collection of enzyme inhibitors which have already demonstrated their commercial success with others on the horizon.

Just a few years ago, in the previous edition of this book, we commented that the medicinal chemist will seldom be so fortunate as to have an X-ray crystal structure of their particular target enzyme. In fact it is becoming more and more apparent that with increasing sophistication of X-ray structure determination of proteins, and understanding of protein folding, and energetics, the way of the future will increasingly find the chemist working from the actual or hypothetical protein structure. Even with this information, chemists have, to date, typically progressed through incremental improvements in inhibitors found by other routes rather than by *de novo* design in which a molecule has been 'custom built' to fit the detailed co-ordinates of an active site. However, there are encouraging signs of progress being made on this front. Thus Ripka and colleagues (1987) have reported an interesting example of 'custom building' inhibitors of phospholipase A_2. This approach will stretch understanding of molecular interactions beyond where they stand now and will force the theoretical understanding of these interactions to be distilled to the point that medicinal chemists can take advantage of the information in inhibitor design.

Theoretical calculation and empirical observation are making inroads on the process of creating three-dimensional protein structures from primary sequence information (Blundell *et al.*, 1987). Protein structure data banks are providing the opportunity to observe the frequency at which residues find themselves in particular elements of secondary structure and scientists can begin to predict these secondary structural elements from sequence data. NMR studies of peptide fragments can add detail to these predicted secondary structural elements. Further refinements are possible through the use of molecular dynamics calculations to minimize the energetics of the fragments. Nuclear magnetic resonance data on peptide fragments are more readily available than crystal structures and new techniques such as two-dimensional NMR are proving very powerful for the elucidation of structure and conformation. If fragment model building becomes reliable, perhaps this route to protein structure determination will become more common than X-ray studies.

Less progress has been made in the folding of these fragments into a tertiary structure but this field is seeing intense research activity, as well. What has been more successful is structure prediction through making substitutions of primary sequence into the determined structure of one protein in order to approximate the structure of a new protein. Again, macromolecular dynamics can then permit local relaxation of this new structure in order to attain more closely the minimum energy conformation of the new protein. Initial attempts required very high degrees of sequence homology (80%) but more recent attempts are pushing this limit to below 50%.

Protein X-ray crystallography itself is speeding up. The availability to crystallographers of new high intensity X-ray synchotron sources, improved array detectors, and faster calculations promise to considerably shorten the time from the availability of crystals to a determined structure.

All of these advances threaten to make the laboratory processes of isolation, purification, sequence determination and crystallization rate limiting in protein structure determination. Here too, however, advances in the science are changing the way we go about the process. Molecular biology permits isolation of a copy of the messenger RNA, or even the gene coding for a specific protein, through screening with oligonucleotide probes designed from just a few amino acids in the primary sequence. The isolated DNA then provides the entire primary sequence through translation of the genetic code. Generally useful expression vectors and the development of several host bacteria, yeast, and mammalian cell lines now make it reasonably straightforward to produce the genetically engineered protein in quantities useful for such activities as crystallization, detailed enzyme kinetics, and other characterization. For the protein biochemist, the problem reduces to the purification of only the small amount of protein required to provide definitive sequence of only a few amino acids. The molecular biologist can then provide relatively large quantities of crude protein for purification, refolding, characterization and crystallization. If the determination of structure from primary sequence advances as we have described above, perhaps even this will become unnecessary or simply confirmatory.

Techniques (especially site-directed mutagenesis) permit point changes in amino acid sequence to be engineered into the gene. These mutant proteins can provide important information on protein folding, active-site residues and interactions occurring during the binding of substrates or inhibitors.

In order to take advantage of this additional information, the medicinal chemist needs to be able to display and manipulate effectively the structures of the enzymes of interest. Advances in powerful, yet relatively inexpensive computers, and high-speed, high-resolution graphical display systems have come to the rescue (Loftus *et al.*, 1987). The medicinal chemist can now readily construct computer graphic models of enzymes from atomic co-ordinates. These models can be manipulated in space, unimportant areas of protein structure can be carved away at will and colour-coded three-dimensional active sites can be probed with potential inhibitors. The structures can be viewed as space-filling models to represent steric interactions most accurately, or as Dreiding-type models to detail specific molecular interactions most effectively. Interatomic distances and steric congestion, as well as other parameters can be simply quantitated.

Of course, computer modelling systems can also be used effectively for construction of hypothetical enzyme active sites when no actual structural data is available. Iterative modification of such models as biological data on each new inhibitor or substrate becomes available should lead to an active-site model as valuable, if not as detailed, as the actual structure.

Not only do modern graphics systems provide the medicinal chemist with an efficient means of visualizing target enzymes, small molecule inhibitors and enzyme–inhibitor complexes, they also serve as a basis for a much more detailed examination of molecular interactions. Quantum chemical calculations, even at the *ab initio* level, are now possible on fairly large systems. These calculations must be used with care since they are still subject to major limitations, for example, the inability to describe solvent effects accurately. They do permit, however, quantification of interactions such as electrostatic attraction, dispersive forces, polarizability, and induced dipolar attractions, which are as important as size and shape in determining the strength and specificity in molecular interactions. State-of-the-art graphics systems provide the chemist with a means to visualize the results from these calculations. Such displays as solvent accessible surfaces and electrostatic-isopotential surfaces can significantly broaden the chemist's view of molecules.

On the horizon as computing speed continues to increase, we see the potential of interactive docking of small-molecule inhibitors with their enzyme hosts. These docking manoeuvres will be able to take advantage of molecular properties beyond just size and shape and will permit the chemist to 'see and feel' the molecule in the active site and to be able to maximize positive interactions and minimize destabilizing interactions in real time.

One of the major advantages of these modelling systems is that they permit the chemist to evaluate selected parameters of potential inhibitors without actually synthesizing each and every compound. A matrix of many inhibitors can be examined for those which most effectively address particular aspects of design such as steric size and shape, dipole moment alignment, acidity or basicity. If used with caution, modelling, coupled with the medicinal chemist's synthetic expertise, can afford highly efficient enzyme active-site exploration which should expedite the discovery of potent inhibitors of optimal design.

In the past, medicinal chemists have shown their capability to develop potent and specific enzyme inhibitors in remarkable ignorance of the host enzyme. They will only improve their skills for this approach with time. The new challenge lies in assimilating and exploiting the unprecedented level of information which will become available for the structure of the target enzyme and the nature of interactions of small and large molecules. This understanding should aid the efficient introduction of new enzyme inhibitors of novel structural types for clinical evaluation and therapy.

REFERENCES

Baker, D. J., Beddell, C. R., Champness, J. N., Goodford, P. J., Norrington, F. E. A., Smith, D. R., and Stammers, D. K. (1981) *FEBS Lett.*, **126**, 49.
Bertino, J. R., Srimatkandada, S., Carman, M. D., Jastreboff, M., Mehlman, L., Medina, W. D., Mimi, E., Moroson, B. A., Cashmore, A. R., and Dube, S. K.

(1986) in *New Experimental Modalities in the Control of Neoplasia*, Plenum Publishing Corp., New York, pp. 183–193.

Blundell, T. L., Sibanda, B. L., Hemmings, A., Foundling, S. F., Tickle, I. J., Pearl, L. H., Wood, S. P. (1986) in *Topics in Molecular Pharmacology Vol 3: Molecular Graphics and Drug Design* (ed. A. S. V. Burgen, G. C. K. Roberts and M. S. Tute), Elsevier, Amsterdam, pp. 323–334.

Blundell, T. L., Sibanda, B. L., Sternberg, M. J. E. and Thorton, J. M. (1987) *Nature*, **326**, 347.

Boger, J., Payne, L. S., Perlow, D. S., Lohr, N. S., Poe, M., Blaine, E. H., Ulm, E. H., Schorn, T. W., LaMont, B. I., Lin, T.-Y., Kawai, M., Rich, D. H. and Veber, D. F. (1985) *J. Med. Chem.*, **28**, 1779.

Brown, A. G. (1981): J. Antimicrob. Chemother., 7, 15. Brown, M. S., Kovanen, P. T. and Goldstein, J. L. (1981) *Science*, **212**, 628.

Brown, M. S., and Goldstein, J. L. (1979) *Proc. Natl. Acad. Sci. USA*, **76**, **7**, 3330.

Brown, M. S., and Goldstein, J. L. (1986) *Science*, **23**, 34.

Brown, W. E. and Rodwell, V. W. (1980) *Experientia Suppl*, **36**, 232.

Burger, A. (1980): in *The Basis of Medicinal Chemistry*, Part I, 4th edn (ed. M. E. Wolf), John Wiley, and Sons, New York, pp. 1–54.

Byers, L. D. and Wolfenden, R. (1973) *Biochemistry*, **12**, 2070.

Calvert, A. H., Jones, T. R., Dady, P. J., Grzelakowska-Sztabert, B., Paine, R., and Taylor, G. A. (1980) *Eur. J. Cancer*, **16**, (**5**), 713.

Christianson, E. W., Lipscomb, W. N. (1986) *J. Am. Chem. Soc.*, **108**, 4998.

Cocco, L., Groff, T. P., Temple, C., Jr., Montgomery, J. A., London, R. E., Matwiyoff, N. A., and Blakely, R. L. (1981) *Biochemistry*, **20**, 3972.

Cody, R. J., Burton, J., Evin, G., Poulsen, K., Herd, J. A. and Haber, E. (1980) *Biochem. Biophys. Res. Commun.*, **97**, 230.

Cody, V. (1985) in *Molecular Basis of Cancer, Part B: Macromolecular Recognition, Chemotherapy and Immunology*, Alan R. Liss, Inc., pp. 275–284.

Cody, V., Welsh, W. J., Opitz, S., and Zakrzewski, S. F., (1984) in *QSAR Des. Bioact. Compd.* (ed. Kuchar, M.), Barcelona, Spain, pp. 241–252.

Cushman, D. W., Cheung, H. S., Sabo, E. F. and Ondetti, M. A. (1977) *Biochemistry*, **16**, 5484.

Dellaria, J. F., Makai, R. G., Bopp, B. A., Cohen, J., Kleinert, H. D., Luly, J. R., Merits, I., Plattner, J. J. and Stein, H. H. (1987) *J. Med. Chem.*, **30**, 2137.

Doherty, J. B., Ashe, B. M., Argenbright, L. W., Barker, P. L., Bonney, R. J., Chandler, G. O., Dahlgren, M. E., Dorn, C. P., Jr., Finke, P. E., Firestone, R. A., Fletcher, D., Hagmann, W. K., Mumford, R., O'Grady, L., Maycock, A. L., Disano, J. M., Shah, S. K., Thompson, K. R. and Zimmerman, M. (1986) *Nature*, **322**, 192.

Dugan, R. E. and Katiyar, S. S. (1986) *Biochem. Biophys. Research Commun.*, **141**, (**1**), 278–284.

Endo, A., Tsujita, Y., Kuroda, M., and Tanzawa, K. (1977) *Eur. J. Biochem.*, **77**, 31.

Endo, A. (1985) *J. Med. Chem.*, **28**, (**1**), 401–405.

Engstrom, R. G., Weinstein, D. B., Kathawala, F. G., Scallen, T., Eskesen, J. B., Rucker, M. L., Miserendino, R. (1986) *Ninth International Symposium on Drugs Affecting Lipid Metabolism*, Florence, Italy, 26 (Abstract).

Fisher, J., Belasco, J. G., Charnas, R. L., Khosla, S. and Knowles, J. R. (1980) *Phil. Trans. R. Soc. London*, Ser. B, **289**, 309.

Frère, J.-M., Ghuysen, J.-M., Degelaen, J., Loffet, A. and Perkins, H. R. (1975) *Nature* (London), **258**, 168.

Fry, D. W., Gewirtz, D. A., Yalowich, J. C., and Goldman, I. D. (1983) *Adv. Exp. Med. Biol.*, **163**, 215–234.

Gelb, M. H., Svaren, J. P., Abeles, R. H. (1985) *Biochemistry*, **24**, 1813.

Grundy, S. M. (1988) *New Engl. J. Med.*, **319**, (1), 24–32.

Heathcock, C. H., Hadley, T. R., Theisen, P. D. and Hecker, S. J. (1987) *J. Med. Chem.*, **30**, 1858.

Hitchings, G. H. and Roth, B. (1980) in *Enzyme Inhibitors as Drugs* (ed. M. Sandler), Macmillan, London, pp. 263–267.

Hitchings, G. H. and Smith, S. L. (1980) *Adv. Enzyme Regul.* **18**, 349.

Hoeg, J. M. and Brewer, H. B. (1987) *JAMA*, **258**, 3532–3536.

Hoffman, W. F., Alberts, A. W., Cragoe, Jr, E. J., Deana, A. A., Evans, B. E., Gilfillan, J. L., Gould, N. P., Huff, J. W., Novello, F. C., Prugh, J. D., Rittle, K. E., Smith, R. L., Stokker, G. E. and Willard, A. K. (1986a) *J. Med. Chem.*, **29** (2), 159.

Hoffman, W. F., Alberts, A. W., Anderson, P. S., Chen, J. S., Smith, R. L., and Willard, A. K. (1986b) *J. Med. Chem.*, **29** (5), 849.

Illingworth, D. R. (1987) *Drugs*, **33**, 259–279.

Karanewsky, D. S., Badia, M. C., Cushman, D. W., DeForrest, J. M., Dejneka, T., Loots, M. J., Perri, M. G., Pettrillo, E. W., Jr. and Powell, J. R. (1988) *J. Med. Chem.*, **31**, 204.

Kelly, J. A., Moews, P.C., Knox, J. R., Frère, J.-M. and Ghuysen, J.-M. (1981) *Science*, **218**, 479.

Kelly, J. A., Knox, J. R., Moews, D. C., Hite, G. J., Bartolone, J. B., Zhao, H., Joris, B., Frère, J.-M. and Ghuysen, J.-M. (1985) *J. Biol. Chem.*, **260**, 6449.

Kelly, J. A., Dideberg, O., Charlier, P., Wery, J. P., Libert, M., Moews, P. C., Knox, J. R., Duez, C., Fraipoint, C.L., Joris, B., Dusart, J., Frère, J.-M. and Ghuysen, J.-M. (1986) *Science*, **231**, 1429.

Kleinsek, D. A., Dugan, R. E., Baker, T. A. and Porter, J. W. (1981) *Methods Enzymol.*, **71**, 462.

Knowles, J. R. (1982) in *Antibiotics, Vol. VI: Modes and Mechanisms of Microbial Growth Inhibition* (ed. F. E. Hahn), Springer Verlag, Berlin, pp. 90.

Knowles, J. R. (1985) *Acc. Chem. Res.*, **18**, 97.

Kokubu, T. and Hiwada, K. (1987) *Drugs of Today*, **23**, 101.

Laragh, J. H. (1978) *Prog. Cariovasc. Dis.*, **21**, 159.

Liang, T.-C. and Abeles, R. H. (1987) *Biochemistry*, **26**, 7603.

Loftus, P., Waldman, M. and Hout, R. F., Jr. (1987) in *Drug Discovery and Development* (eds Williams, M. and Malick, J. B.), Humana Press, Clifton, New Jersey, pp. 73–96.

Luly, J. R., BaMaung, N., Soderquist, J., Fung, A. K. L., Stein, H., Kleinert, H. D., Marcotte, P. A., Egan, D. A., Bopp, B., Merits, I., Bolis, G., Greer, J., Perun, T. J. and Plattner, J. J. (1988) *J. Med. Chem.*, **31**, 2264.

Marciniszyn, J., Hartsuck, J. A. and Tang, J. (1976) *J. Biol. Chem.*, **351**, 7088.

Marquet, A., Frère, J.-M., Ghuysen, J.-M. and Loffet, A. (1979) *Biochem. J.*, **177**, 909.

Marshall, G. R. (1976) *Fed. Proc., Fed. Am. Soc. Exp. Biol.*, **35**, 2494.

Matthews, D. A., Alden, R. A., Bolin, J. T., Filman, D. J., Freer, S. T., Hamlin, R., Hol, W. G.-J., Kisliuk, R. L., Pastore, E. J., Plante, L. T., Xuong, N.-L. and Kraut, J. (1978) *J. Biol. Chem.*, **253**, 6946.

Matthews, D., (1987) in *New Advances in Development Cancer Chemotherapy*, Bristol Meyer's Cancer Symposium, No. 18, Academic Press, pp. 65–81.

Matthews, D. A., Bolin, J. T., Burridge, J. M., Filman, D. J., Volz, K. W., Kraut, J. (1985) *J. Biol. Chem.*, **260** (**1**), 339–399.

Nakamura, C. E. and Abeles, R. H. (1985) *Biochemistry*, **24**, 1364.

Navia, M. A., Springer, J. P., Lin. T.-Y., Williams, H. R., Firestone, R. A., Pisano, J. M., Doherty, J. B., Finke, P. E., and Hoogsteen, K. (1987) *Nature*, **327**, 79.

Ondetti, M. A., Rubin, B. and Cushman, D. W. (1977) *Science*, **196**, 441.

Ozaki, Y., King, R. W., and Carey, P. R. (1981) *Biochemistry*, **20**, 3219.

Patchett, A. A., Harris, E., Tristram, E. W., Wyvratt, M. J., Wu, M. T., Taub, D., Peterson, E. R., Ikeler, T. J., ten Broeke, J., Payne, L. G., Ondeyka, D. L., Thorsett, E. D., Greenless, W. J., Lohr, N. S., Hoffsommer, R. D., Joshua, H., Ruyle, W. V., Rothrock, J. W., Aster, S. D., Maycock, A. L., Robinson, F. M., Hirschmann, R., Sweet, C.S., Ulm, E. H., Gross, D. M., Vassil, T. C., and Stone, C. A. (1980) *Nature* (London), **288**, 280.

Peach, M. J., (1977) *Physiol. Rev.*, **57**, 313.

Quiocho, F. A. and Lipscomb, W. N. (1971) *Adv. Protein. Chem.*, **25**, 1.

Ripka, W. C., Sipio, W. J. and Blaney, J. M. (1987) *Lectures in Heterocyclic Chemistry*, **IX**, 95.

Roberts, G. C. K., Feeney, J., Burgen, A. S. V., and Daluge, S. (1981) *FEBS. Lett.*, **131**, 85.

Roth, B. (1986) *Federation Proc.*, **45** (**12**), 2765–2772.

Sato, A., Ogiso, A., Noguchi, H., Mitsul, S., Kaneko, I. and Shimada, Y. (1980) *Chem. Pharm. Bull.*, **28**, 1509.

Sham, H. L., Bolis, G., Stein, H. H., Fesik, S. W., Marcotte, P. A., Plattner, J. J., Rempel, L. A. and Greer, J. (1988) *J. Med. Chem.*, **31**, 284.

Sielecki, G. E., Ayakawa, K., Fujinaga, M., Murphy, M. E. P., Fraser, M., Muir, A. K., Camilli, C. T., Lewicki, J. A., Baxter, J. D., James, M. N. G. (1989) *Science*, **243**, 1346.

Stammers, D. K., Champness, J. N., Beddell, Z. C. R., Dann, J. G., Eliopoulous, E., Geddes, A. J., Ogg, D., and North, A. C. T., (1987) *FEBS. Lett.*, **218** (**1**), 178.

Stein, R. L., Strimpler, A. M., Edwards, P. D., Lewis, J. J., Mauger, R. C., Schwartz, J. A., Stein, M. M., Trainor, D. A., Wildonger, R. A. and Zottola, M. A. (1987) *Biochemistry*, **26**, 2682.

Stokker, G. E., Hoffman, W. F., Alberts, A. W., Cragoe, Jr, E. J., Deana, A. A., Gilfillan, J. L., Heff, J. W., Novello, F. C., Prugh, J. D., Smith, R. L., and Willard, A. K. (1985) *J. Med. Chem.*, **28**, 347.

Stokker, G. E., Alberts, A. W., Anderson, P. S., Cragoe, Jr, E. J., Deana, A. A., Gilfillan, J. L., Hirschfield, J., Holtz, W. J., Hoffman, W. F., Huff, J. W., Lee, T.-J., Novello, F. C., Prugh, J. D., Rooney, C. S., Smith, R. L., and Willard, A. K., (1986) *J. Med. Chem.*, **29**, 170.

Strominger, J. L. (1970) *Harvey Lect.*, **64**, 179.

Suguna, K., Padlan, E. A., Smith, W. C., Carlson, W. D., Davies, D. R. (1987) *Proc. Natl. Acad. Sci. USA*, **84**, 7009.

Szelke, M., Tree, M., Leckie, B. J., *et al.* (1985) *J. Hypertension*, **3**, 13.

Takahashi, L. H., Radhakrishnan, R., Rosenfeld, R. E. Jr., Meyer, E. F. Jr., Trainor, D. A. and Stein, M. M. (1988) *J. Mol. Biol.*, **201**, 423.

Tang, J. (ed) (1977) *Adv. Exp. Med. Biol.*, **95**, 1.

Thaisrivongs, S., Pals, D. T., Kati, W. M., Turner, S. R. and Thomasco, L. M. (1985) *J. Med. Chem.*, **28**, 1554.

Tipper, D. J. (1979) *Rev. Infect. Dis.*, **1** (**1**), 39.

Tobert, M. B. (1987) *Circulation*, **76** (**3**), 534.

Tomasz, A. (1979) *Annu. Rev. Microbiol.*, **33**, 113.

Trainor, D. A. (1987) *Trends in Pharmacological Sciences*, **8 (8)**, 303.

Yocum, R. R., Amanuma, H., O'Brien, T. A., Waxman, D. J. and Strominger, J. L. (1982) *J. Bacteriol.*, **149**, 1150.

Wyvratt, J. J. and Patchett, A. A. (1985) *Med. Res. Rev.*, **5**, 483.

Veber, D. F., Bock, M. G., Brady, S. F., Ulm, E. H., Cochran, D. W., Smith, G. M., LaMont, B. I., DiPardo, R. M., Poe, M., Freidinger, R. M., Evans, B. E. and Boger, J. (1984) *Biochem. Soc. Trans.*, **12**, 956.

Volz, K. W., Matthews, D. A., Alden, R. A., Freer, S. T., Hansch, C., Kaufman, B. T., and Kraut, J. (1982) *J. Biol. Chem.*, **257**, 2528.

Yasutake, A. and Powers, J. C. (1981) *Biochemistry*, **20**, 3675.

6 | The impact of metal ion chemistry on our understanding of enzymes

Donald H. Brown and W. Ewan Smith

6.1 INTRODUCTION AND GENERAL CHEMICAL PRINCIPLES

The study of the biological role of metal ions has a long history in medicine, in pharmacology and in toxicology, but it is only recently that the extent and variety of metal ion involvement has been appreciated. For example, among the transition metals, the elements V, Cr, Mn, Fe, Co, Ni, Cu, Zn and Mo have been shown to be essential to life and the elements Au, Ag, Pt, Pd, Ir, Os, Ti, and others have either been used in therapy or claimed to be of therapeutic value. In recent years the toxic nature of the complexes of many transition metals and of main group metals, such as cadmium, lead, mercury and aluminium, has evoked much worry and stimulated research into the effect of these metals on terrestial and marine biochemical systems. The ecological and health effects of these elements are often exacerbated by environmental transformation of less hazardous forms to extremely toxic compounds. An example of the latter is the biosynthesis of methylmercury cations from inorganic mercury salts by, amongst others, sulphate-reducing bacteria in marine sediments (Campion and Barthia, 1985).

An essential feature of evolution is that the chemistry of living systems including inorganic and organic constituents tends to be optimally adapted to function. However, as the uptake of an element depends on its availability and on the availability of a potential substrate for it, sometimes replacement can occur involving the alteration of biological activity. Thus the effects of manmade redistribution of metals (i.e. pollution) are generating socially sensitive problems which are of particular concern to the inorganic biochemist. A corollary to this is the increasing awareness of the frequency and economic importance of trace element deficiency induced disease, which can arise, for example, from high intensity agriculture, from the geochemistry of a particular region, from inherent biochemical defects or from medical practices such as intravenous feeding.

Thus, since inorganic biochemistry is a socially meaningful area with many interesting problems, an ever increasing number of chemists with a wide range of specialities have been attracted into the field. This has triggered off research into the preparation and reactivity of many model compounds such as square-planar cobalt complexes (cf vitamin B_{12}) and iron–sulphur clusters (cf ferredoxin). The interest of electronic spectro-scopists has been stimulated by such problems as the very high extinction coefficients found in the 'blue' copper proteins, while kineticists have been involved in studying the catalytic activities of many metalloenzymes and theoretical chemists have been examining the reasons for the specific reactivities of their metal centres. Thus, the very variety of problems presented by *in vivo* chemistry has provided a stimulating climate for increased research activity. However, the complexity of the chemistry of metal containing proteins means that the present understanding of the action of metal ions in their protein environment is patchy with some processes being more understood than others. In part this is due to the low concentration of some of the reactive metal containing species, for which the detection and quantification of the metal is a current research problem in its own right, and in part it is due to the uncertainty inherent in much of the biochemistry concerning the true *in vivo* role of species with defined *in vitro* activities.

There are two different approaches to the application of inorganic chemistry to biology. First, it is possible and in some cases essential to consider the way in which the chemistry of the metal ion affects complete mammalian systems, whole organs or intact cells. For example, some platinum(II) complexes are effective anticancer agents. The effective ones are *cis* complexes with specific ligand labilities and it is now generally believed that these complexes work by bridging between two nitrogen bases on DNA by substitution of the more labile pair of ligands. *In vitro* studies confirm that reactions of this type are likely to take place within the cell. Secondly, the chemistry of the metal ion held in a well defined environment in specific proteins has been investigated successfully *in vitro* in many instances. There are numerous examples of such proteins, but what is more difficult to discover are good examples of studies in which well characterized enzymes have been restudied in the environment in which they must function *in vivo*. Both approaches are discussed here with reference to individual elements, with two of the elements which have been more exhaustively researched, namely iron and copper, providing most of the examples of studies of individual enzymes.

One aspect of the behaviour *in vivo* of metals which cannot be over emphasized is that their chemistry is essentially that of the complexed ion, irrespective of whether more polar ions such as Na^+ or K^+ or more covalent species such as Au(III) or Pt(II) are being considered. Properties such as the effective size and solubility of a metal ion *in vivo* are a function of ligand and solvent present as well as of the metal ions themselves. Further, the correct metal ion balance in various *in vivo* compartments is important for the functioning of specific metal containing sites in many enzymes and proteins. For example, if the concentrations of some metal ions are raised considerably above the norm, blocking of transport sites can occur and symptoms more normally attributed to depletion of certain metal ions can appear. A corollary to this is that metal ion distribution can be affected by alterations in the *in vivo* concentration of naturally occurring low molecular weight ligands or of complexing sites in proteins. An example where this may be the case is in rheumatoid arthritis, where serum histidine levels and albumin thiol levels are significantly lower than in normal subjects. However, as yet little is known about the effect this has on the disease process.

In biological fluids, there are large numbers of complexing species for metal ions. For example, in serum there are many more complexing sites for iron or copper in amino acids, low molecular weight peptides and proteins than there are metal ions, so that the chemistry of these ions *in vivo* is that of ions present in an excess of competing complexing groups. Thus, considerable impetus has been given recently to the study of complexing in determining likely distributions and reactivities of transition metal ions in such fluids.

A major problem which emerged in considering *in vivo* activities is that even if the most likely complex species in terms of concentration can be predicted and the relative abundance of these species can be determined, this information may or may not directly relate to physiological activity. There are a wide range of microenvironments present *in vivo* and each would be expected to affect the chemistry of the metal in much the same way as would a change in the solvent or the absorption of the metal ions on a surface. For example, when transition metals such as copper react with amino acids, quite different complexes are formed in ethanol to those which are formed in water, both because of changes in the solubility of the complexes and because of the different strengths of water and ethanol as ligands. The *in vivo* microenvironments are often separated by membrane barriers which can be penetrated by specific complexes rather than by the ion or other complexes of different size, charge etc. An example of a situation in which a minor component of the metal ion concentrations present *in vivo* is believed to be the most active species is in the use of platinum drugs in cancer therapy. Milligram quantities of drugs are administered, but, although the action of platinum is believed to be on DNA, fewer platinum ions reach the nucleus than there are DNA molecules present. The remaining platinum is complexed to other cellular and extracellular fractions or is rapidly excreted.

These problems have often been viewed in the past as reasons for ignoring inorganic biochemistry, but a better understanding of inorganic chemistry and the advent of more powerful physical techniques means that many more problems can be investigated with a reasonable chance of success. Thus, before looking at the individual metal systems, it seems worthwhile to outline some of the basic principles developed particularly with reference to transition metal ions.

The classification of metal ions into the class I (hard) and class II (soft) acids, based on their *in vitro* reactions is useful in discussing their *in vivo* chemistry. 'Hard' metal ions are small, and are either not easily oxidized or reduced, or have a relatively high positive charge, e.g., Na^+, K^+, Mg^{2+}, Cr^{3+}. 'Soft' metal ions, on the other hand, are large with a low positive charge, e.g., Cu^+, Au^+, Hg^+, Cd^{2+}, Pt^{2+}. Intermediate between those extremes lie the divalent first-row transition-metal ions. In general, 'hard' metal ions favour complexing with oxygen and nitrogen donors and 'soft' metal ions with sulphur and phosphorus donors. This concept is difficult to apply absolutely, but it is useful in a relative sense. For example, copper(II) is 'harder' than copper(I). Therefore, the latter will complex more readily with thiol ligands and a redox enzyme such as caeruloplasmin in which both oxidation states of copper are required uses a mixture of sulphur and nitrogen or oxygen donor atoms to complex the copper. The distinction between 'hard' and 'soft'. metal ions is reflected in the biochemical properties of the essential metal ions listed in Table 6.1.

By definition, transition metals have partly filled d orbitals. Thus, strictly speaking zinc(II) (d^{10}) is not a transition metal ion, but, it is convenient, if not strictly accurate, to include it in this section. The properties of transition metal ions which make them particularly suitable for the types of reaction mentioned above are:

1. they are good Lewis acids forming wide ranges of complexes with nitrogen, oxygen and sulphur donor ligands;
2. because of their unfilled d orbitals, they have a readily accessible range of oxidation states available for oxidation reduction reactions.

Property 2 is perhaps the major difference between the main group and transition metal ions *in vivo*, but the more covalent nature of the bonding of these ions also produces more directional bonding and quite specific variations in chemical properties which are a function of the nature of the complex rather than of the ligand alone.

Table 6.1 Classification of metal ions

$Na^+ K^+$	$Mg^{2+} Ca^{2+}$	Transition metal ions
Charge carriers	structure formers triggers	redox catalysts and Lewis acids
Mobile	semi-mobile	static
Oxygen anion binding	oxygen anion binding	O/N/S ligands
Weak complexes	slightly stronger complexes	strong complexes
Very fast exchange	moderately fast exchange	slower exchange

Most first-row transition-metal ions prefer octahedral or, to a lesser extent, tetrahedral geometries. However, where there is a single d electron in the electronic structure, distorted structures may be preferred. For example in superoxide dismutase, copper(II) which is a d^9 ion is found in a distorted tetragonal site whereas zinc(II), a d^{10} ion, favours a tetrahedral site.

Square planar configurations and other specific geometries determined largely by the nature of the protein environment are quite common with the consequence that variations in size of the metal can critically affect metal ion reactivity. The divalent ions of the first row have a maximum ionic radius at Mn^{2+} (1.00 Å) and a minimum at Ni^{2+} (0.80 Å). An increase in oxidation number usually produces a corresponding decrease in size (Mn^{3+} (0.66 Å)). Structural variations can affect other properties, such as redox potentials, and catalytically active sites in enzymes are often formed by producing geometries which would be regarded as less common in *in vitro* chemistry such as the ferridoxin cluster or the planar haem system.

The ligands present in these complexes can affect the reactions of the complex by virtue of their electronic structure as well as their steric preferences. For example the substitution of $[Co^{(III)}(NH_3)_6]^{3-}$ by halide ligands gives the stability order $Cl^- > Br^- > I^-$, whereas in $[Co^{(III)}(CN)_6]^{3-}$ the order is $I^- > Br^- > Cl^-$. The reason for this is that the NH_3 ligand is relatively hard compared to CN^-, and the hardness of the complex is affected by the ligands, altering the affinity of the cobalt for hard or soft ligands. A similar argument may be advanced to explain the relatively soft behaviour of zinc in carboxypeptidase.

There are two ways to consider the stability of complexes, namely kinetic and thermodynamic. The kinetic aspect can often be explained in terms of the ligand field stabilization energies of the initial complexes and the transition states produced during reaction. This is why Co^{3+}, Cr^{3+} and Ni^{2+} react much more slowly than the other commonly occurring metal ions. The thermodynamic aspect is related to the alteration in size of the metal ion and the bonding characteristics of their complexes. This results in a sequence of stability constants for the respective complexes formed between one ligand and various metal ions. This series is known as the Irving Williams order and is $Cr^{2+} > Mn^{2+} < Fe^{2+} < Co^{2+} < Ni^{2+} < Cu^{2+} > Zn^{2+}$. This effect is of obvious importance, since the difference in the stability constants of, say, copper(II) and manganese(II) complexes with a ligand may be of the order of 10^3 to 10^5, and copper will substitute for manganese if the availability of the ion is not controlled.

One advantage that the study of transition metal ions has over that of the alkali metals is that a wider range of physical techniques is available, due to the presence of unpaired electrons. These techniques include UV–visible spectroscopy, ESR spectroscopy, and magnetic susceptibility. By means of these and other methods, such as NMR and EXAFS, considerable progress has been made in elucidating the role of the transition-metal ions in many enzymatic reactions.

6.2   THE TRANSITION ELEMENTS IRON AND COPPER

6.2.1   Iron

Iron is the most abundant transition metal in the body with approximately 3–4 g present in adult humans. About 60% of it circulates as haemoglobin, 10% as myoglobin, and 1% is distributed between transferrin, other enzymes and the iron storage proteins ferritin and haemosiderin.

The aqueous chemistry of iron mainly involves two oxidation states – ferrous (iron(II)) and ferric (iron (III)). The former is a weak reducing agent and is readily oxidized by dioxygen in aqueous solution. Iron(III) in aqueous solution will precipitate out as a polymeric hydroxide at biological pH unless complexed with oxygen, nitrogen or sulphur donor ligands. Iron(II) is a d^6 ion and iron(III) is a d^5 ion. Both can exist as spin-paired or spin-free species. Six-coordinate complexes are known. In addition to the insoluble nature of iron(III) hydroxide, some iron(III) complexes with oxygen ligands such as catechols are inert to substitution and, in general, high-spin iron(II) complexes tend to be more labile than their iron(III) counterparts. This chemistry is used in the transfer and storage of iron *in vivo*.

Iron is absorbed into the body mainly from the gut. With excess iron, the path becomes partly blocked producing an accumulation of ferritin. It is thought that since most iron is likely to be absorbed as iron(II), dietary iron has to be reduced and dispersed before absorption takes place (Forth and Rummel, 1973). For example, ascorbic acid enhances the uptake of iron apparently by reducing and complexing it. Powerful complexing ligands such as EDTA have the opposite effect, probably because they compete strongly with the uptake sites for the available iron. The poor uptake of iron from vegetable sources is probably due to strong complexing by phytic or oxalic acids. Iron from animal sources is better absorbed because a large part of it is haem iron which is readily absorbed unchanged. The presence of peptides and amino acids from protein digestion helps in the absorption, presumably, by the formation of complexes of the correct stability.

The iron is transported from the gut by transferrin (a β-globulin). It has been suggested that the uptake of iron by transferrin involves the oxidation of iron(II) to iron(III) by the copper-containing protein caeruloplasmin and this is thought to be the reason why copper deficiency causes anaemia (Mareschal *et al.*, 1980). The transferrin then delivers the iron to bone marrow, iron stores and other tissues. The further transfer in these organs is again thought to involve the reduction of iron(III) to the more labile iron(II).

Transferrin also cycles iron from the catabolism of haem in the reticulo-endothelial system to erythoblasts in the bone marrow. Iron exerts a negative feedback effect on the biosynthesis of transferrin so that in iron deficiency more protein circulates (measured as higher total iron-binding capacity) with a reduced relative and absolute iron content. Thus, in determining the type of anaemia, both total iron and iron-binding capacity are measured for the patient. Transferrin is not thought to have any other role but metal-ion transport. It is related to ovalbumin and lactoferrin in that all three proteins have similar amino acid sequences and bind two ferric ions strongly.

The main iron-storage protein in the body is ferritin. Such a reservoir is needed, both to provide a source for iron for the many molecules that need it and also to limit the concentration of free iron. These requirements explain the presence of ferritin in most organisms. Iron entering a cell often appears to stimulate the synthesis of apoferritin. The protein is in the form of a spherical ball of external diameter *ca.* 125 Å with an internal cavity of about 80 Å diameter. The structure is made from a symmetrical assembly of 24 polypeptide chains arranged in pairs, each pair lying on the face of a rhombic dodecahedron (Banyard *et al.*, 1978). At six of the apices there is a small space about 10 Å across which provides a passage for iron transport. The central cavity has a capacity for over 4000 ferric ions which are present as iron(III) oxide/hydroxide/phosphate aggregates of general formula $(FeO . OH)_8(FeO . H_2PO_4)$. The units vary in size and more than one may be present per molecule. Their orientations do not appear to be related to the protein shell. EXAFS measurements suggest that each iron is coordinated by six oxygen atoms. Polymeric iron(III) hydroxides form readily at biological pH and if unrestricted would become too large for a protein shell. To prevent this, the apoferritin incorporates ferrous iron which is then oxidized within the protein shell and polymerizes. The release of iron probably involves reduction of iron(II). The reducing ligands must approach the iron(III) core through the same passages as did iron(II) on incorporation. A number of reducing agents such as thioglycolic acid, cysteine, ascorbic acid and reduced riboflavin have been found experimentally to remove iron (Crichton, 1973). Ferritin is also degraded in lysosomes. Here the protein shell is partly digested, leaving a microcrystalline iron(III) residue of similar composition to that found originally in the ferritin centre. This substance is known as haemosiderin and is the second principal storage form of iron found in animals. In iron deficiency, ferritin and haemosiderin levels drop. As the stores of iron disappear, transferrin concentrations and saturation change. If excess iron is absorbed or if erythrocyte lifetime is reduced, ferritin and haemosiderin appear in many tissues where damage occurs eventually. There does not appear to be any natural method for excess iron removal.

The single most important iron coordination geomery *in vivo* is that associated with porphyrin ligands (Fig. 6.1). The characteristics of the porphyrin ring confer specific properties on the reactive iron at its centre and the unique chemistry which arises has led to many studies attempting to explain both the functions in biology and the physical properties of iron porphyrin groups (haems) *in situ* in proteins.

For example, the role of haemoglobin circulating in human blood is to

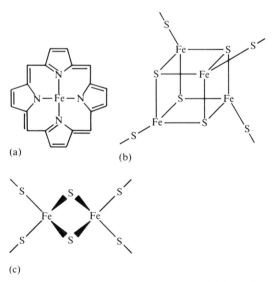

(a)

(b)

(c)

Fig. 6.1 Geometries of some entities containing iron which are found in biological systems.
(a) Haem; (b) Ferredoxin; (c) Rubredoxin.

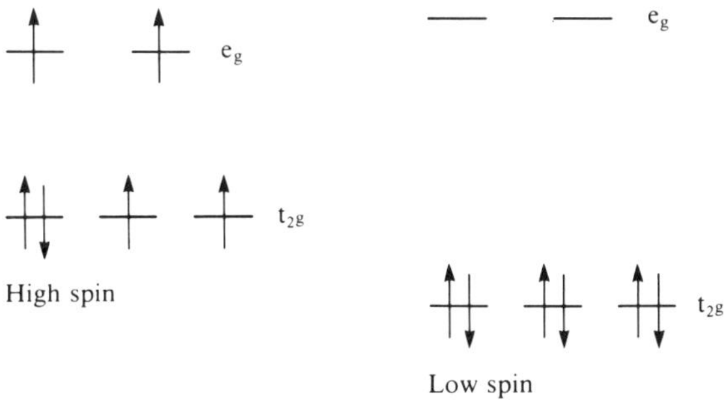

High spin

Low spin

provide an efficient uptake of oxygen in the lungs together with an efficient unloading in the body where it is needed. The cooperative properties of the molecule are such that, while saturation of the molecule with oxygen in the lungs is not much affected by variation in loading pressure, unloading of oxygen in the tissue is sensitive to local anoxia. Haemoglobin contains ferrous iron which reacts with oxygen to form the dioxygen complex. It can be oxidized to the less active ferric form methaemoglobin. From magnetic measurements it is thought that the deoxy-ferrous complex is a high-spin complex whereas the oxygenated form is low-spin. e_g orbitals are slightly larger than t_{2g} orbitals and so the high-spin form of iron(II) is too large to fit into the haem ring, and lies slightly out of plane. On complexing with oxygen to form a metal oxygen bond believed to be angled thus, the iron moves into the ring, causing a slight puckering of the ring.

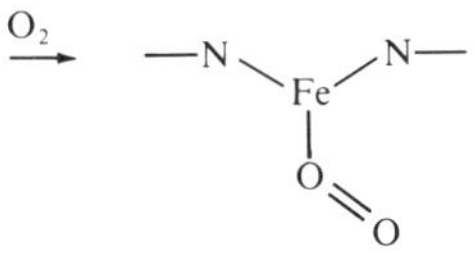

The ability of haemoglobin to load and unload oxygen with only a minimum change in partial pressure arises from both the structure of the protein ligand and from controlling substances such as DPG (diphosphoryl-glycolic acid), carbon dioxide and hydrogen ions. There are four subunits in haemoglobin, 2 α units and 2 β units. Each half of the molecule contains one α and one β subunit. Most oxygen transport involves the take up and release of the second and third oxygen molecules observed. Between second and third oxygen site absorption, the two halves of the protein swing by about 7° giving a change from a so-called tense (T) to relaxed (R) state. The ease with which this occurs is controlled by the DPG, CO_2 and H^+ receptors, which by this means control oxygen absorption and release. Thus the haemoglobin molecule is a good example of an extended ligand system in which the local environment round the ion, the macrostructure of the ligand and specific non-haem-binding receptor sites all play a part in the careful control of an important biological process.

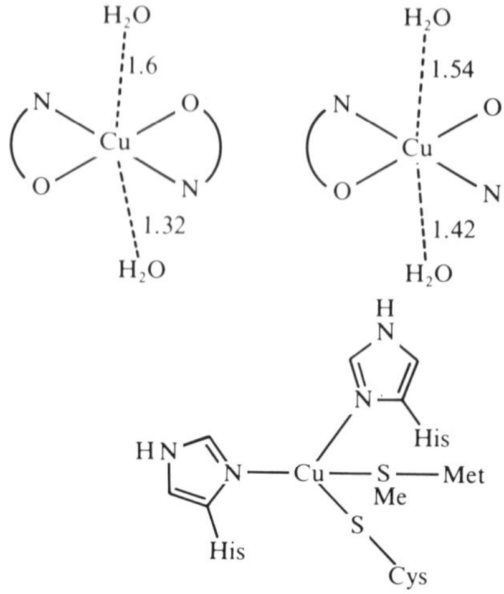

Fig. 6.2   The geometries of copper (II) in complexes can be quite varied. Illustrated are the different bond lengths and angles exhibited by water molecules in *cis* and *trans* bis alanine copper (II) complexes and in a model of the site in copper blue proteins such as plastocyanin.

There are a large number of proteins containing haem groups which are of importance in biology. These include such systems as cytochrome C, cytochrome oxidase and cytochrome-P450 which are involved in the respiratory chain. The last type of protein has been the subject of some considerable interest recently because its action is selectively to oxidize or hydroxylate key biological compounds which could also be of value in commerical synthesis. Model reactions for this protein have been developed to achieve high stereoselectivity. The reactions are quite complex, involving oxygen uptake, reduction and substrate binding (see section 4.6.5).

P450 proteins have developed into many subfamilies, each with a different selective reaction. In some cases the protein sequences are quite close (*ca.* 97% in one pair) and in some they are quite different with only certain regions conserved. A significant biological function of P450 proteins is that they convert many environmental toxinogens to their toxic form *in vivo*, making their role in environmentally produced cancer an important one. In some cases, it is believed that they also have a protective role in causing further reaction. The key active site in this protein is a haem ring with an iron–sulphur bond in the fifth position. Reactivity of the enzyme depends on the subunits on the ring and on the protein sequence, but the effect is very subtle. The two proteins with 97% sequence homology illustrate this: one is active and the other only weakly active with substrates. The most effective criterion for activity changes other than the measurement of substrate reactivity are changes in the iron oxidation spin states, and properties that can be measured by surface enhanced resonance Raman spectroscopy. This protein is less well understood than haemoglobin but gradually it, and other haem containing proteins, are being investigated to a stage where a full picture of the sophisticated ligand control of iron chemistry can be appreciated (Wolf *et al.*, 1988).

There are many other important iron containing enzymes, some of which do not have haem groups. For example, the protein haemoerythrin, despite its name, has two iron atoms bound in a single polypeptide chain about 3.5 Å apart. The dioxygen binding affinity of each pair of iron atoms is similar to that in haemoglobin, but, in contrast, there is little co-operative action between the different complexing sites. It is thought that dioxygen bridges the two iron atoms (initially in the high-spin iron(II) state) to give what can formally be described as two iron(III) atoms with briding peroxide group. Both forms can be easily oxidized to a deoxygenated methaemrythrin protein which contains iron(III) pairs (Klotz *et al.*, 1976).

Another important group of iron containing proteins are the so-called 'iron–sulphur' proteins. These generally contain non-haem iron sulphide ion in equal proportions. They include the ferredoxins and other high-potential iron proteins and they function by electron transfer rather than substrate conversion (Fig. 6.1). They take part in many reactions in bacteria including nitrogen fixation, ATP formation and photosynthesis electron transfer. In addition they have a role in higher organisms involving the enzymes xanthine oxidase and nitrate and sulphite reductase. The sulphide ions are labile and can be displaced as H_2S by acid treatment. This test distinguishes them from the thiol groups bound to the iron atoms. Three different types of iron–sulphur clusters are commonly found

1. two iron atoms with two bridging sulphurs
2. a six-membered ring with alternate iron and sulphur atoms
3. a cube with alternate iron and sulphur atoms at the corners.

In each case the iron–sulphur cluster is linked to the polypeptide chain by cysteine residues – the thiol groups being complexed to the iron atoms. The encapsulation of an iron–sulphur group into a polypeptide brings other atoms into close proximity, resulting often in hydrogen bonding from protein amide groups to sulphide groups. Hydrogen bonding appears to be significant in controlling the redox potential at which the cluster operates. In each of the iron–sulphur centres there are two accessible redox states, the difference between them corresponding to a one-electron change, thus permitting electron transfer.

Many studies of iron complexes have been stimulated by the very specific *in vivo* chemistries which arise from specific environments. Perhaps the greatest attention has been paid to haem systems with their unusual redox and coordination properties, but lately the ability of the ferredoxins to reduce nitrogen has led to more extensive investigation of cluster systems of this type. These studies have produced novel results and probably form some of the best examples of the way in which biochemistry may be expected to have an impact on inorganic chemistry as our knowledge of the coordination and mode of action of other metal ion species increases.

6.2.2 Copper

Copper is a member of the group 1B elements along with gold and silver. Thus although in some ways a typical transition metal, it is distinct in its properties. Copper can exist under normal conditions in four oxidation states, 0, 1, 2, 3. Copper(0) (i.e., copper metal) is relatively inert and is used, for example, in water piping. However, in the presence of oxygen and a strong complexing ligand, copper(0) dissolves easily – an example of this is the ready solubility of copper metal in aqueous solutions of amino acids. This is reputedly the therapeutic mechanism of copper bangles – the dissolution of copper(0) by the amino acids in sweat and the subsequent absorption of the copper–amino acid complexes.

Copper(I) forms mainly 4- or 6-coordinate species with essentially tetrahedral or octahedral structures. It is unstable in aqueous solution unless complexed to soft ligands such as thiols or thiourea, but it is stable in some non-aqueous solvents (acetonitrile) and insoluble copper(I) complexes are readily prepared from aqueous solution. As a d^{10} ion, there are no unpaired electrons, with the result that the detection of copper(I) is more difficult than copper(II) and its *in vivo* role has received much less attention. Nevertheless, soft thiol complexing sites and non-aqueous environments are suitable for copper(I) and it has been identified in specific proteins such as caeruloplasmin and metallothionein.

The most common oxidation state of copper is copper(II) and being a d^9 ion, its complexes are often distorted resulting in unequal bond lengths and interbond angles and a large variety of possible geometries. Thus, biologically, copper(II) will occupy unique sites. Because of the relative stability of its complexes, labile copper concentrations in natural fluids will need to be kept low, otherwise Cu^{2+} will displace other transition metal ions such as Mn^{2+} or Zn^{2+}. Copper(II) also forms dimeric species, the classic example of this is copper acetate, a dimer with bridging carboxylate groups but dimeric complexes also occur in nature. For example, in haemocyanin, oxygen is carried as a bridged species between two copper ions and not by a haem group as the name would imply.

Copper(III) is relatively rare and is only stabilized by hard ligands such as O^{2+} or F^-, as in $NaCuO_2$ and K_3CuF_6. Whether copper(III) has any role biologically is as yet unknown, but it is possible to form copper(III) *in vitro* in a copper albumin complex in which the hard ligand is believed to be an N^- atom in the protein chain.

Copper(I) usually oxidizes rapidly in aqueous solution to copper(II) and thus copper(I) is most likely to be found complexed to thiols or in lipid-rich tissue or other non-aqueous environments. However, copper(I) can be stabilized in aqueous solution with soft ligands such as thiourea, and the thiols *in vivo* in combination with the non-aqueous nature of proteins may also lead to copper(I), circulating in aqueous compartments. An example of this is the storage protein metallothionein. Copper(II), on the other hand, is stable in aqueous environments complexed to chloride, water, carboxylate groups, amino acids, or amines. However, in enzymes copper is involved in oxidation–reduction systems and the coordination spheres are often composed of a mixture of hard and soft ligands in which both oxidation states are relatively stable. CNDO calculations on thiol–copper complexes suggest that on going from copper(II) to copper(I), the added electron is distributed over molecular orbitals associated with the ligands as well as the copper and therefore to speak of formal copper(I) and copper(II) in such complexes is somewhat misleading.

Copper is thought to be absorbed mainly in the gut. There appear to be two routes. The first involves diffusion of low molecular weight copper complexes and transport by albumin to the liver. The second envisages a specific transfer involving the high sulphur containing protein metallothionein. The latter mechanism is probably the major one. It is known that high zinc(II) dietary content can block copper absorption and zinc(II) is thought to be transported via metallothionein. The ability of zinc to block copper transport has been used clinically to control copper absorption. The antagonism of molybdenum (particularly thiomolybdates) for copper absorption in ruminant animals is less easy to understand. Again, metallothionein could be involved (Bremmer and Mills, 1981). Thiomolybdates(VI) are easily reduced to a dimeric sulphur-bridging molybdenum(V) species which is known to form strong complexes with cysteine – the main amino acid in metallothionein. However, present work on the dietary absorption of copper is not sufficiently definitive to determine whether species of this type are involved.

Copper enzymes are involved in a wide range of metabolic pathways and some examples are listed in Table 6.2. The functions given in the table can perhaps be described as indicative rather than comprehensive, because many of the enzymes appear to be multifunctional. For example, metallothionein is also a storage protein for zinc and perhaps also a protective protein against the poisonous effect of some heavy-metal ions such as cadmium(II). Caeruloplasmin, which contains up to six copper atoms per molecule, has been variously described as a storage protein, a peroxidase in iron mobilization, a transport protein, a serum antioxidant, a low molecular weight thiol oxidant, and an acute phase reactant (Frieden, 1980). Whether one of these functions is of primary importance and the rest are incidental is not known. On the other hand, the role of the intracellular protein

Table 6.2 Copper proteins

Function	Enzyme
Storage, transport	Metallothionein
	Caeruloplasmin
Dioxygen transport	Haemocyanin
Electron transfer	Azurin
	Plastocyanines
	Stellacyanin
Monooxygenation	Dopamine-β-monoxygenase
Dioxygenation	Indole-2,3-dioxygenase
Superoxide dismutation	Superoxide dismutase
Substrate oxidation with $O_2 \rightarrow H_2O_2$	Amine oxidases

superoxide dismutase, (Michelson *et al.*, 1977) is much better understood; it catalyses very efficiently the reaction

$$2O_2^- + 2H^+ \rightarrow O_2 + H_2O_2$$

This protein contains two copper and two zinc ions and each copper and zinc ion are linked by a single complexing group. The arrangement at the copper site consists of two histidine imidazole groups and two cysteines arranged in a distorted 4-coordinate geometry. O_2^- can readily complex with the copper site and so aid in catalysing further reduction.

The enzyme appears to help to protect the cell when superoxide is produced. The most likely example of its use is in macrophages, where during the process of phagocytosis a chemical bleach containing O_2^- is produced. Both copper/zinc and manganese containing dismutases are known. The copper/zinc one is present in high concentrations in the red cell as well as being present in various white cells. The reason for the presence of high levels in the red cell is probably that it prevents damage caused by the formation of methaemoglobin.

$$Fe(II)O_2 \rightarrow Fe(III) + O_2^-$$

The erythrocyte contains a much larger quantity of the general radical scavenger glutathione and this would dominate any radical scavenging activity including those which produce O_2^- by a radical chain. However, glutathione does not react rapidly with O_2^- and, consequently, the reaction above could require the additional superoxide dismutase scavenger. The balance between these radical scavenging processes is altered in some disease states.

Although some of the specific functions of the enzymes are still open to question, the role of copper in most of them seems to be concerned with redox reactions. Recently, the determinations of the structure of a number of these enzymes have been of sufficient precision to reveal the immediate copper environment. This has produced a greater understanding of the redox functions of copper in different protein environments. For example, the structure of plastocyanin shows that the copper is present in a flattened tetrahedron which can be considered as intermediate between common geometries of copper(I) and copper(II). The four ligands are two histidines complexed through imidazole nitrogens and methionine and cysteine complexed through their sulphur atoms. Stellacyanin is thought to have a similar structure but with two sulphur atoms arising from cysteines. The immediate environment of the copper in both of these proteins is broadly similar to that of superoxide dismutase, but they do not contain the zinc. Thus, slight changes in structure and subtle changes in complexing ligands confer different redox properties on enzymes. The copper sites are close to the surface of the enzymes, but they do not appear on the surface, enabling the protein to exert steric control of complexation to the copper. In some cases separate anion and cation complexing sites have been identified close to the copper, whose role seems to be essentially one of electron transport.

Another type of mechanism is shown by haemocyanins. The structures of these compounds are varied depending on source and their functions are complex. However, in each case the copper seems to function as a dioxygen carrier with the ratio of O_2:Cu of 1:2. In the iron containing oxygen transporting enzyme, haemerythrin, the O_2:Fe is also 1:2. The reason for the apparent duplication of roles of iron and copper in oxygen transport is as yet unexplained.

The complexity of copper reactions and environments is perhaps best illustrated by the copper in the multifunctional enzyme caeruloplasmin (Frieden, 1980). This enzyme accounts for over 70–80% of the circulating serum copper. The remaining copper is distributed between the proteins albumin, and transcuprein and low molecular weight copper(II) complexes with histidine and other amino acides. Since caeruloplasmin concentration fluctuates in a number of diseases, it has stimulated considerable research work, both chemical and medical. Its structure has not been completely elucidated although its 1064 amino acids have been sequenced and some specific receptor sites have been identified in responsive tissues (Takahashi *et al.*, 1984). Some concept of the environment of the six copper atoms has been derived mainly from electron spin resonance (ESR) and electronic spectra measurements. The copper atoms have been classified in different groups. Type I is a copper(II) ion which is associated with a detectable ESR signal and an intense blue colour. Its environment is possibly similar to that described above for plastocyanin (Fig. 6.2). There are thought to be two of these present in each caeruloplasmin molecule. Type II, also copper(II), is ESR-detectable, but is not associated with the strong blue colour. It appears to be able to complex anions readily. In Type III, copper is again thought to be copper(II), but since it is not ESR-detectable, it is probably in a coupled pair with bridging ligands. Type IV copper is required for the total copper content, but little is known about it. Since our four copper atoms seem to be essential for the basic properties of caeruloplasmin, this last copper could be associated with copper transport. There is evidence that the copper content of the caeruloplasmin molecule falls in some disease states such as acute rheumatoid arthritis, but the activity of the molecule is unaltered.

The impact of biology on copper chemistry has been largely to stimulate more specific studies of areas where adequate knowledge seemed to exist until the more specific questions relating to particular types of biological activity were posed. For example, the geometry of copper(II) complexes is expected to be distorted from octahedral, but the question of the manner in which the different geometries of amino acids would affect this distortion is less clear. One good example of the empirical classification of this effect is amino acid complexes of type $CuL_2(H_2O)_2$ for which a detailed structural analysis of a wide range of complexes is available (Freeman 1967). The ligand has a considerable effect on the position of the water molecules and each complex studied is different, but broadly they fall into two types as illustrated for alanine (Fig. 6.20).

As in the case of iron, increasing understanding of the biological role of the metal has led to a large increase in the number of studies of model complexes designed to mimic some aspect of the role of copper *in vivo*. For example, copper Schiff-base complexes are believed to be key intermediates in some pyridoxal dependent enzyme processes and possibly in the cross linking of collagen by lysyl oxidase (Harris *et al.*, 1982). The use of sulphur containing amino acids produces a more complex series of compounds than that found with alkyl amino acids and suggests that, as is the case with most of these systems, much remains to be done before *in vitro* knowledge is adequate to begin an interpretation of *in vivo* behaviour (MacDonald *et al.*, 1982). The discovery of the nature of the site in blue copper proteins such as stellocyanin has led to studies of model systems based on this group and the role of copper complexes as oxygen carriers has also been studied. Further, the catalytic role of the copper site can sometimes be effectively studied *in vitro*. For example, copper amino acid complexes can behave as superoxide dismutases. In medicine, ligands specifically designed to complex and remove copper in Wilson's disease, a condition involving the accumulation of excess copper, have been synthesized and the discovery that copper aspirinate is a more effective and less ulcerogenic anti-inflammatory agent has led to the reinvestigation and extension of the chemistry of complexes of this type (Sorensen, 1982).

6.3 OTHER TRANSITION METAL IONS

Some other metals such as zinc and cobalt could equally well be considered in the same detail as iron and copper, whereas in the case of others, such as vanadium, the chemical understanding of the role of the metal is still in its infancy. However, enough has been said to establish the principles governing the effect of metal ion chemistry on enzyme chemistry and so only the main points relating to the action of other metals will be considered. For elements such as vanadium and chromium, the main emphasis will be on giving a plausible explanation of their chemical action in mammalian systems, whereas for cobalt and zinc, much more specific studies of particular enzymes can be considered.

6.3.1 Vanadium

Interest in vanadium chemistry arose from the experimental observation that ATP supplied by different companies resulted in differences in the activity of the Na–K ATPase required for the action of the sodium pump (Karlish *et al.*, 1979). One difference was the presence of vanadium in the less active ATP. Further experiments showed that vanadium selectively inhibits Na–K ATPase. The vanadium seems to be transported into the cell by the anion transport mechanism and acts at the inner surface of the cell membrane by inducing a conformational change in the sodium pump. Vanadate appears to act synergistically with potassium to stimulate binding of the potassium ion. In so doing it alters the response of the sodium pump to external potassium which becomes inhibitory instead of stimulatory in the presence of vanadium. Thus, vanadium may control the response of the sodium pump to potassium. In view of this fairly dramatic effect, vanadium complexes are surprisingly non-toxic. Experiments in mice, rats and humans have shown few side effects. However, in common with most trace elements, if large excesses are involved, as is the case with workers in vanadium mines, widespread inhibition of metabolic processes seems to occur. A possible reason for this (Cantley and Aisen, 1979) seems to be that vanadium has two stable oxidation states in normal aqueous solutions. Vanadium(V) exists as the simple vanadate ion $[VO_4]^{3-}$ or as a polymeric form of it, depending on pH. However, in the presence of weak reducing agents, a complexed vanadyl ion is easily formed. This contains vanadium(IV) bonded to an oxygen giving the ion $(VO)^{2+}$, which then forms five- or six-coordinate complexes with a wide range of other ligands. For example, the structure of the vanadyl(IV) complex with L-cysteine methyl ester is square pyramidal with the complexing nitrogen and sulphur atoms in the *trans* positions (Sakurai *et al.*, 1988). It is thought that there is present on cell membranes a NADH-dependent vanadate reduction enzyme which controls the concentration of vanadium(IV). Thus, alteration of the redox status or the pH of cells could have a greater effect on vanadium metabolism than small variations in dietary intake, but it is also possible that vanadium deficiency or excess can result in the circumvention of normal body control mechanisms.

However, the explanation may be simpler in that the obvious structural parallel between the tetrahedral vanadate and phosphate could account for the inhibition of several natural processes, in particular phosphohydrolases (Chasteen and Weiss, 1983). Another property of vanadate ion that could be of biological significance is its ability to complex with disulphides. The observed increase in the level of membrane peroxides after vanadate exposure has been emphasized by the complexation of vanadate ion with glutathione disulphide thus inhibiting the action of glutathione reductase (Cohen *et al.*, 1987). Recently, it has been suggested that vanadium has a role in the aetiology of manic–depressive illness. In this disease, whole body changes in water and electrolyte metabolism have been reported, probably because of the increase in Na^+/K^+ ATPase activity observed in patients with the disease. This latter activity has been found to correlate with plasma vanadium. The plasma vanadium concentrations were high in the manic depressive stage and fell to normal on recovery.

Although an essential role for vanadium has been identified in certain primitive biological species, such as some species of algae and mushrooms, an understanding of the physiological functions of vanadium in higher species has remained an elusive goal.

6.3.2   Chromium

Chromium is an essential element apparently required for normal carbohydrate metabolism. It is closely associated with insulin metabolism; normal chromium nutrition levels decrease the requirement for insulin, whereas experimental animals on a chromium deficient diet have impaired growth, decreased longevity, elevated serum cholesterol and related cardiovascular maladies (Anderson and Koglovsky, 1985). Many of these symptoms have also been observed in humans thought to be suffering from chromium deficiency and have been removed by chromium supplementation. Most fresh foods are good sources of dietary chromium with one of the highest concentrations being found in brewer's yeast. From this source a chromium complex has been isolated but not yet identified (Schwarz and Mertz, 1959). This complex has been found to be capable of potentiating insulin action.

The structure of the active chromium complex is thought to involve nicotinic acid, and glutathione or its constituent amino acids, glycine, cysteine and glutamic acid. Synthetic insulin-potentiating chromium complexes have been prepared with glutathione, but, again, the exact nature of this active species is unknown as the complexes dissociate in aqueous solution. Part of the problem in this work has been to establish the correct amount of chromium present in biological fluids. Contamination from stainless-steel equipment and technical problems in chromium analysis are only now being overcome and, for example, the accepted value for chromium in serum has fallen in the past few years from about 1000 parts per billion (p.p.b.) to below 1 p.p.b. With such low concentrations, the isolation and characterization of complexes is very difficult and most of the successful work has depended on the monitoring of biological functions using bioassays. For example, chromium potentiating factors are being sought using an *in vitro* insulin assay in the presence of different concentrations of chromium.

The application of inorganic chemistry to this problem can at least help to define the likely types of complex which may be found and the reactions which may occur. Under normal conditions with most ligands such as amino acids, the stable oxidation state of chromium is III. It is kinetically and thermodynamically stable and in almost all complexes formed the metal is six-coordinate. Chromium(III) can be reduced to chromium(II), but it is only when complexed to ligands such as cyanide, dipyridyl and thiocyanate, that water-soluble and stable complexes can be formed and it is unlikely that if nicotinic acids and amino acids are complexed, a chromium(III)–(II) redox system is involved. Thus, with ligands such as nicotinic acid and glutathione, six-coordinate chromium(III) complexes are the most likely species *in vivo*. A stable chromium(V) glutathione complex has been described (O'Brian, 1985), and it is possible that chromium(III)–chromium(V) redox equilibria could be involved

$$2RS^- = RSSR + 2e \text{ and } Cr(V) + 2e = Cr(III)$$

This would fit with the suggestion that the role of the chromium complex is as a catalyst in thiol disulphide interchange between insulin and sulphydryl membrane acceptor sites. *In vitro*, the slow kinetics of chromium(III) substitution reactions would make this unlikely, but it is always possible *in vivo* that a specific energized reaction does occur. On the other hand, chromium(VI) is known to cause both carcinogenic and mutagenic reactions (Wetterhahn and Connett, 1983). This is thought to be caused by chromate ion passing into the cell *via* anion channels and being reduced by cellular components ultimately to chromium(III) species *via* chromium(V) species in some cases. There are several potential intracellular reducing agents – one of the most important being the tripeptide glutathione. It is formed in millimolar concentrations in many mammalian cells and is thought to be a major defence component in cells challenged by oxidizing agents. However in the case of chromate toxicity, cells with elevated glutathione have been shown to have increased susceptibility to damage perhaps indicating that a chromium–glutathione complex is an active intracellular toxin. Whether this is due to a stable chromium(III) or chromium(V)–glutathione complex, is unknown.

It is interesting that a recent dietary survey (Anderson and Koglovsky, 1985) suggests that many western diets based on processed foodstuffs are low in chromium and could therefore give rise to marginal deficiency in large segments of the population. However, as yet there is no medical evidence to support this contention.

6.3.3   Manganese

Although the essential nature of manganese in some biological systems has been known for many years, information about the effects of managenese deficiency in humans and of the biochemical functions of the elements is still sparse (Hurley 1984). This is possibly due to the low levels of manganese present in the serum (approximately $0.5 \rightarrow 0.6 \ \mu g \, l^{-1}$) and to the difficulty of analysis and speciation at some low concentration levels. The *in vivo* chemistry of manganese seems to be mainly associated with oxygen or one of its reduced forms. Manganese in aqueous solution can exist in a range of oxidation states from (II) to (VII). Of these the most common is the divalent and in that state manganese has been used as a probe element in the replacement of zinc(II) from zinc metalloenzymes. In plants, manganese is involved in the respiration and photosynthesis chain. A manganese protein is involved in the abstraction of electrons from water to give dioxygen and where this enzyme is missing, as in certain bacteria, an exogenous reducing agent such as a sulphide is required to feed electrons into the chain (Harrison and Hoare, 1980).

Another important manganese enzyme is a superoxide dismutase found in bacteria and often in the mitochondria of more elaborate cells (Michelson *et al.*, 1977). The structures of these enzymes are unknown, but they contain several manganese atoms with the actual number depending on where they are found. Their role is to protect tissue from the superoxide ion O_2^-. However, in doing this they produce peroxide ion (O_2^{2-}), which itself must be removed presumably by an adjacent enzyme such as glutathione peroxidase or catalase. The manganese superoxide dismutase is not the only body defence to superoxide ion and indeed the copper–zinc superoxide dismutase has been more widely studied. It is of interest that in Down's syndrome, where the copper–zinc superoxide dismutase concentration is increased, there is a decrease in the manganese dismutase concentration.

6.3.4   Cobalt

Although cobalt occurs in only a limited number of metal containing proteins, its role in vitamin B_{12} is vital for a wide range of life forms from micro-organisms to man (Wood and Brown, 1972). Cobalt forms stable complexes with nitrogen donor ligands, both in the di- and trivalent states, and both are thought to be implicated in the catalytic cation of the vitamin. The cobalt is complexed in a corrin ring system by four equatorial nitrogen atoms. One axial position is occupied by a donor atom of a side chain of the ring system. The other axial position can be occupied by a variety of ligands such as water, cyanide or a carbon chain, depending on the type of vitamin B_{12}. The discovery of a cobalt–carbon bond in this molecule was unexpected at the time and had an immediate impact in stimulating research into the production of organometallic cobalt complexes.

Two of the main catalytic reactions associated with vitamin B_{12} enzymes are hydrogen and methyl transfers. In the former, the mechanism generally involves abstraction of the hydrogen from the substrate by the coenzyme. Substrate rearrangement then takes place and the hydrogen returns to its

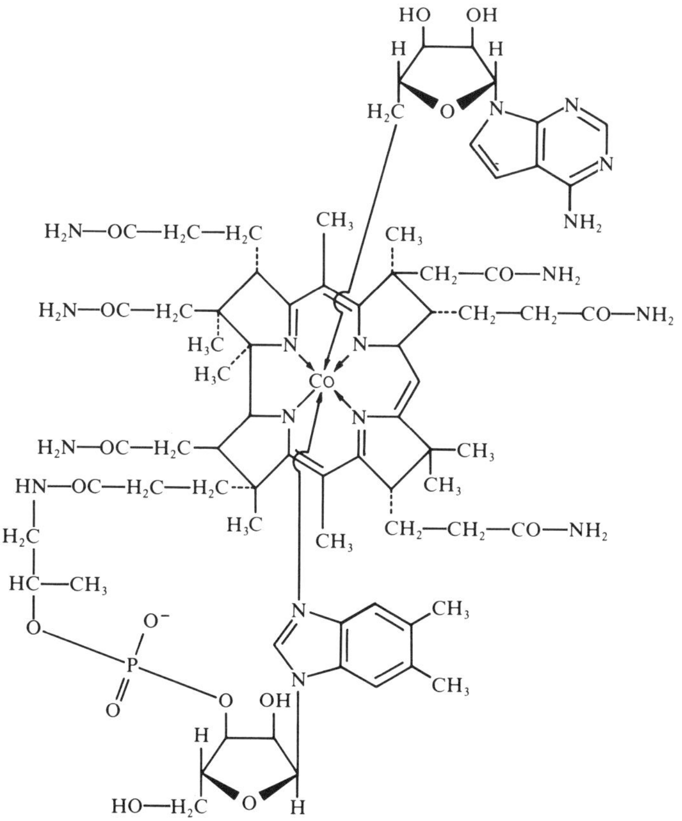

Coenzyme B_{12}

new position. The cobalt–carbon bond is split prior to reaction with the substrate and is reformed at the end of the reaction. In the latter group of B_{12} enzymes, methyl transfer occurs, examples being in the synthesis of acetate and methionine. These use methylated corrin derivatives as intermediates. Methionine synthetase is one example which is found in birds and mammals as well as bacteria. It action is to transfer a methyl group from N-methyltetrahydrofolate to homocysteine. The absence or inactivity of this enzyme is thought to contribute to the pathology of pernicious anaemia.

6.3.5 Zinc

Zinc exists in solution mainly as a divalent cation. It occurs in a wide range of enzymes and is involved in the transfer of information by RNA. It is similar in size to copper(II) and can compete with it in complex formation. However, unlike copper(II), it is not easily reduced. Thus its role in enzymes appears to be as a structure former and as a Lewis acid in catalytic reactions. Zinc(II), being a d^{10} ion, prefers regular tetrahedral or octahedral geometries. However, in biological systems zinc(II) is often found in distorted environments necessary for its catalytic or structural role (Bertini *et al.*, 1982).

The zinc in carboxypeptidase (Dunn 1975) lies in a shallow hole in the protein which can also accommodate the peptide substrate. The peptide is so orientated that the carbonyl oxygen is positioned in the co-ordination sphere of the zinc(II). This polarizes the carboxyl bond and facilitates nucleophilic attack. Another example of ligand bond polarization occurs in carbonic anhydrase, where the ligand is water. Substrate carbon dioxide reacts with the polarized water molecule to produce the bicarbonate ion.

The role in zinc in superoxide dismutase, on the other hand, appears to be mainly structural. Catalysis is carried out at the copper site, but binding of the zinc(II) ion to the apoprotein results in the enzyme taking up a specific structure, resulting in the protein in this region being less accessible to water. Quite recently, a new structural role for zinc has been discovered. Regions of the RNA protein depend for their function on protein conformation; the base of each of the loops of this section are held together by complexed zinc ions producing the so called 'zinc fingers'. Thus, zinc is essential for RNA function.

6.3.6 Molybdenum

Molybdenum is unique in that it currently seems to be the only essential second-row transition metal. It occurs in a number of enzymes – one of which, nitrogenase, has received much attention, because of the obvious economic advantages of low-cost nitrogen fixation (Coughlan, 1980). The peculiar factors that appear to be important in molybdenum chemistry are: it has a wide range of oxidation states which are physiologically available (III–VI); it can form complexes with a range of co-ordination numbers and geometries; and it can break bonds easily to give oxygen, nitrogen, sulphur or hydrogen transfer. For example, in xanthine oxidase the reaction probably follows the following mechanism

$$\begin{array}{c}\diagdown \\ \diagup\end{array}\!\!Mo^{VI}\!\!\begin{array}{c}\diagup\!\!\!\diagup O \\ \diagdown\!\!\!\diagdown S\end{array} + RH \longrightarrow Mo^{IV}\!\!\begin{array}{c}\diagup OR \\ \diagdown SH\end{array}$$

followed by cleavage of OR^- and the intramolecular reoxidation of molybdemun(IV) to molybdenum(VI) initially involving dioxygen.

The molybdenum(VI) ion can exist at biological pH in polymeric forms. One of these, a heptamolybdate has been found to have antitumour activity (Yamase *et al.*, 1988). These isopolyacids can be reversibly reduced and oxidized and still retain the same basic structure.

$$[Mo_7O_{24}]^{6-} + e + H^+ \rightarrow [HMo_7O_{24}]^{6-}$$

Thus the destruction of cells could be due to a reversible redox reaction interfering with cell metabolism rather than to a complexing reaction as is the case with platinum(II). This mechanism could also apply in the use of 21-tungsto-9-antimonate $[Sb_9W_{21}O_{86}]^{19-}$ which has been tested in the treatment of AIDS. It has been reported to inhibit the reverse transcriptase of both lymphocytopathic retrovirus and simian AIDS virus (Rosenbaum *et al.*, 1985). The fact that these polyoxoanions adhere to biological surfaces has been utilized in the staining of microscopic slides of biological samples. There have been many different examples of these polyoxoanions and their reactions *in vivo* could prove a productive line of research.

In nitrogenase it has been assumed without any definitive evidence that the molybdenum ion is the site at which the N_2 molecule is complexed and also reduced directly to ammonia. This nitrogen fixation enzyme is important in that it provides a major path for the reduction of atmospheric nitrogen to ammonia, thus making nitrogen available to plants and eventually to animals. The enzyme consists of two protein fractions – one containing only iron and the other both iron and molybdenum. Both parts of the protein are very sensitive to oxidation by dioxygen and so cells which respire aerobically prevent this by a variety of means generally involving either rapid metabolism of oxygen or the presence of a protein such as leghaemoglobin, which has a very high affinity for oxygen.

There is still some ambiguity about the nature of the molybdenum present. However, recently EXAFS (extended X-ray absorption fine structure) measurements have shown that the molybdenum is bonded to three or four sulphur atoms at a distance of 2.36 ± 0.01 Å (Cramer *et al.*, 1978). There is also evidence for one or two sulphur atoms at a distance of 2.49 Å. However, further information is still required for an understanding of the mechanics of the electron transfer system.

Thus, present knowledge of the chemistry of each transition metal *in vivo* is insufficient to obtain a complete picture of their biological activities. In some cases such as iron, copper, cobalt or zinc, much work has been done and model studies of specific enzyme sites, such as those found for example in vitamin B_{12} or carbonic anhydrase, have been extensive and useful. In others, such as vanadium or chromium, much remains to be done before even the nature of the ion sites involved in *in vivo* processes can be characterized.

6.4 MAIN GROUP ELEMENTS

6.4.1 Sodium, potassium, magnesium, and calcium

As mentioned previously, these ions are class I (hard) in nature and will favour complexing with oxygen anions. They are strongly electropositive and not easily oxidized or reduced. Thus, biologically, they have a major role in neutralizing the charge on anions and in maintaining isotonic pressure in natural fluids, which accounts for their relative abundance in living systems. Chemically, Na^+ and K^+ ions are very similar, as are Mg^{2+} and Ca^{2+} ions. However, in biological chemistry, small differences are magnified and lead to selective roles for the different cations. For example, K^+ and Mg^{2+} are concentrated in the cytoplasm, whereas Na^+ and Ca^{2+} are found mainly in serum. These cation concentration gradients are maintained across cell walls by the expenditure of considerable energy through the hyrolysis of ATP (Racker, 1979). An efficient separation is important in that a rapid influx of Na^+ or Ca^{2+} ions into cells often acts as a trigger. Magnesium and calcium ions are also involved in stabilizing certain conformations of enzymes, proteins and membranes. Thus, again, efficient separation is critical. For example, with calcium the intracellular level is 10^{-7} mol dm^{-3} and the extracellular level 10^{-3} mol dm^{-3}. During stimulation, the Ca^{2+} acts as a second messenger amplifying the original signal for appropriate cell action. The calcium must act at a molecular level by binding proteins or enzymes which then modulate the activity of others. Two of the more important calcium binding proteins are calmodulin and troponin C (Wasserman, 1977). The former, as an example, has a molecular weight of 17 000 and contains four sites for Ca^{2+} ions. The extent of occupancy varies for different reactions. There are also calcium storage proteins such as calsequestrin, which can bind up to 43 calcium ions per mole of protein.

Efflux of Ca^{2+} from these storage proteins can take place in milliseconds to trigger specific reactions. Subsequently, the calcium will be pumped back out of the cell by energy supplied by the hydrolysis of ATP. Although many calcium modulated reactions are fairly specific, other divalent cations can interfere – for example, Zn^{2+} inhibits platelet aggregation and phagocytosis, both of which require Ca^{2+} ions. Thus, the overall balance of metal ions *in vivo* is of considerable importance. A range of 'calcium antagonist' cardioactive drugs has been developed (Brewer *et al.*, 1982). These apparently function by controlling the influx of calcium ions into cells, with the subsequent inhibition of ATP synthesis and oxygen uptake.

Thus, although there are larger concentrations of main group ions than of transition metal ions *in vivo*, their environment is still often specific and tightly controlled. One example of the synergism between biochemistry and inorganic chemistry which has arisen from these studies has been the continuing development of specific complexing ligands such as crown ethers. The main criterion has been to produce ligands with the correct hole size to accept one ion in preference to another. More recently, the use of complexing groups other than oxygen has been investigated, thereby changing the chemical affinity of the site for specific metal ions.

6.5 SOME TOXIC METALS

6.5.1 Cadmium

Occupational and environmental exposure to cadmium is a known health hazard. Cadmium accumulates in body tissues and organs and in particular in the liver and kidneys where it has a long biological half life (Friberg *et al.*, 1985). It is present in the blood of non-exposed people at a concentration of about $1\mu g\,ml^{-1}$, in cigarette smokers up to $3\mu g\,ml^{-1}$ and in industrial workers exposed to cadmium dust or fumes at much higher levels. The proposed threshold for industrial workers is now about $10\mu g\,ml^{-1}$ (Rogenfelt, 1984). It is interesting that the body has its own defence against a limited amount of cadmium poisoning in that it synthesizes more of the protein metallothionein which contains a large number of sulphydryl groups. It is normally involved in copper and zinc storage *in vitro*, but cadmium binds strongly to the sulphydryl group sites thus removing the cadmium from circulation. From investigations of the effects of cadmium toxicity in the rat, simultaneous determination of plasma cadmium and plasma metallothionein show that these are related and if picked up early enough coincide with early pathological changes in the kidney which may still be reversible (Aughey 1984).

6.5.2 Lead

The poisonous effects of ingested lead have been known for many years, although it is only in recent times that the possible harmful effects of low levels of lead on the behaviour and intelligence of children have been appreciated. Lead is ingested mainly from food and to a lesser extent from water with very little being absorbed by inhalation. Industrial workers using lead or its complexes, those people living in soft water areas where there is extensive lead piping involved in the water supply, or children in areas where there is a high traffic density are examples of groups exposed to higher than normal lead intake. Evidence suggests that the primary route for the intestinal uptake of lead is by diffusion through the extracellular cation-specific channels.

6.5.3 Aluminium

The biochemistry of aluminium has in recent years become a subject of increasing interest. The reason for this is twofold: it is now belived to have a role in the death of fish populations resulting from acid rain and it is believed to have a key role in certain disease processes such as Alzheimer's disease. In the former, aluminium appears to be retained in the gills of fish so preventing sodium ion replacement. The effect can be reversed by calcium ions. Thus, the addition of calcium carbonate to lochs, lakes and ponds affected by acid rain can help to preserve fish life as well as to neutralize the pH. In the human disease, possible toxic effects of aluminium were first seen in patients undergoing kidney dialysis. Whereas aluminium cannot pass through the normal process of absorption in the gut, it is possible for it to pass through the dialysis membrane and consequently it is concentrated in the body of patients on long-term dialysis in areas where there is a high aluminium content in the water. This resulted in several cases of encephalopathy in dialysis (dialysis dementia). The correspondence between this and the presenile dementia of Alzheimer's disease is interesting particularly because higher than normal aluminium levels have been found in certain parts of the brain of some patients with Alzheimer's disease. A major problem in this work appears to be the question of whether the name Alzheimer's disease covers one or several disease states, not all of which may be equally sensitive to aluminium. However, no causal link has been demonstrated between aluminium and Alzheimer's disease and it is not known whether the metal plays a significant role in the disease or whether it merely concentrates in specific damaged cells. On the other hand, it has been found that aluminium can affect the action of certain enzymes in the brain such as acetyl cholinesterase (Patocka and Bajgar, 1987). Thus, it would seem that the general population is perhaps not at risk from aluminium present in the environment in reasonable quantities and if there is a connection between aluminium and brain disease, it may be possible to identify those sections of the population who are specifically at risk.

6.6 METAL IONS AS DRUGS

Many compounds of metal ions have been used as drugs, with compounds of calcium, iron, cobalt or zinc being prime examples. Zinc, for example, can be used in the treatment of wounds and burns, for nutritional deficiency, for the treatment of rashes and as a replacement for sulphur compounds in the treatment of Wilson's disease. Recently, there has been considerable interest in metal ion compounds of non-essential metals and the use of platinum compounds has already been mentioned. One example of a transition metal (gold) and one of a main group metal (lithium) are discussed. One other metal (platinum) has been much discussed recently. Its complexes are the basis for a series of drugs which are effective anticancer agents. This area is very active and has been well reviewed. Essentially *cis* complexes, mainly of platinum(II), are believed to complex with DNA and prevent cell replication. The basic compound used is *cis*-diaminodichloro-platinum(II), but a range of more active and less toxic materials based on it is now available. It is a particularly effective agent against testicular cancer, but has effects on other cancers such as ovarian cancer.

6.6.1 Gold

Gold is a non-essential element but its metabolism is of interest in that several of its complexes have found pharmacological use at various times. The beneficial effects of gold complexes in the treatment of rheumatoid arthritis was confirmed in 1961 by the report of the Empire Rheumatism Council. Since then, gold complexes have been quite widely used (Shaw, 1979; Brown and Smith, 1980). There are problems, however, in that some patients do not respond at all and others suffer various toxic side effects. Despite fairly extensive literature on the clinical aspects of gold therapy, only recently have attempts been made to determine the mechanism of gold action. The gold complexes most commonly used are gold(I) sodium thiomalate (Myocrisin), gold(I) thioglucose (Solganol) and, more recently, S-2,3,4,6-tetraacetyl-1-β-D-thioglucose triethylphosphine gold(I) (Auranofin) (Fig. 6.3). The first two are injected intramuscularly and the last given orally.

Experiments with animals have shown that after intramuscular injection gold is widely distributed throughout the body although higher concentrations are found in organs such as the liver. Gold is retained in identifiable

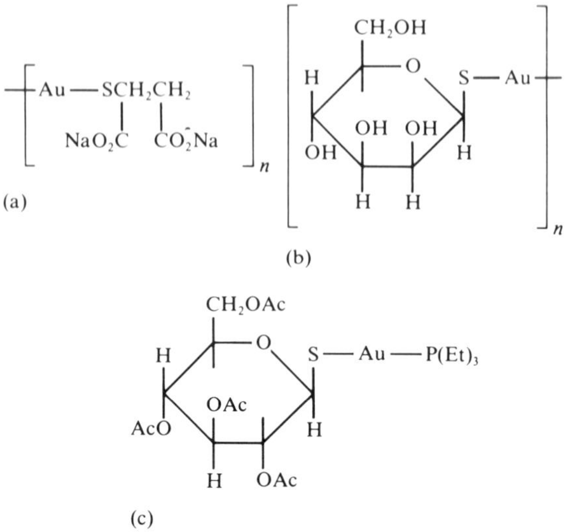

(a)

(b)

(c)

Fig. 6.3 The structures of the gold drugs Myocrisin (a), Solganol (b) and auranofin (c). They are used in the treatment of rheumatoid arthritis. The gold is believed to exchange with other thiol ligands in vivo. It is probably the active species rather than the released thiol.

quantities in humans up to twenty years after gold injections have been stopped. This long-term storage raises questions as to its nature. Experimental evidence has suggested that the gold aggregates found in certain cells are probably gold(I)–sulphur species and this seems the most likely type of complex by comparison with the *in vitro* chemistry of gold.

Gold has two stable oxidation states in aqueous solution, gold(III) and gold(I). The former is stabilized by 'hard' ligands such as chloride or hydroxide and the latter by 'soft' ligands such as thiols, cyanide and thiosulphate. Gold(III) is a fairly strong oxidizing agent and is unlikely to be stable for long in the presence of organic material *in vivo*. Thus the most likely gold species *in vivo* will be gold(I) and the most common ligand the thiol group. Other gold(I) complexing ligands which occur naturally are cyanide and thiosulphate and there is some evidence that the former may have a part to play *in vivo*.

The role of thiols appears to be of considerable importance in the aetiology of rheumatoid arthritis. Many of the drugs used in its treatment are themselves thiols (D-penicillamine), metabolized to other thiols (levamisole or azothioprine); react with thiols (gold complexes); or alter thiol levels on proteins (chloroquine). Thus, the reactivity of different gold drugs with naturally occurring thiols *in vivo* could determine the gold metabolites and, hence, the effectiveness of the drug. Myocrisin and Auranofin give different gold distributions in blood, and produce forms of gold with different *in vivo* behaviour such as the ability of the circulating gold to cross cell membranes (Table 6.3).

Gold(I) forms 1:1 complexes with a wide range of thiols including cysteine and glutathione. The full structure of these complexes is unknown, but they are likely to be linear – similar to gold(I) thiosulphate, but with bridging sulphur atoms. Gold(I) is unusual amongst the heavy metals in that most of its complexes whose structures have been described are linear. This could be due to the small contribution to bonding of the gold *p* orbitals, leaving the two bonds mainly *s* in character and hence linear. Only when the ligands have more diffuse *p* orbitals, such as with phosphines, does greater mixing occur and three- and four-coordinate gold(I) compounds are obtained. Gold(III), on the other hand, forms mainly square-planar complexes, as is common for a d^8 metal ion. As mentioned above, gold(III) is a good oxidizing agent and quickly oxidizes thiols to the disulphide, the gold being reduced to gold(I) and stabilized as a thiol complex by excess thiol. If there is no excess of thiol, metallic gold is formed. An interesting exception to this is D-penicillamine (dimethylcysteine). In contrast to cysteine it forms a stable complex with gold(III). The reason for this is probably steric. Because of the restrictions of the two methyl groups on the carbon atoms adjacent to the sulphur donor atoms, the sulphur atoms will probably be *trans* to each other in the complex, whereas with cysteine a *cis* form is possible and any two sulphur atoms *cis* to each other could easily be oxidized to the disulphide causing reduction of the gold.

Such is the affinity of gold(I) for thiols that the widespread distribution of gold probably reflects the widespread distribution of thiols *in vivo*. The other likely ligands mentioned above, *viz.* cyanide and thiosulphate, are present in much lower concentrations in biological fluids. There is some

Table 6.3 The different transporting effect of the gold drugs Myocrisin and Auranofin. After 12 weeks therapy on either the injected gold drug Myocrisin or the oral gold drug Auranofin, a sample of blood was taken from each patient and separated into a cellular fraction and plasma. Each fraction was incubated with a placebo fraction of the opposite type. The samples were re-separated and gold concentrations determined by atomic absorption spectrometry. A different fraction of the plasma gold was transferred to the placebo cells indicating different metabolites in the plasma. Gold concentrations are in µg/ml.

Cells	Plasma	Cells	Plasma
Auranofin	Placebo	0.72	0.0
Myocrisin	Placebo	0.28	0.0
Placebo	Auranofin	0.68	0.05
Placebo	Myocrisin	0.05	3.0

evidence that cyanide in the blood of heavy smokers increases the gold concentration inside cells and hence affects the balance of gold concentrations across cell membranes. However, no work has been published yet on the possible effects of gold on the metabolism of cyanide or thiosulphate or even possibly thiocyanate.

Thus, in conclusion the therapeutic effect of gold is probably due to a wide range of reactions of which the thiol metabolism appears to be the main target. A more specific point of action of the gold could be on the structurally and metabolically important thiol groups on cell membrane. The reaction of gold with the hexose transport protein of the erythrocyte has recently been reported. The reaction is likely to affect whole cell metabolism and therefore a range of enzyme processes, but as yet the only target investigated is the red cell. It is likely that similar proteins exist in white cells but the effects have yet to be evaluated. If one enzyme is to be chosen, the most likely is glutathione peroxidase which has an active selenium site. Class II or soft gold ions would be expected to complex with selenium in preference to sulphur and there is animal and some human evidence to support this. However, the relationship between glutathione peroxidase and disease is rather tenuous at present.

6.6.2   Lithium

Lithium, the lightest of the alkali metals, is not, as far as is known, an essential element, but lithium carbonate has been found to be effective in the treatment of manic–depressive psychosis (Johnson, 1980). It is thought to function by competing with the naturally occurring metal ions and, because of the different stability of its complexes, it alters the existing reaction pathways. The question then arises as to which metal ions it will most likely compete with. In aqueous solution, lithium will exist as the hydrated Li^+ cation. Therefore, it will have the same charge as Na^+ and K^+ ions. However, it is smaller than these two monovalent cations and is roughly the same size as the Mg^{2+} ion. Using another criterion for comparison, its surface charge density is similar to that of the Ca^{2+} ion. Thus, depending on the conditions, the Li^+ cation could compete with any of the four naturally occurring cations Na^+, K^+, Mg^{2+} and Ca^{2+}. A complicating factor is that the lithium ion is heavily hydrated in aqueous solution and this affects its complexing ability. For example, with some ion-exchange resins the order of strength of binding has been shown to be $Li^+ < Na^+ < K^+$. Thus, lithium can have a wide range of effects, several of which could have some influence on the same physiological system.

Lithium is the only other metal ion which, *in vitro*, is able to replace sodium in surrounding fluids in the maintenance for any significant time of the resting and active potentials in isolated nerve cells (Birch and Sadler,

1982). The resting potential is mainly due to the potassium potential across the nerve cell membrane, and the lithium ion does not interfere significantly with potassium ion transport. However, the active potential relates to the sodium gradient suggesting that lithium ions may be transported through sodium channels. In general it is thought that lithium ions enter cells in a similar manner to sodium ions but are transported out much more slowly, resulting in intra- and extracellular lithium concentrations being of the same order. In the more specific sodium sites, as in the sodium pump, lithium does not compete. Thus in animal experiments, the addition of lithium ions does not greatly affect sodium and potassium ion concentrations.

Lithium ions appear to have an effect on a wide variety of processes, making an interpretation of its biological role somewhat complicated. For example in rats, serum magnesium and calcium levels increase after lithium injections. Also, lithium often competes directly with calcium and magnesium in the stimulation of enzymes. For example, lithium can replace calcium and stimulate the secretion of acetylcholine.

6.7 MODERN PHYSICAL METHODS

The development of chemistry during the past two decades has produced a battery of powerful techniques which can be applied to problems in inorganic biochemistry. X-ray techniques for example have improved to the point where the environment of a metal in a protein may be elucidated much more quickly and even in a system which cannot be crystallized some information can be obtained using EXAFS.

However, other information on such points as rates of reaction, structure in solution and the effect of changing microenvironments is required, and for this purpose the chemist often requires a specific reaction. It is only possible to mention a selection of such techniques here, and we have chosen as examples techniques which we believe will have an increasing impact on inorganic biochemistry.

Modern nuclear magnetic resonance spectrometers are much better equipped to obtain spectra of direct value in inorganic biochemistry. For example, nucleii such as Na and Pt can readily be detected and spin echo techniques have been used to identify the ^1H spectrum of potential ligands such as glutathione in whole cells (Fig. 6.4). ^1H and ^{13}C spectra of metal ion protein sites in enzymes such as the Cu/Zn superoxide dismutase have been successful in predicting the metal ion environment in advance of X-ray structural determination and have the inherent advantage that they do not require single crystals and can be used in principle to determine structure *in situ* or at least in solution in model studies. In general the use of 2D NMR to determine structure in relatively low molecular weight species is a growing field with an increasing number of successful investigations being reported.

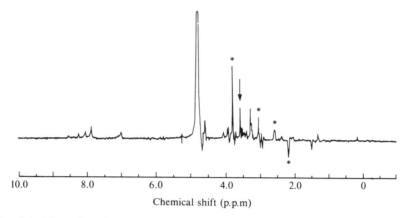

Chemical shift (p.p.m)

Fig. 6.4 The spin echo NMR spectrum of intact red cells. This technique utilizes the difference in resolution times between protons in small molecules and in proteins to produce a relatively simple spectrum of the small molecules. Peaks due to glutathione are marked *.

Circular dichroism (CD) can be used to define protein configuration in such problems as the changes caused in the α and β side chains in haemoglobin in sickle cell anaemia (Woody, 1978). Chiral transition metal complexes give good CD spectra, but good spectra are not restricted to such complexes. The metal ligand bond is partly covalent and, if there is a chiral centre on the ligand, then there is often sufficient interaction for the formally $d \rightarrow d$ transition on the metal to give intense circular dichroism in the readily accessible visible region. For example in six-co-ordinate amino acid chromium(III) complexes, two ligand field bands

$$^4A_{2g} \rightarrow {}^4T_{2g}(^4F) \text{ and } {}^4A_{2g} \rightarrow {}^4T_{1g}(^4F)$$

appear in the visible near IR (Fig. 6.5) spectrum. In complexes of lower symmetry such as those with amino acid ligands, these bands are each split into three components which are much more easily identified. Circular dichroism has the advantage that its signals can be obtained from optically inactive species and more specific electronic information can be obtained from both optically active and optically inactive species. There have been a number of significant studies of metal ion containing enzymes using this method.

Raman spectroscopy has the advantage in biological systems that measurements of vibrational structure can be taken in aqueous solution. However, it is relatively insensitive and sophisticated equipment is required. Extensions of the Raman technique, such as CARS (coherent antistokes Raman scattering), FT Raman, stimulated Raman and resonance Raman,

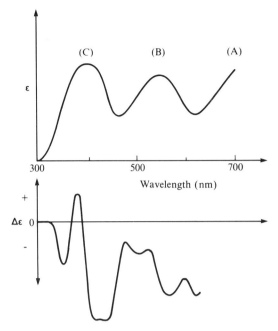

Fig. 6.5 The UV-visible spectrum and circular dichroism of a chromium (III) serine complex indicating the increased resolution which can be obtained from the latter technique. The UV-visible peaks are due to d → d transitions (a) $^4A_g \rightarrow {}^4T_{2g}$ (b) $^4A_g \rightarrow {}^4T_{1g}(^4F) {}^4T_{1g}$, (c) $^4A_g \rightarrow {}^4T_{1g}(^4F)$.

can be used to overcome this disadvantage. The most extensively used of these methods at present is resonance Raman spectroscopy, in which a laser beam of wavelength similar to that of an absorption band of the molecule is used as the exciting source. An enhancement of intensity of up to about 10^5 compared to Raman spectroscopy can be achieved in favourable cases and since the effect is specific for the molecule with the electronic structure which causes the resonance, it is selective.

An example of the use of this technique is the determination of glutathione concentration (Banford *et al.*, 1982) (-2000 μm) in a red cell lysate in the presence of haemoglobin (Fig. 6.6). Among the many compounds investigated in this way are several haem containing proteins, blue copper proteins and retinal pigments. More recently, this technique has been extended using surface-enhanced resonance Raman spectroscopy to enable very small concentrations ($10^{-8} - 10^{-9}$M) of enzyme to be detected. Substrate interactions involving mammalian cytochrome-P450 have been studied in this way (Wolf, 1988).

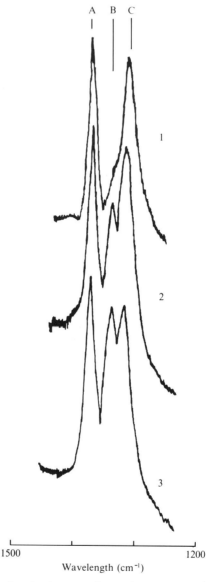

Fig. 6.6 Raman signals from a haemolysate treated with 5,5-dithiobis (2-nitrobenzonic acid)(ESSE). The vibrations are produced by haemoglobin and the coloured $ES^{(-)}$ ions produced by reaction of ESSE with intracellular thiol. The electronic spectra of the two overlap but the vibrational spectra shown here are separate and the signal is enhanced so that it can be observed in biological systems at 10^{-5} to 10^{-6}M. A = haemoglobin, B = ESSE, C = $ES^{(-)}$.

Electron spin resonance can produce good signals with metal ions in biological systems in favourable cases. For example, copper in serum ultrafiltrate (1 to 5% of serum copper) and caeruloplasmin can be observed, as can the high spin paramagnetic forms of haemoglobin (Fig. 6.7). The main problem with this technique is that the line widths are very sensitive to the metal ion used and its environment. For example in caeruloplasmin, much of the copper is copper(II) and thus should be detected, but it is 'silent' giving no signal as a result of exchange interactions. Thus, although ESR has proved a useful and sensitive probe the results must be interpreted with caution.

Finally, one advantage of the study of metal ions is that it is possible to analyse for them in very dilute solutions using techniques that have become rapid, reliable and semiroutine enabling quite precise pharmacokinetic experiments to be carried out. Perhaps neutron activation analysis is the most reliable method and the ability to scan electron micrographic sections and map metal ion distributions is the most impressive, but the most widely used technique because of its simplicity and availability is atomic absorption spectrometry. Figure 6.8 shows an example of the use of this method applied to the uptake of cuprous oxide by guinea pigs following oral administration. That most of the copper is initially in albumin can be demonstrated by a combination of electrophoresis and atomic absorption. Later most of the copper is present bound to caeruloplasmin.

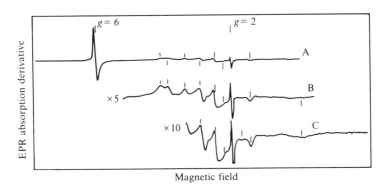

Fig. 6.7 The use of ESR spectra to identify oxidation states and closely related species illustrated in these spectra of haemoglobin in packed red cells. They have been incubated with nitrate to produce a prominent signal due to high spin ferri haemoglobin (A) and a variety of low spin forms (B and C). The total amount of paramagnetic haem is *ca.* 5% of haem present.

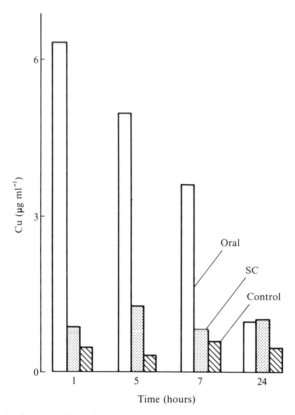

Fig. 6.8 A pharmocokinetic study of the oral absorption of Cu_2O by guinea pigs, illustrating the ease of analysis possible with atomic absorption (s.c., subcutaneous injections). The samples were diluted 10 times and injected into the furnace within a few hours of sampling, thus minimizing delays and contamination from sample handling, two of the most important causes of error in this type of analysis.

6.8   CONCLUSIONS

This chapter has aimed to show that although biochemistry was originally just an applied form of organic chemistry, now as our knowledge increases the chemistry of metal ions has an increasingly important part to play. Since many biochemists have little formal training in transition metal chemistry, this has resulted in inorganic chemists moving into the field. Their contributions in the main have been in the elucidation of the environment and reactions of specific metal ions in certain enzymes and the production of an accompanying array of model systems. In a sense this is a logical step for traditional metal complex chemistry, with the same principles applying but in a rather more complicated environment.

Parallel with this are the investigations into *in vivo* chemistry – absorption, transport and general metabolism. This work involves a more interdisciplinary approach particularly if 'in health' and 'in sickness' variables are considered. Inevitably this work has lagged behind the more orthodox chemistry, but ultimately should prove the most rewarding. Some impetus for this approach has also come from the now recognized pollution hazards of certain metals. Acceptable levels of non-essential metals have still to be rationalized in most cases. Medical treatment involving chelating agents is effective, but has disadvantages as labilizing the chelated metal for excretion may also redistribute the complex within the body with perhaps other unwanted effects.

REFERENCES

Anderson, R. A. (1981) *Sci. Total Environ.*, **17**, 13.

Anderson, R. A. and Koglovsky, A. S. (1985) *Am. J. Clin. Nat.*, **41**, 1177.

Aughey, E., Fell, G. S., Scott, R. and Black, M. (1984) *Environ. Health Prospect.*, **54**, 153.

Banford, J. C., Brown, D. H., McConnell, A. A., McNeil, C. J., Smith, W. E., Hazelton, R. A. and Sturrock, R. D. (1982) *Analyst*, **107**, 195.

Banyard, S. H., Stammers, D. K. and Harrison, P. M. (1978) *Nature (London*, **271**, 282.

Bertini, I., Luchinat, C. and Scozzafa, A. (1982) *Struct. Bond.*, **48**, 45.

Birch, N. J. and Sadler, P. J. (1982) in *Specialist Periodical Reports in Inorganic Biochemistry*, Vol. 3 (ed. H. A. O. Hill), Royal Society of Chemistry, London, pp. 372–396.

Bremner, I. and Mills, C. F. (1981) *Philos. Trans R. Soc. London. Ser. B.*, **294**, 75.

Brewer, G. J., Bereza, U., Kretzchmar, P., Brewer, L. F. and Aster, J. C. (1982) in *Inflammatory Diseases and Copper* (ed. J. R. Sorenson), Humana Press, New Jersey, pp. 532–535.

Brown, D. H. and Smith, W. E. (1980) *Chem. Soc. Rev.*, **9**, 217.

Campion, G. C. and Barthia, R. (1985) *Applied and Environmental Microbiology*, 498.

Cantley, L. C. and Aisen, P. (1979) *J. Biol. Chem.*, **254**, 1781.

Chasteen, N. D. and Weiss, R. (1983) *Struct. Bonding*, **53**.

Cohen, M. D., Sen., A. C and Wei, C. I. (1987) *Inorg. Chim. Acta*, **138**, 91.

Coughlan, M. (ed.) (1980) *Molybdenum and Molybdenum-Containing Enzymes*, Pergammon Press, Oxford.

Cramer, S. P., Gillum, W. O., Hodgson, K. O., Moretensen, L. E., Stiefel, E. I., Chisnel, J. R., Brill, W. J. and Shah, V. K. (1978) *J. Am. Chem. Soc.*, **100**, 3814.

Crichton, R. R. (1973) *Struct. Bond.*, **17**, 67.

Dunn, M. F. (1975) *Struct. Bond.*, **23**, 61.

Forth, W. and Rummel, W. (1973) *Physiol. Rev.*, **53**, 724.

Freeman, H. C. (1967) *Adv. Protein Chem.*, **22**, 257.

Friberg, L., Elinder, C. G. and Jell Strom T. K. (1985) in *Cadmium and Health. A Toxicological and Epidemiological Appraisal*, (ed. G. F. Norberg), Vol. 1, CRC Press, Florida.

Frieden, E. (1980) *CIBA Found.*, **79**, 93.

Harris, E. D., DiSilvestro, R. A. and Balthrop, J. A. (1982) in *Inflammatory Disease and Copper* (ed. J. R. J. Sorenson), Humana Press, New Jersey, pp. 183–199.

Harrison, P. M. and Hoare, R. J. (1980) *Metals in Biochemistry*, Chapman and Hall, London.

Hurley, L. S. (1984) *Trace Elements in Analytical Chemisty in Medicine and Biology* (Eds., P. Bratter and P. Schamel), Walter de Grayter & Co, Berlin, **2**, 239.

Johnson, F. N. (ed.) (1980) *Handbook of Lithium Therapy*, University Park Press, Baltimore.

Karlish, S. J. D., Beauge, L. and Glynn, I. M. (1979) *Nature (London)*, **282**, 333.

Klotz, I. M., Klippenstein, G. L. and Hendrickson, W. A. (1976) *Science*, **192**, 335.

Lippard, S. J. (ed.) (1983) *Biochemistry of Platinum, Gold and other Therapeutic Agents* (American Chemical Society. symposium, no. 209), American Chemical Society, Washington D.C.

MacDonald, L. G., Brown, D. H. and Smith, W. E. (1982) *Inorg. Chim. Acta*, **67**, 7.

Mareschal, J. C., Rama, R. and Crichton, R. R. (1980) *FEBS Lett.*, **110**, 268.

Michelson, A. M., McCord, J. M. and Fridovich, I. (eds.) (1977) *Superoxide and Superoxide Dismutase*, Academic Press, New York.

O'Brian, P., Barnett, J. and Swanson, F. (1985) *Inorg. Chim Acta* **108**, C19.

Patocka, J. and Bajgar, J. (1987) *Inorg. Chim. Acta*, **135**, 161.

Racker, E. (1979) *Acc. Chem. Res.*, **12**, 338.

Rosenbaum, W., Dormont, D., Spiro, B., Vilmer, E., Gentibini, M., Griskelli, C., Montagener, L., Barre-Simoussi, F. and Cherman, J. C. (1985) *Lancet*, 450

Rogenfelt, A., Elinder, C. G. and Jarup, C. (1984) *Anch. Occup. Environ. Health*, **55**, 43.

Sakurai, H., El Taira, Z. and Sakai, M. (1988) *Inorg. Chim. Acta*, **151**, 85.

Schwarz, K. and Mertz, W. (1959) *Anal. Biochem. Biophys.*, **85**, 292.

Shaw, C. F. (1979) *Inorg. Persp. Biol. Med.*, **2**, 287.

Sorenson, J. R. J. (ed.) (1982) *Inflammatory Diseases and Copper*, Humana Press, New Jersey.

Takahashi, M. *et al.* (1984) *Proc. Nat. Acad. Sci.*, **81**.

Wasserman, R. H. (1977) *Calcium Binding Proteins and Calcium Function*, Elsevier, Amsterdam.

Wetterhahn, K. E. and Connett, P. H. (1983) *Struct. Bonding*, **54**, 93.

Wolf, C. R., Miles, J. S., Seilman, S., Burke, M. D., Rospendowski, B., Kelly, K., and Smith, W. E. (1988) *Biochemistry*, **27**, 1597.

Wood, J. M. and Brown, D. G. (1972) *Struct. Bond.*, **11**, 47.

Woody, R. (1978) in *Biochemical and Clinical Aspects of Haemoglobin Abnormalities*, Academic Press, New York, pp. 279–298.

Yamase, T., Fusith, H. and Fukashima, R. (1988) *Inorg. Chim Acta*, **151**, 15.

7 | The enzymology of the biosynthesis of natural products

David E. Cane

7.1 INTRODUCTION

Over the last fifteen years, the study of the biosynthesis of natural products has undergone a revolution in both methodology and scope. The widespread use of stable isotope nuclear magnetic resonance methods has provided a firm experimental base for the widely accepted precursor–product relationships between the simple compounds of central metabolism – acetate, amino acids and carbohydrates and the seemingly endless variety of organic natural products, while establishing many of the key details of the pathways which link precursor to product (Vederas, 1987). While the vast majority of such incorporation experiments have been carried out with intact organisms, an ever increasing number of investigations have begun to focus on the individual enzymes of secondary metabolic pathways (Hutchinson, 1986). In concert with the powerful new tools of molecular genetics and recombinant DNA methodolgy, the study of the enzymology of biosynthesis of natural products has begun not only to reveal the detailed biochemical mechanisms of numerous biosynthesis tranformations, but has opened up new opportunities to study the regulation as well as the evolution of secondary metabolic pathways.

For many years, the biosynthesis of natural products has been studied primarily by feeding isotopically labelled precursors to growing cells or otherwise intact tissues followed by isolation of the resulting labelled metabolites and analysis by chemical or spectroscopic methods to determine the sites of labelling (Brown, 1972). Incorporation experiments of this type, many of which have achieved the pinnacle of intellectual elegance and experimental finesse, have led not only to the identification of the primary precursors of individual polyketides, terpenoids,and alkaloids, but have allowed plausible deductions as to the probable pathways by which the simple precursors are transformed to the more complex natural product. Guiding the design and interpretation of all these incorporation experiments has been a small number of unifying mechanistic concepts which form the core of biogenetic theory and which are thought to account for the biochemical formation of carbon–carbon and carbon–heteroatom bonds (Mann, 1978; Herbert, 1981). The enormous success of these central generalizations in predicting as well as rationalizing such a massive body of experimental results, together with the host of biomimetically modelled

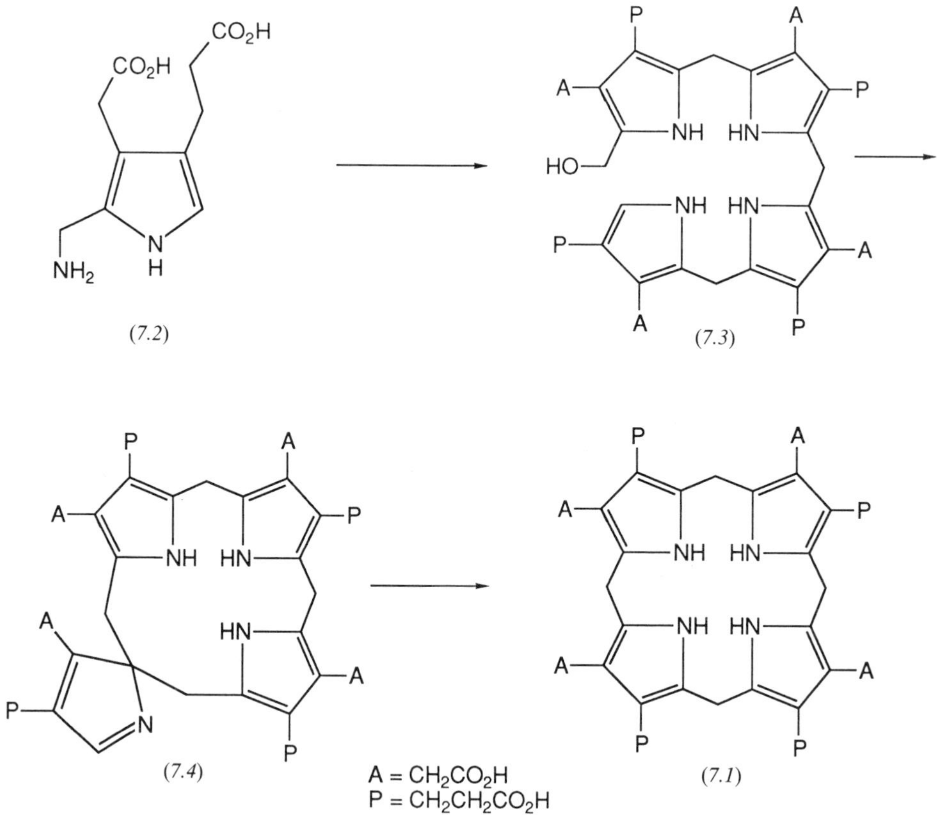

Fig. 7.1 Formation of urogen III (*7.1*) from PBG (*7.2*) by way of hydroxymethyl-bilane (*7.3*).

syntheses inspired by these paradigms, has left little doubt as to the essential correctness of these biogenetic hypotheses.

In spite of the impressive successes of these classical precursor–product experiments, there have been powerful incentives to carry out investigations at the cell-free level. Not the least of these have been the difficulties frequently encountered in achieving efficient incorporation of labelled precursors due to the failure of these substrates to traverse cellular barriers or reach the appropriate sites in the metabolizing cell. The problem can be particularly severe with highly reactive substrates which may not be sufficiently long-lived to allow efficient incorporation or detection, or with highly polar substances which, in the absence of any active transport system, are unable to traverse the normally lipophilic barriers of the cell. Attempts to establish the sequences of intermediates in a metabolic pathway have often been frustrated by differential uptake of exogenously administered precursors leading to misleading conclusions based on the relative efficiency of utilization or rate of incorporation of competing substrates. While whole-cell experiments can provide a wealth of details about the net consequences of a biosynthetic pathway, true mechanistic information, including an understanding of the precise timing of bond-forming and bond-breaking processes as well as the identification of the cofactor requirements for biosynthetic transformations, can only come from a study of the enzymes catalysing the key biosynthetic events which lie at the heart of biogenetic theory. Several recent examples will serve to illustrate the impressive power of biosynthetic studies at the cell-free level.

7.2   RECENT ADVANCES IN THE STUDY OF BIOSYNTHETIC ENZYMES

7.2.1   Porphobilinogen Deaminase

Uroporphyrinogen III (7.1) (urogen III) is the universal precursor of the tetrapyrrolic porphyrin, chlorin, and corrin pigments which form the nucleus of the organometallic cofactors haem, chlorophyll, and vitamin B_{12} respectively (Battersby and McDonald, 1979) (Fig. 7.1). Extensive investigations carried out in several laboratories have established that the formation of urogen III from porphobilinogen (7.2) is mediated by two enzymes (Leeper, 1985). The first, porphobilinogen deaminase (PBG deaminase), catalyses the conversion of porphobilinogen to a linear tetrapyrrole, hydroxymethylbilane (7.3), which in turn serves as the substrate for uroporphyrinogen III synthase (cosynthetase) (Battersby et al., 1980a; Battersby et al., 1982; Scott et al., 1980). Extremely elegant experiments utilizing multiple-[^{13}C]-labelled substrates have conclusively demonstrated that the cyclization of (7.3) involves an intramolecular rearrangement of the terminal pyrrole (Battersby *et al.*, 1978; Battersby *et al.*, 1980b), presumably through a spiro intermediate such as (*7.4*), so as to generate the characteristic ring D substitution pattern of urogen III (Stark *et al.*, 1986).

By the early 1980s it had been recognized that porphobilinogen becomes covalently attached to PBG deaminase in the formation of hydroxymethyl-bilane. Initial attempts to identify the nature of the nucleophilic enzyme X-group had led to controversy, with speculation centering on a lysine–NH_2 (Battersby *et al.*, 1983) or cysteine–SH residue (Evans *et al.*, 1984). A major breakthrough in the understanding of PBG deaminase came from the cloning of the corresponding structural gene, first reported by Jordan (Thomas and Jordan, 1986). By carrying out subcloning experiments on restriction fragments of an *E. coli* plasmid which had been shown to carry the *hem*C gene locus, the Jordan group located the PBG deaminase gene within a 1.69 kb B*am*H1-*Sal*1 fragment and succeeded in overexpressing deaminase activity in host strains 50-fold. Using an analogous approach, the Battersby group was able to obtain 50–100 mg quantities of *E. coli* PBG deaminase. Incubation of the recombinant enzyme with [11-^{13}C]PBG and examination of the resultant mono- and dipyrrole complexes (*7.5*) and (*7.6*) by high-field difference ^{13}C NMR, led to the startling discovery that PBG is covalently attached to the enzyme by an enzymic *pyrrole* moiety which extensive chemical and spectroscopic investigations have conclusively shown to be an enzyme-bound dipyrromethane, itself derived from two equivalents of PBG (Hart *et al.*, 1987) (Fig. 7.2). This discovery, for which there is no enzymic precedent, has been independently confirmed, and the amino acid residue to which the novel cofactor is attached identified as a cysteine (Hart *et al.*, 1988; Jordan *et al.*, 1988a; Scott *et al.*, 1988a; Alefounder *et al.*, 1989). In the meantime, the PBG deaminase gene has been sequenced (Thomas and Jordan, 1986; Jordan *et al.*, 1987; Alefounder *et al.*, 1988) and the site of attachment of the dipyrromethane cofactor has now been identified as Cys-242 by site-directed mutagenesis (Scott *et al.*, 1988b; see also Miller *et al.*, 1988). The experiments on PBG deaminase provide a dramatic illustration of the power of the combined use of recombinant DNA methodology, enzymology, high-field NMR spectroscopy, and organic synthesis in probing the intimate details of biosynthetic transformations.

In related work, three groups have now cloned and sequenced the *E. coli* *hem*D gene, corresponding to uroporphyrinogen III synthase (Sasarman *et al.*, 1987; Jordan *et al.*, 1988b; Alefounder *et al.*, 1988). The latter gene is contained in an open reading frame which overlaps the last four bases of the *hem*C gene and there is evidence that both enzymes are transcribed under the control of a common promoter as part of a putative *hem* operon which may also code for two additional structural genes. The way is now open not

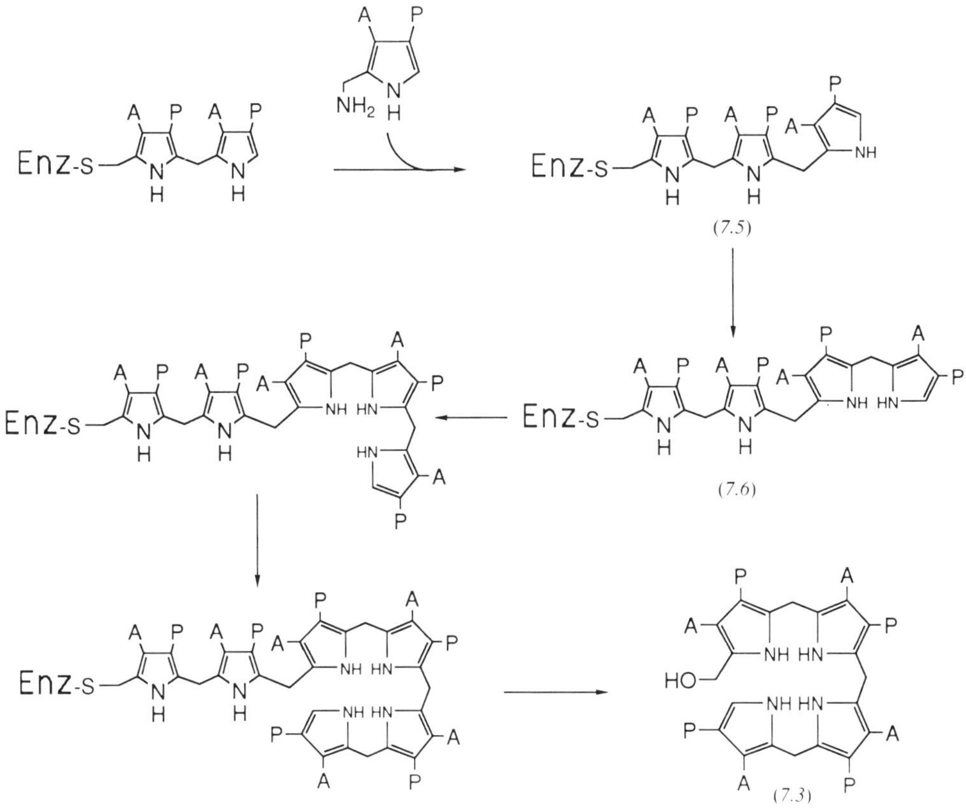

Fig. 7.2 Conversion of PBG to hydroxymethylbilane (*7.3*) by PBG deaminase, involving covalent attachment to an enzyme-bound dipyrromethane cofactor.

only for detailed mechanistic studies of cosynthetase itself, but for studies of the regulation of the porphyrin biosynthetic pathway at the genetic level.

7.2.2   Isopenicillin N Synthase

The biosynthesis of the enormously important penicillin and cephalosporin families of β-lactam antibiotics has been the subject of intensive investigations over the last 20 years. By 1976, incorporations of both radioactive and ^{13}C-labelled substrates had firmly established the amino acids valine, cysteine, and α-aminoadipic acid as the common precursors of penicillin N (*7.7*) and cephalosporin C (*7.8*) (Aberhart, 1977) and implicated a likely intermediate in the form of the Arnstein tripeptide, δ-(L-α-aminoadipyl)-L-cysteinyl D valine (*7.9*) (*L I D*-ACV) (Arnstein and Morris, 1960 (Fig. 7.3).

Nonetheless all attempts to demonstrate intact incorporation of the simple tripeptide into penams or cephems using cultures of *Penicillium chrysogenum* or *Cephalosporium acremonium* met with universal failure until Abraham discovered that viable cell-free preparations of the key cyclase could be obtained by osmotic lysis of protoplasts of *C. acremonium* (Fawcett *et al.*, 1976). Since that time, dramatic progress has been made and isopenicillin N synthase (IPNS) has also been obtained from two other β-lactam producers, *P. chrysogenum* (Ramos *et al.*, 1985) and *Streptomyces clavuligerus* (Wolfe *et al.*, 1985). The purified cyclase has been shown to be a single polypeptide, M_r 32 000, requiring Fe^{2+}, ascorbate, and one equivalent of molecular oxygen for the conversion of *LLD*-ACV to isopenicillin N (*7.10*) (Pang *et al.*, 1984).

Over the last several years, Baldwin has produced a substantial corpus of work which has contributed enormously to our understanding of the

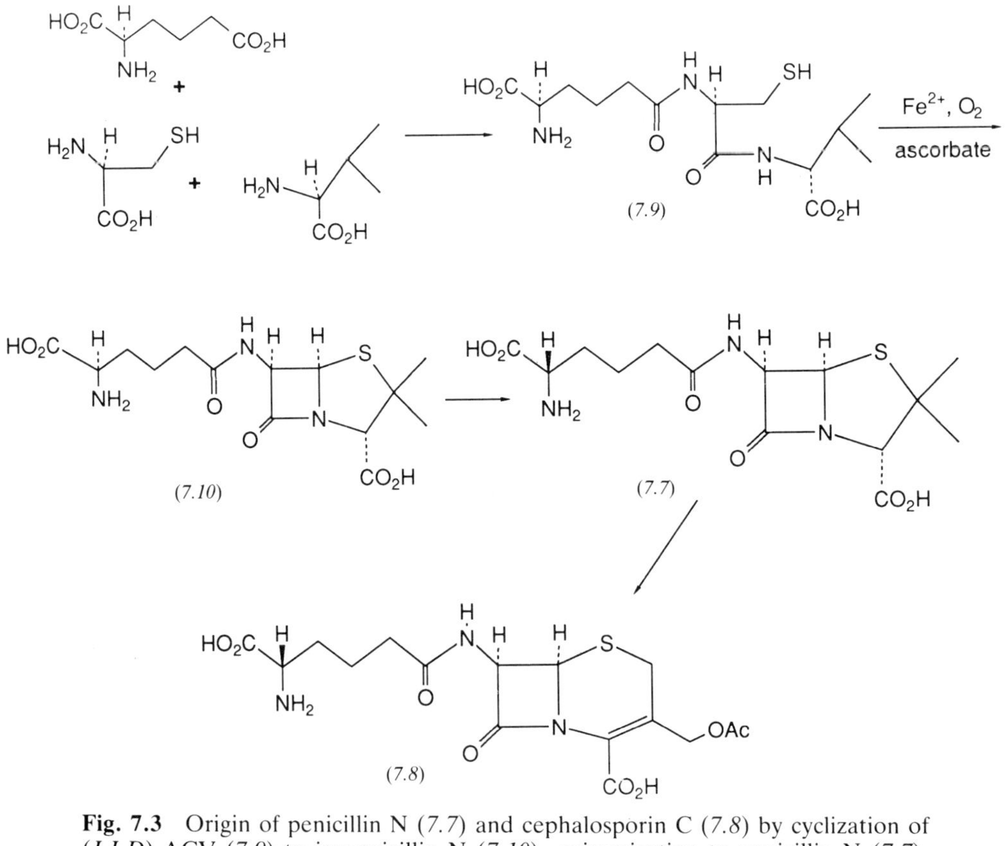

Fig. 7.3   Origin of penicillin N (*7.7*) and cephalosporin C (*7.8*) by cyclization of (*LLD*)-ACV (*7.9*) to isopenicillin N (*7.10*), epimerization to penicillin N (*7.7*), followed by oxidative ring expansion and methyl hydroxylation.

mechanism of formation of isopenicillin N (Baldwin and Abraham, 1988). In a particularly penetrating series of experiments, the Oxford group has made ingenious use of isotope effects to sort out the order in which the characteristic β-lactam and thiazolidine rings of the penam skeleton are generated. Earlier work had established the stoichiometry and net stereo-chemical course of the bond-forming reactions at the relevant cysteine and valine carbon centres, but had failed to identify any enzyme-free intermedi-ates in the cyclization. By incubating a 1:1 mixture of A-[3-^{2}H$_2$]-CV (7.9a) (Fig. 7.4) and unlabelled tripeptide with IPNS and determining the ratio of deuteriated to non-deuteriated species in recovered substrate as a function of the extent of turnover, it was found that the cyclase exercised a significant discrimination in favour of unlabelled ACV, corresponding to a V_{max}/K_m isotope effect in the removal of the 3-H$_{si}$ hydrogen of the cysteine residue (Baldwin et al., 1984). Such V/K isotope effects can only result from isotopically sensitive steps up to and including the first irreversible step in an enzyme-catalysed transformation (Ray, 1983). When the corresponding competition experiment was carried out with AC-[3-^{2}H]-V (7.9b) and un-labelled ACV, however, no such isotopic discrimination was observed in the breaking of the valine C–H bond. A separate series of experiments, carried out with the individual deuteriated tripeptides under conditions of substrate saturation, revealed significant deuterium isotope effects on V_{max} for both labelled substrates, indicating comparable energy barriers for the breaking of both the cysteine and valine C–H bonds (Monroney, 1985; Baldwin and Abraham, 1988). Furthermore, the V_{max}
the tripeptide with deuterium at both sites, A-[3-^{2}H$_2$]-C-[3-^{2}H]-V (7.9c), was within experimental error identical with the value predicted for two con-secutive processes involving an enzyme-bound intermediate. Baldwin has explained these results in terms of the reaction profile and mechanism illus-

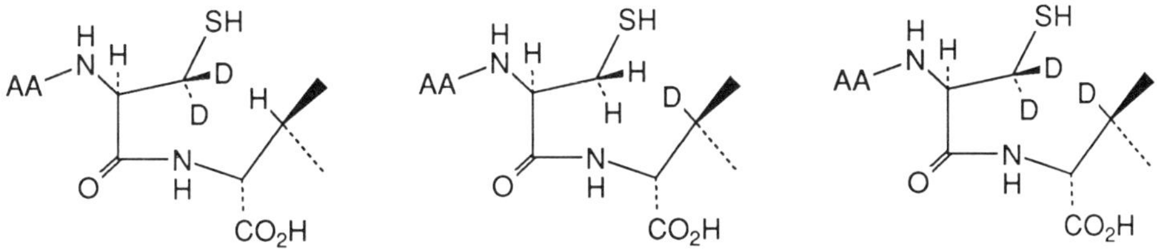

$(7.9a)$ $(7.9b)$ $(7.9c)$

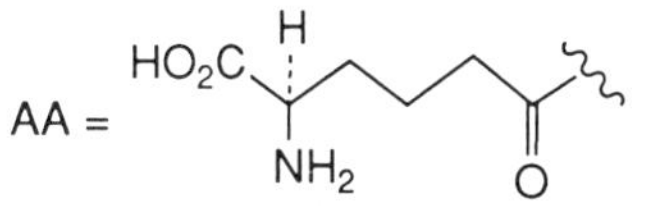

AA =

Fig. 74.   Deuterated ACV substrates for isopenicillin N synthase.

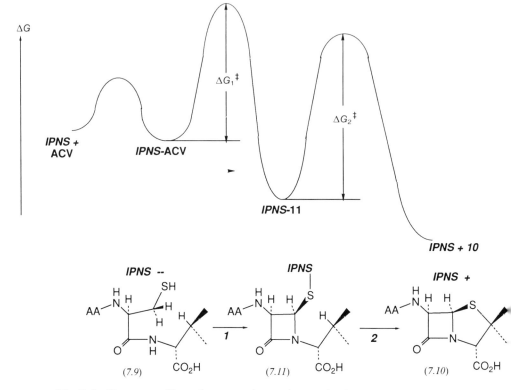

Fig. 7.5 Energy profile and proposed stepwise mechanism of ACV cyclization by isopenicillin N synthase (IPNS) involving initial formation of the β-lactam ring of (7.11).

trated in Fig. 7.5 in which removal of the cysteine hydrogen atom and formation of the enzyme-bound β-lactam ring in (7.11) represent the committed step of penam ring formation.

Further insight into the mechanism of formation of isopenicillin N has come from the study of a wide variety of ACV analogues (Baldwin and Abraham, 1988). Thus substitution of D-α-aminobutyrate for valine and incubation of modified tripeptide (7.12) with IPNS led to formation of a 7:1 mixture of the normethyl stereoisomers (7.13) and (7.14) along with an additional three parts of the unexpected cephem product (7.15) (Baldwin et al., 1983a). (Fig. 7.6). Remarkably, incubation of the corresponding (3S)- and (3R)-deuteriated tripeptides, (7.12a) and (7.12b), led to formation of the identical monodeuteriated product (7.13a) in each case, implying net retention and net inversion of configuration, respectively, in the formation of the C–S bond of the thiazolidine ring (Baldwin et al., 1983b) (Fig. 7.7).

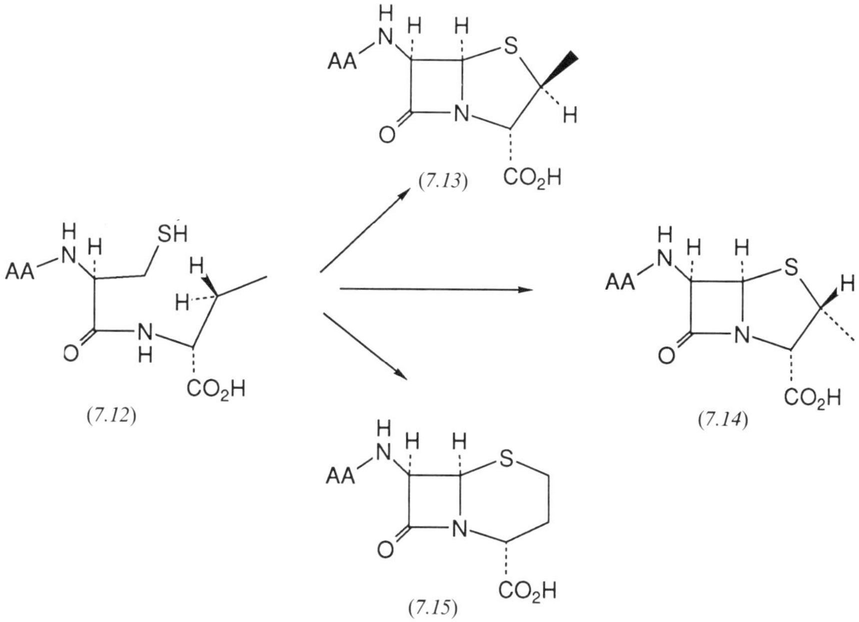

Fig. 7.6 Conversion of the α-aminobutyrate analogue of ACV to normethyl-isopenicillin N stereoisomers (7.13) and (7.14), as well as modified cephem (7.15).

This result was completely unexpected since earlier experiments had conclusively shown that the cyclization of the natural ACV substrate involves exclusively retention of configuration at C-3 of the valine moiety (Kluender *et al.*, 1973; Neuss *et al.*, 1973; Aberhart and Lin, 1974; Baldwin *et al.*, 1986). These observations establish that C–H bond cleavage and C–S bond formation are not concerted and are readily explained by a postulated radical mechanism, shown in Fig. 7.8, a proposal which is consistent with the results of a substantial body of subsequent experiments by the Oxford group.

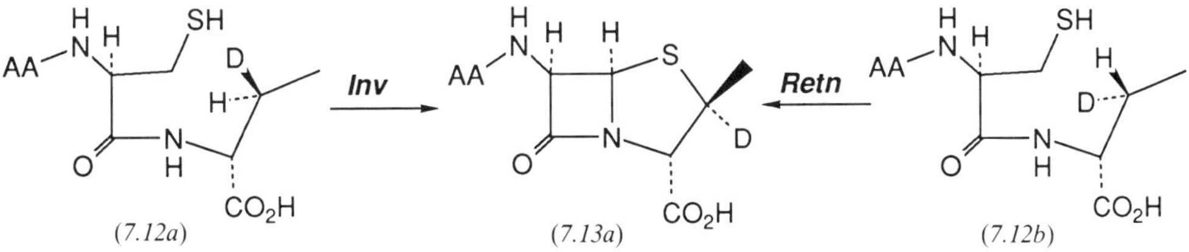

Fig. 7.7 Cyclization by IPNS of stereospecifically deuterated α-aminobutyrate ACV analogues to a common product by stereodivergent pathways.

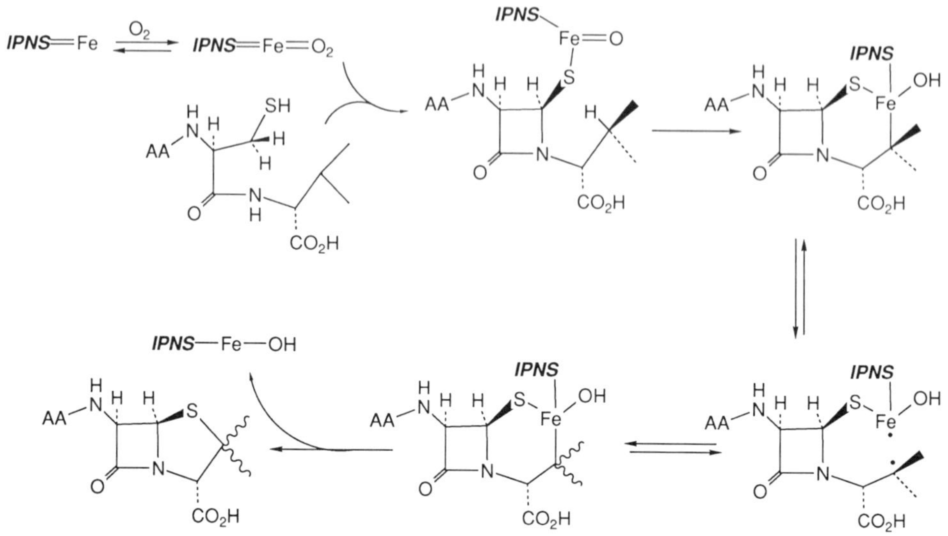

Fig. 7.8 Proposed mechanism of IPNS reaction involving ferryl(IV) species and competing equilibration of free radical intermediates.

In the meantime, there has been substantial progress on the enzymological and molecular biological front. To isolate the IPNS gene, researchers at Eli Lilly Laboratories and at Oxford used oligonucleotide probes based on the N-terminal amino acid sequence of the cyclase to screen an *E. coli* cosmid library of *C. acremonium* DNA (Samson *et al.*, 1985). The cloned IPNS gene has now been sequenced and over-expressed and recombinant cyclase is now available in significant quantities (Baldwin *et al.*, 1987). The corresponding *P. chrysogenum* (Carr *et al.*, 1986) and *S. clavuligerus* (Leskiw *et al.*, 1988) genes have subsequently been cloned and sequenced as well. Most recently, site-directed mutagenesis has demonstrated that neither of the two cysteine residues (Cys-106 and Cys-225) of *C. acremonium* IPNS is essential for cyclase activity (Samson *et al.*, 1987a). At the same time, there have been several reports of the cloning of the deacetoxy-cephalosporin C synthase (expandase) gene coding for the enzyme which catalyses the conversion of penicillin N (*7.7*) to the parent cephalosporin ring system (Samson *et al.*, 1987b; Chen *et al.*, 1988). While the *C. acremonium* enzyme, which also catalyses the hydroxylation of deace-toxycephalosporin C (*7.16*) to yield deacetylcephalosporin C (*7.17*) (Fig. 7.9), is clearly a bifunctional protein (Scheidegger *et al.*, 1984), the corresponding reactions in *S. clavuligerus* are catalysed by separate enzymes (Jensen *et al.*, 1985). Both the ring-expanding and hydroxylating enzymes require α-ketoglutarate as well as Fe^{3+} and molecular oxygen.

Finally, in related areas, cell-free preparations catalysing key steps in the formation of clavulanic acid (Elson *et al.*, 1987) and nocardicin (Wilson *et al.*, 1988) have been reported, opening the way for further progress in understanding the biosynthesis of these novel β-lactams.

7.2.3 Bialaphos Biosynthesis

The herbicide bialaphos (*7.18*) produced by *Streptomyces hygroscopicus* is a tripeptide formed from two equivalents of alanine and the novel amino acid phosphinothricin, containing a rare methylphosphonyl residue. Incorporation experiments with [^{13}C]-labelled glucose and acetate have established the basic building blocks for the phosphinothricin skeleton (Seto *et al.*, 1982) while extensive investigations with a series of mutants blocked in antibiotic production have led to the identification of the major intermediates of the bialaphos biosynthetic pathway (Seto *et al.*, 1983a, 1983b; Ogawa *et al.*, 1983; Imai *et al.*, 1984, 1985; Seto *et al.*, 1984), illustrated in Fig. 7.10. Among the most intriguing steps in this sequence is the formation of phosphinopyruvate (*7.19*) by the reaction of phosphonoformate (*7.20*) with phosphoenolpyruvate (PEP, *7.21*). Although the mechanism of this reaction is still unclear, it is known that acid-catalysed decomposition of (*7.20*) can lead to generation of phosphorous acid with evolution of CO_2 (Warren and Williams, 1971). The enzyme catalysing this step has apparently been isolated and characterized (Murakami *et al.*, 1986) although no details have as yet been reported. Phosphonoformate is itself derived in several steps by a process initiated by the rearrangement of PEP to phosphonopyruvate (*7.22*). The latter reaction has in fact been implicated as an early step in the biosynthesis of several naturally occurring phosphates (Trebst and Geike, 1967; Horiguchi, 1972; Rogers and Birnbaum, 1974). Recently, a Mg^{2+} dependent phosphomutase catalysing the PEP/phosphonopyruvate conversion has been purified from *Tetrahymena pyriformis* and the thermodynamics of the rearrangement have been studied (Bowman *et al.*, 1988). Seto

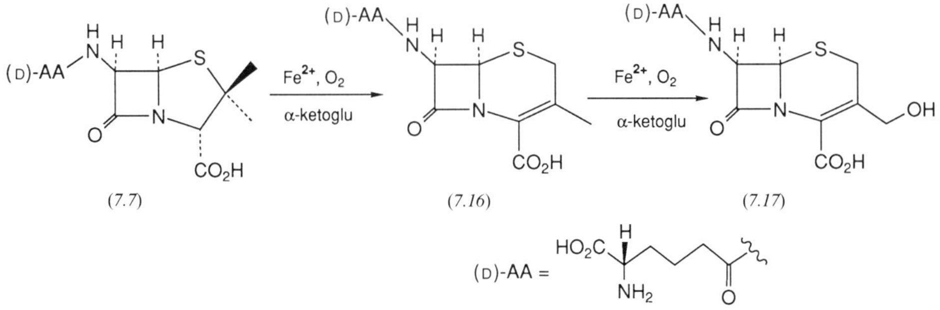

Fig. 7.9   Conversion of penicillin N (*7.7*) to deacetyl cephalosporin C (*7.17*).

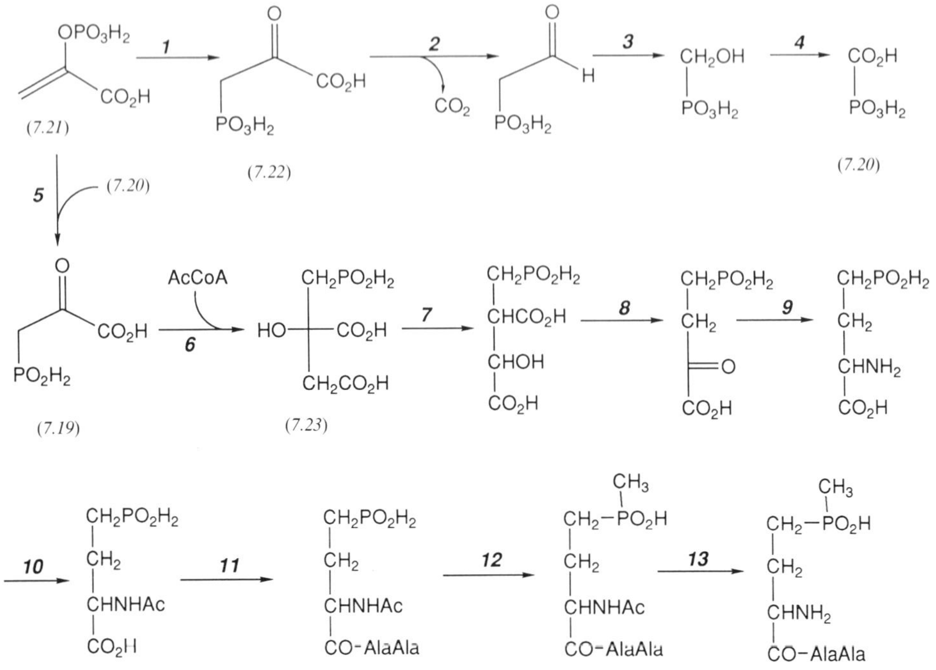

(7.21) (7.22) (7.20)

(7.19) (7.23)

Fig. 7.10   The biosynthesis of bialaphos (7.18) from PEP (7.21).

and his collaborators have also purified to apparent homogeneity the phosphinomethylmalate (7.23) synthase which catalyses the condensation of phosphinopyruvate, an oxaloacetate analog, and acetyl-CoA (Shimotohno *et al.*, 1988). The latter enzyme is a dimer of subunit M_r 48 000 requiring a divalent cation for activity and shown to be distinct from either (R) or (S)-citrate synthase.

The availability of blocked mutants has also allowed the cloning of most or all of the bialaphos biosynthetic genes which appear to be clustered on a 35 kb chromosomal DNA fragment (Murakami *et al.*, 1986; Hara *et al.*, 1988). (Fig. 7.11) Several structural genes were initially isolated by complementation of *S. hygroscopicus* mutants blocked at various steps of the bialaphos biosynthetic pathway. Additional production genes were identified by subcloning experiments and several of the resulting subclones were used as hybridization probes to screen *E. coli* cosmid clones, leading to identification of the bialaphos gene cluster. Although these clones have not yet been used for cloning and over-expression of individual enzymes of the bialaphos pathway, a 1.3 kb *Bam*H1 fragment has been shown to complement mutations in the phosphomutase reaction (step *1*) which a 2.0 kb

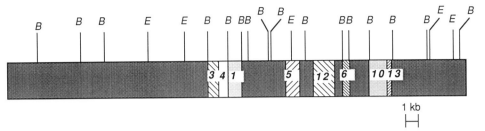

Fig. 7.11 The bialaphos (*bap*) gene cluster of *Streptomyces hygroscopicus*. Numbered regions correspond to steps shown in Fig. 7.10. B = *Bam*H1, E = *Eco*R1.

*Bgl*II/*Bcl*I fragment restored productivity to a mutant defective in the late stage methylation of the phosphinic acid (step *12*). DNA fragments related to the phosphinomethylmalate synthase reaction have also been cloned, but in the latter case restoration of bialaphos biosynthesis may have taken place by low-efficiency homologous recombination between cloned DNA and the corresponding defective chromosomal sites rather than by complementation in *trans* by vector DNA. In any case, one can anticipate the expression and sequencing of several key enzymes of this fascinating pathway within the near future as well as further elucidation of the regulation of the biosynthetic pathway itself.

7.2.4 Polyketide Biosynthesis

Polyketides are among the most widely studied groups of natural products. These metabolites which include a wide array of aromatic, macrolide, and polyether antibiotics, are all derived from simple metabolic building blocks acetate, propionate, and butyrate (Mann, 1978; Herbert, 1981). There is now extensive evidence supporting the idea that the formation of the characteristic carbon skeletons of these diverse metabolites takes place by mechanisms closely analogous to the well-understood chain-elongation steps of fatty acid biosynthesis (Volpe and Vagelos, 1973; Bloch and Vance, 1977; Wakil and Stoops, 1983). For example, there is good evidence that the key condensation reactions in the biosynthesis of the macrolide erythromycin and the polyethers monensin and lasalocid take place by a decarboxylative acylation of the methylmalonyl-CoA substrate with net inversion of configuration (Cane *et al.*, 1986; Sood *et al.*, 1984; Sherman and Hutchinson, 1987) (Fig. 7.12), directly analogous to the β-ketoacyl-CoA synthase reaction of fatty acid biosynthesis (Sedgwick and Cornforth, 1977; Sedgwick *et al.*, 1977, 1978; Arnstadt *et al.*, 1975). In spite of an impressive array of isotopic incorporation experiments, the study of polyketide biosynthesis has been hampered by the near absence of detectable intermediates of polyketide chain-elongation in both antibiotic-producing cultures as well as

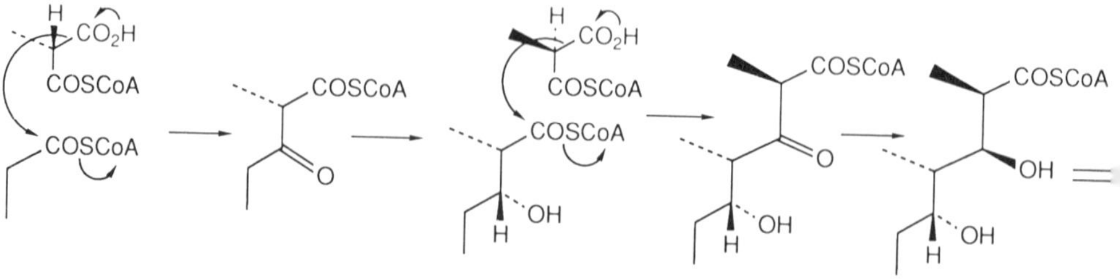

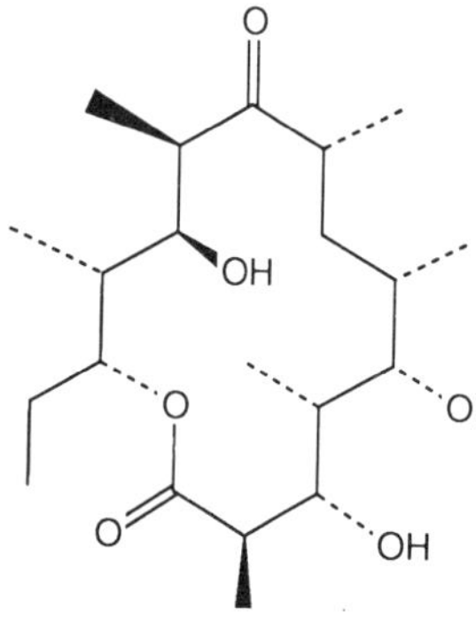

Fig. 7.12 Formation of 6-deoxyerythronolide B (*7.25*), the precursor of erythromycin A (*7.24*), by polyketide chain elongation involving stepwise condensations of methylmalonyl-CoA with acyl-CoA intermediates. The β-ketoacyl synthase reactions have been shown to take place with net inversion of configuration by a presumed decarboxylative acylation process.

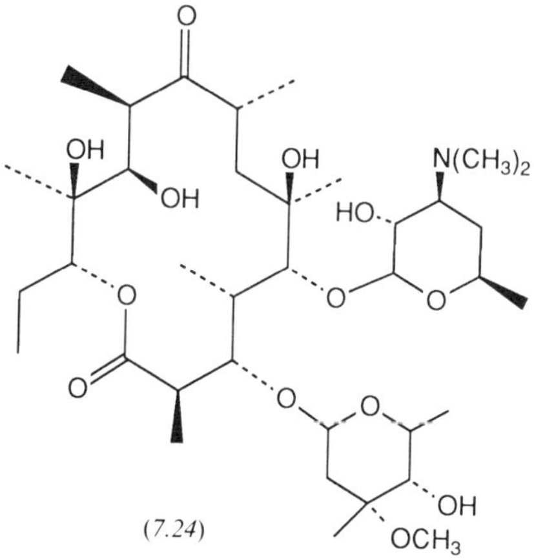

in blocked mutants (see however, Kinoshita *et al.*, 1988). Although some progress has recently been made in the intact incorporation of presumed intermediates of polyketide chain elongation (Yue *et al.*, 1987; Cane and Yang, 1987; Cane and Ott, 1988; Cane *et al.*, 1988a), the details of the chain elongation process have remained obscure. The difficulty in detecting intermediates can be ascribed, in part, to the probability that the actual physiological intermediates are covalently bound to the relevant polyketide synthase complex. Unfortunately, little is known about the synthases themselves, which as a group have proved resistant to persistent attempts at isolation. Nonetheless, recent advances, at both the enzymological and the molecular genetic level hold out the promise of significant advances in this critical area within the next several years.

Methylsalicylate (*7.26*) synthase, which catalyses the condensation of acetyl-CoA with three equivalents of malonyl-CoA, is one of the few polyketide synthases to have been isolated and one of only two such proteins which have been purified to date (Scott *et al.*, 1974; Dimroth *et al.*, 1976). The enzyme isolated from *Penicillium patulum*, M_r 1.1–1.5 × 10⁶, requires one equivalent of NADPH and shows strong mechanistic and enzymological similarities to yeast fatty acid synthase. One of the few pieces of information about the chain-building sequence comes from the observation that omission of NADPH from the incubation medium leads to formation of the derailment product triacetic acid lactone (*7.27*), suggesting the reasonable but still tentative conclusion that reduction of the diketoester (*7.28*) precedes condensation with the last malonate unit (Dimroth *et al.*, 1976) (Fig. 7.13). It will be noted, of course, that while the formation of triacetic acid lactone under anomalous conditions is certainly reasonable from a mechanistic point of view, the corresponding diketoester intermediate must be processed without incident by orsellinate synthase. What is not clear is why the omission of the reduction step is fatal in one case but causes no problems in the other.

The formation of reduced polyketides such as macrolides and polyethers presents an experimental problem which is an order of magnitude more complicated than that for methyl salicylate. It would appear that at each step of the chain elongation process the relevant synthase must make the appropriate choice among the available malonyl-CoA, methylmalonyl-CoA, and ethylmalonyl-CoA building blocks and then adjust the oxidation level and stereochemistry of the growing polyketide chain prior to the next condensation reaction. The factors governing the order of steps or the choice of substrates are completely unknown. There is considerable indirect evidence from the results of isotopic labelling experiments, however, that the microscopic mechanisms of the individual chain-building reactions may bear a close resemblance to the reactions catalysed by eucaryotic and procaryotic fatty acid synthases (Cane *et al.*, 1982, 1983; Hutchinson *et al.*, 1981; O'Hagan *et al.*, 1983). In addition, it has been found that the fungal metabolite cerulenin, a potent inhibitor of the β-ketoacyl thioester component (condensing enzyme) of fatty acid synthases from a variety of sources, will also inhibit the biosynthesis of a variety of aromatic and macrolide polyketides. For example, Leadlay has reported that addition of cerulenin to cultures of *Saccharopolyspora erythraea* (formerly *Streptomyces erythreus*) inhibits production of erythromycin and has found that [³H]tetrahydocerulenin binds irreversibly to a 35–37 kDa polypeptide which is detectable only in cultures actively producing antibiotic (Roberts and Leadlay, 1983, 1984). The relationship between this inactivated polypeptide and the so far elusive 6-deoxyerythronolide B synthase has yet to be established.

Several groups have attempted to detect the synthase mediating the

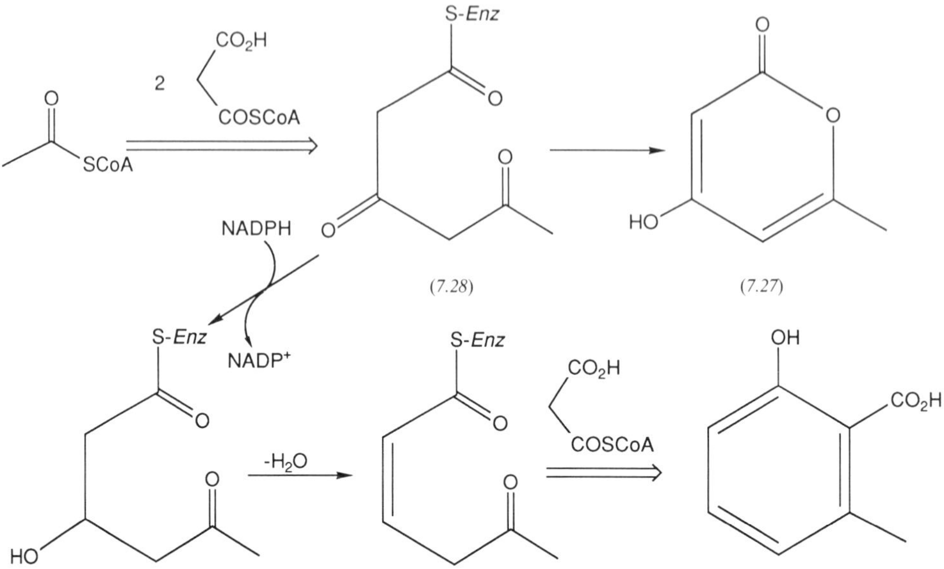

Fig. 7.13 Conversion of acetyl-CoA and malonyl-CoA to 6-methylsalicylate (7.26) and formation of triacetic acid lactone (7.27) in the absence of NADPH.

formation of 6-deoxyerythronolide B (7.25), the parent macrolide aglycone of the erythromycin family of antibiotics (Corcoran, 1981) (Fig. 7.12). Early experiments in the laboratories of both Lynen (Wawszkiewicz and Lynen, 1964) and Corcoran (Corcoran et al., 1967) were without success and no further details have appeared in the literature concerning a claim from Kolattukudy's laboratory involving the observation of in vitro 6-deoxyeryth-ronolide B synthase activity (Hunaiati and Kolattukudy, 1984). There is still no definitive evidence establishing whether the relevant macrolide synthases are true multienzyme polypeptides, analogous to Type I fatty acid synthases of yeast and high animals, or whether the polyketide synthase exists as a dissociable multienzyme complex comparable to the Type II fatty acid synthase of E. coli and other bacteria. It is interesting to note in this regard, however, that Leadlay has purified a discrete acyl carrier protein (ACP) from Sac. erythraea (see Fig. 7.15) which appears to be required for in vitro fatty acid biosynthesis, an observation consistent with the inference that the fatty acid synthase, at least, of Sac. erythraea is in fact a Type II synthase consisting of discrete subunits (Hale et al., 1987).

    In the biosynthesis of aromatic polyketide and macrolide antibiotics, the key chain-elongation and ring-forming events are often followed by a series of late stage modifications including one or more oxidations, methylations,

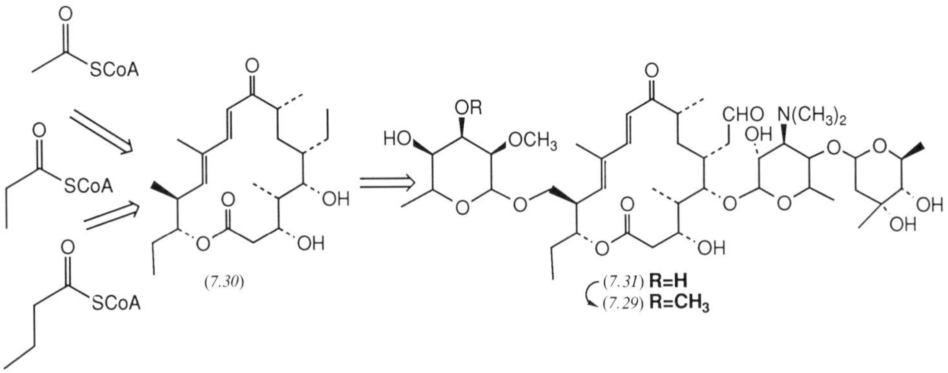

Fig. 7.14   Derivation of tylosin (*7.29*) from acetate, propionate, and butyrate, by way of tylactone (*7.30*) and macrocin (*7.31*).

and glycosylations. Based on extensive work with blocked mutants, Baltz and his collaborators have identified the sequence of intermediates lying between the antibiotic tylosin (*7.29*) and the parent macrolide aglycone tylactone (*7.30*) (Baltz and Seno, 1981; Baltz *et al.*, 1981; see also Omura *et al.*, 1980). Macrocin-*O*-methyl transferase, catalyses the last step in this sequence, involving the methylation of macrocin (*7.31*) by *S*-adenosylmethionine to yield tylosin (Seno and Baltz, 1981, 1982) (Fig. 7.14) The enzyme has been purified to homogeneity and the corresponding *N*-terminal amino acid sequence was used to construct a mixed 44-base oligonucleotide probe which utilized either guanosine or cytidine as the third base in each codon, based on the known codon preference of *Streptomyces* (Fishman *et al.*, 1987). Screening of a *S. fradiae* genomic library followed by appropriate subcloning experiments led to the identification of a 2.3 kb fragment apparently corresponding to the macrocin-*O*-methyltransferase structural gene. Thus plasmid vectors containing the relevant insert restored tylosin biosynthesis to *Tyl*F mutants which normally accumulate macrocin due to lack of the native methyltransferase. Furthermore, transformation of the naive host *S. lividans* with an ultrahigh copy number vector containing the same 2.3 kb insert resulted in the expression of low levels of the corresponding macrocin-*O*-methyltransferase. In related experiments, the *Tyl*F gene was mapped into a 70 kb gene cluster which was shown to contain several previously described late stage genes of the tylosin biosynthetic pathway along with the corresponding tylosin resistance genes and a region containing an amplifiable unit of DNA (AUD). Unfortunately, none of the components of this gene cluster was able to complement mutations in the tylactone-forming gene and to date the tylactone synthase gene has not been isolated.

   In the meantime, some progress has been reported in the study of the

Acyl transferases

yeast, acetyl transferase	L K G A T G H **S** Q G L V T A V
yeast, malonyl, palm. transferase	D A T F A G H **S** L G E Y A A L
Sac. erythraea eryAI	V A A I L G H **S** S P D A V G Q
S. glaucescens tcmIa ORF1	R K S M R G H **S** L G A I G S L
S. glaucescens temIa ORF2	R H G M R G H **S** S V F V T I Q

β-Ketoacyl ACP synthases

yeast	G P I K T P V G A **C** A T S V E S V
Sac. erythraea eryAI	V S L A R L W G A **C** G V S P S A V
S. glaucescens tcmIa ORF1	G P V T V V S T G **C** T S G L D A V

Acyl carrier proteins

yeast	L V G G K **S** T V Q N E I L G D L G K
E. coli	D L G A D **S** L D T V E L V M A L E E
Sac. erythraea FAS	D L G M D **S** L D L V E V
Sac. erythraea eryAIb	G S D G F **S** L D L V D M A D G P G E
S. glaucescens temIe	D L G Y D **S** I A L L E I S A K L E Q

Fig. 7.15   Peptide sequence homology of *Sac. erythraea* and *S. glaucescens* gene products with microbial FAS components (From Donadio *et al.*, 1988).

tylactone synthase itself. Hutchinson has isolated two radioactive protein fractions of M_r 10 000 and 65–75 000, obtained after administering [^{14}C]acetate, -propionate, or -butyrate to *S. fradiae* GS14 cells (Hutchinson *et al.*, 1988). These proteins have tentatively been identified as an ACP and either an acyltransferase or a β-ketoacyl-ACP synthase, respectively. Much of the argument implicating these proteins in polypropionate biosynthesis is based on the observation that the corresponding fatty acid synthase components of *E. coli* do not bind propionate or methylmalonate. Consistent with these observations was the isolation and partial purification of a set of acylCoA-pantotheine transferase activities from cell-free extracts of *S. fradiae*. The connection with the actual tylactone synthase is so far only circumstantial. Nonetheless, these preliminary findings do support the notion that the bacterial polyketide synthase may be a dissociable Type II synthase and open up new avenues of inquiry.

In further studies of the erythromycin biosynthetic pathway, Baltz has reported the cloning of the structural genes for erythromycin A biosynthesis (Stanzak *et al.*, 1986). A 35 kb segment of *Sac. erythraea* DNA, isolated by screening a cosmid library for erythromycin resistance, was shown to restore antibiotic biosynthesis to a family of mutants blocked at various stages in macrolide biosynthesis. Transformation of a naïve host also led to the production of small quantities of antibiotically active material presumed to be erythromycin, although the actual chemical products have not yet been rigorously isolated and characterized. It might be anticipated that appropriate subcloning experiments might eventually lead to expression of viable quantities of the elusive 6-deoxyerythronolide B synthase. In related work, Katz and Hutchinson have identified a 3.2 kb region of DNA downstream from the erythromycin resistance gene, *ermE*, which contains *eryA* genes coding for at least a portion of the 6-deoxyerythronolide B synthase. (Donadio *et al.*, 1988). Sequencing data led to the identification of three putative open reading frames (ORFs) which were suggested to correspond to the ACP, acyl transferase, and β-ketoacyl synthase portions of the macrolactone synthase, based on amino acid homology with fatty acid synthases from a variety of sources (Fig. 7.15). Hutchinson has also reported the purification of two forms of the cytochrome P450-dependent 6-deoxyerythronolide B hydroxylase and is actively investigating the cloning of the corresponding structural genes (Shafiee and Hutchinson, 1987).

In 1984, Hopwood reported the first example of the cloning of the structural genes for a polyketide, the octaketide actinorhodin (*7.32*), a blue pigment produced by *Streptomyces coelicolor* (Malpartida and Hopwood, 1984) (Fig. 7.16). The entire gene cluster was shown to lie within a 26 kb region of chromosomal DNA. Although little is still known about the biochemical details of the assembly of the aromatic system, subcloning experiments have resulted in the isolation of a 2.2 kb *Bam*H1 fragment, termed *ActI*, believed to be related to the β-ketoacyl synthase activity, based on its ability to complement mutations in polyketide chain elongation (Malpartida and Hopwood, 1986). Based on this lead, *ActI* DNA has been used as a hybridization probe to search for polyketide biosynthetic genes in a variety of organisms (Malpartida *et al.*, 1987). Recently, Hutchinson has reported the sequencing of a 2.97 kb fragment of *S. glaucescens* DNA which complements defects in the polyketide chain-elongation and cyclization steps of tetracenomycin C (*7.33*) biosynthesis (Hutchinson *et al.*, 1988). Three apparent open reading frames were identified and an attempt was made to assign the biochemical function of the presumptively derived

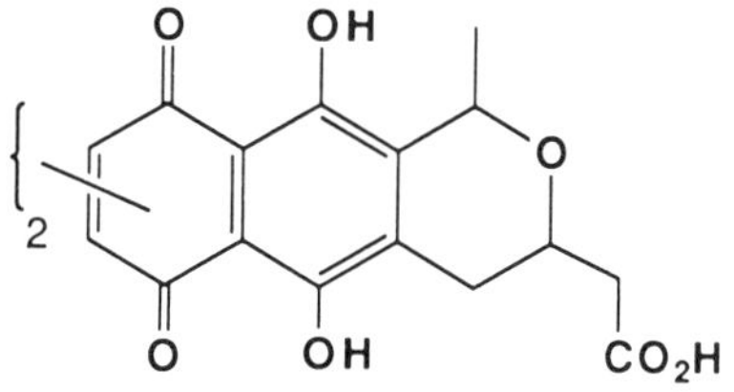

(*7.32*) 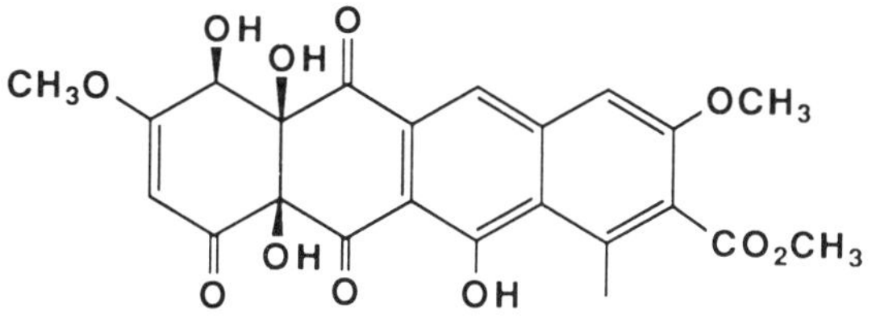   (*7.33*)

Fig. 7.16 Dimeric octaketide actinorhodin (*7.32*) and decaketide tetracenomycin C (*7.33*).

proteins (Fig. 7.15). ORF1 and ORF2, which hybridize to *ActI*, were proposed to code for 426 and 405 amino acid subunits, respectively, of a β-ketoacyl-ACP synthase, while ORF3, corresponding to an 83 amino acid protein, would appear to be an ACP, based on significant amino acid homology to acyl carrier proteins in related species. These are exciting discoveries which are certain to be followed up by work now in progress in several laboratories. There is little doubt that the long-guarded secrets of polyketide enzymology stand ready to be revealed.

7.3   TERPENOID CYCLASES

Interest in the biological origins of terpenoids goes back more than a century. As early as 1887, Wallach had pointed out that terpenes were constructed from apparent isoprene building blocks (Wallach, 1887). This insight eventually led to Ruzicka's proposal of the empirical Isoprene Rule, (Ruzicka and Stoll, 1922) subsequently reformulated in explicitly mechanistic terms as the Biogenetic Isoprene Rule (Ruzicka, 1959, 1963). Ruzicka pointed out that ionization of the commonly occurring sesquiterpene alcohol farnesol and intramolecular electrophilic attack of the resulting allylic cation on either the central or distal double followed by various combinations of rearrangements and further cyclizations could, in principle, account for the formation of all cyclic sesquiterpenes (Fig. 7.17). The actual cyclization substrate has been shown to be farnesyl pyrophosphate (*7.34*), (Lynen *et al.*, 1958, 1959; Chayken *et al.*, 1958) while the original Ruzicka hypothesis has been further elaborated by several groups of authors (Hendrickson, 1959; Parker *et al.*, 1967; Cane, 1985) In the intevening three decades, numerous investigations carried out at the whole cell level have enriched our knowledge of terpene biosynthesis and supported the general correctness of these cyclization schemes. (Cane, 1981; Croteau, 1981, 1987).

   Much of our understanding of isoprenoid biosynthesis has come from the pioneering investigations of the biosynthesis of cholesterol carried out by Lynen, Bloch, Cornforth, and Popjak, among others (Bloch, 1952; Popjak and Cornforth, 1960, 1966; Clayton, 1965; Cornforth, 1973). The study of the biosynthesis of farnesyl pyrophosphate itself has proved to be a particularly rich source of mechanistic, stereochemical, and enzymological information. Farnesyl pyrophosphate synthetase (prenyl transferase) catalyses the synthesis of *trans, trans*-farnesyl pyrophosphate from dimethylallyl pyrophosphate (*7.35*) and two equivalents of isopentenyl pyrophosphate (*7.36*) (Poulter and Rilling, 1981) (Fig. 7.18). The key carbon–carbon bond-forming reaction is the condensation of C-1 of the allylic pyrophosphate substrate with C-4 of the isopentenyl pyrophosphate partner. The resulting intermediate, geranyl pyrophosphate (*7.37*), is itself a substrate for prenyl transferase, undergoing condensation with the second equivalent of

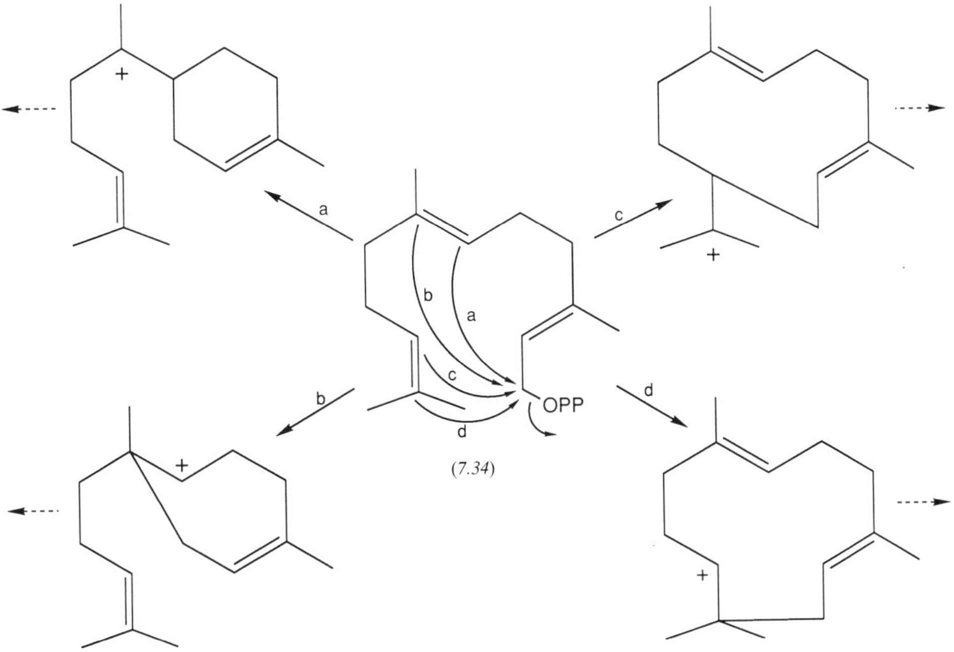

Fig. 7.17 Cyclization of farnesyl pyrophosphate (*7.34*) by intramolecular electrophilic attack on the central (pathways a and b) or distal (pathways c and d) double bonds.

isopentenyl pyrophosphate to generate farnesyl pyrophosphate. The condensation reaction has been shown to involve inversion of configuration at C-1 of the allylic substrate, a result originally interpreted to imply simultaneity in the bond-breaking and bond-forming events (Donninger and Popjak, 1966; Cornforth *et al.*, 1966a; Popjak and Cornforth, 1966). Detailed studies by Poulter and Rilling, however, have conclusively shown that the condensation is in fact a stepwise process, with ionization of the pyrophosphate and generation of the resulting allylic cation preceding electrophilic attack on the isopentenyl pyrophosphate double bond and loss of a proton from C-2 of the resulting intermediate (Poulter and Rilling, 1978; Poulter *et al.*, 1978, 1981; Mash *et al.*, 1981). The latter allylic addition–elimination reaction, corresponding to a formal S_E' process, is known to take place with net *syn* stereochemistry (Cornforth *et al.*, 1966b) and evidence has been adduced supporting a stepwise process which does not involve covalent interaction with a nucleophilic X-group at the enzyme active site (Poulter and Rilling, 1981). Prenyl transferases which have been purified from a variety of microbial, higher plant, and animal sources are homodimers of subunit molecular weight 38 000–43 000 (Poulter and

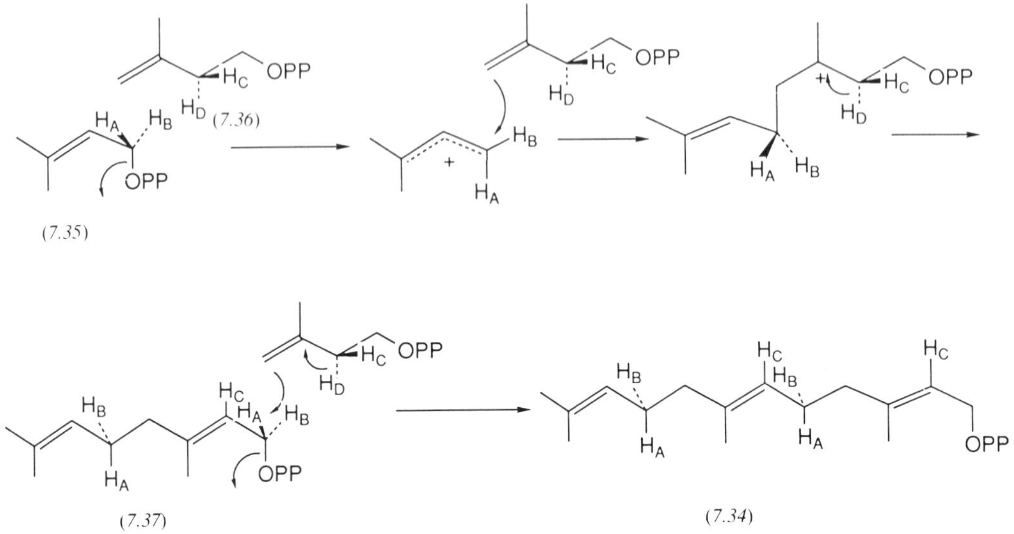

Fig. 7.18 Conversion of dimethylallyl pyrophosphate (7.35) and isopentenyl pyrophosphate (7.36) to farnesyl pyrophosphate (7.34) catalysed by prenyl transferase involving net inversion of configuration at C-1 of the allylic substrate and syn S_E' addition–elimination to IPP.

Rilling, 1981; Rilling, 1984; Barnard, 1984; Ogura *et al.*, 1984) The enzymes all require a divalent cation for activity, either Mg^{2+} or Mn^{2+}. The apparent K_m values for the substrates dimethylallyl pyrophosphate, geranyl pyrophosphate, and isopentenyl pyrophosphate are typically in the 0.5–$2 \mu M$ range. The cyclization of farnesyl pyrophosphate to generate sesquiterpenes can be thought of as the intramolecular analogue of the isoprenoid chain elongation reaction catalysed by prenyl transferase (Saito and Rilling, 1981; Davisson *et al.*, 1985). For this reason, the extensive studies of prenyl transferase have provided both a useful conceptual framework and an experimental model for studies of terpenoid cyclizations.

In our own laboratories, we have been investigating the biosynthesis of a number of terpenoid metabolites produced by a variety of bacterial, fungal, and higher plant systems. In many cases the early stages of this work have been carried out by feedings to intact organisms and analysis of the resulting labelled metabolites by traditional chemical degradative techniques or by application of modern spectroscopic methods including high-field ^{13}C and ^2H NMR. Throughout much of this work we have been concerned with the metabolism of allylic pyrophosphates. (Cane, 1980) Since 1980, our research in this area has concentrated on investigations of terpenoid cyclizations at the cell-free level, in some cases with enzymes already described in the recent literature, and in several others with enzyme preparations which we have developed in our own laboratories. In the early stages of our studies we have been interested in defining as closely as possible the characteristic changes in structure and bonding undergone by each individual substrate in the course of its enzyme-mediated transformation. More recently we have begun to turn our attention to these terpenoid synthases as protein or catalytic entities in their own right, with a view to understanding the relationship between structure and catalytic function. In the following sections are presented some of our own contributions to the understanding of sesquiterpene synthases.

7.3.1   Trichodiene Synthase

Trichodiene synthase, first described by Hanson (Evans and Hanson, 1976), catalyses the cyclization of farnesyl pyrophosphate to trichodiene (*7.38*), the parent hydrocarbon (Nozoe and Machida, 1972) of the trichothecane family of sesquiterpene antibiotics (Tamm and Bretenstein, 1980; Cane, 1981). Incubation of [1-^{3}H, 12, 13-^{14}C]farnesyl pyrophosphate (*7.34a*) with crude preparations of trichodiene synthase isolated from *Trichothecium roseum* and degradation of the resulting labelled trichodiene has established that the cyclization takes place without loss of either of the hydrogen atoms originally attached to C-1 of the allylic pyrophosphate precursor. (Cane *et al.*, 1981a; Cane *et al.*, 1985) To explain these results, we have proposed the cyclization mechanism, illustrated in Fig. 7.19, involving initial isomerization of *trans,trans*-farnesyl pyrophosphate to the tertiary allylic ester nerolidyl pyrophosphate (*7.39*). By rotation about the 2,3 single bond, the intermediate nerolidyl pyrophosphate can adopt a conformation capable of cyclization to the bisabolyl cation (*7.40*). This isomerization–cyclization sequence is necessary in order to overcome the geometric barrier to direct interaction of C-1 of the *trans,trans*-farnesyl pyrophosphate precursor with the C-6,7 double bond which would otherwise result in the formation of a prohibitively strained *trans*-cyclohexene intermediate. Further cyclization of the bisabolyl cation and appropriate rearrangments will then generate the eventual product trichodiene.

  To test this mechanistic scheme further, we have determined the stereochemical course of the cyclization by incubating 1*S*- and 1*R*-[1-^{3}H, 12,13-^{14}C]farnesyl pyrophosphate with trichodiene synthase in separate experiments and subjecting the derived samples of trichodiene to chemical degradation to determine the site and stereochemistry of labelling (Cane *et al.*, 1985). The presence of tritium at C-11 of trichodicne was established in both cases by base-catalysed exchange of the ketone (*7.41*) obtained by acid-catalysed rearrangment of 9,10-epoxytrichodiene (*7.42*) (Fig 7.20). Whereas *syn* elimination of the phenylselenide derivative (*7.43*) obtained

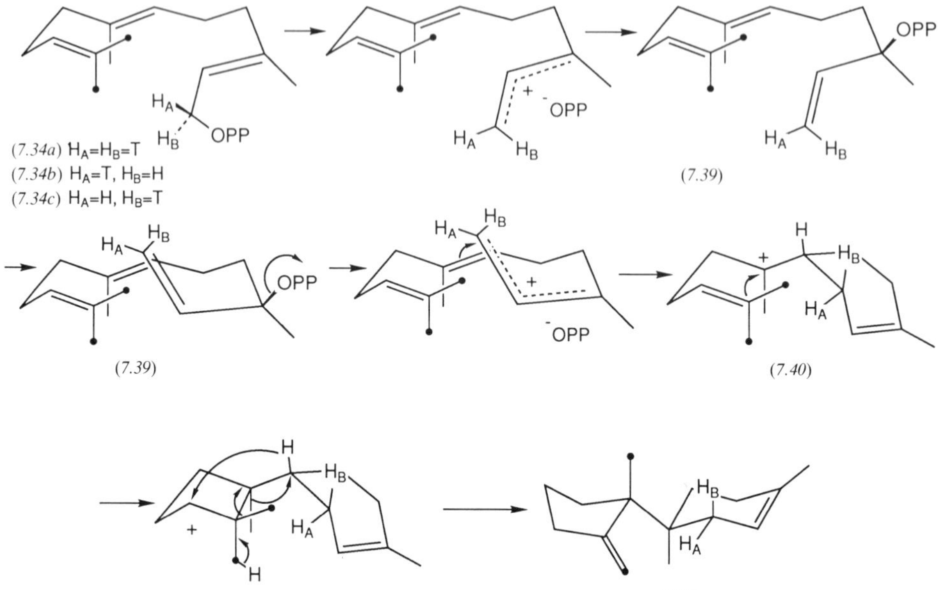

(7.34a) H$_A$=H$_B$=T
(7.34b) H$_A$=T, H$_B$=H
(7.34c) H$_A$=H, H$_B$=T

(7.39)

(7.39)

(7.40)

(7.38)

Fig. 7.19 Cyclization of [1-^{3}H, 12,13-^{14}C]farnesyl pyrophosphate (7.34a), and 1S- and 1R-[1-^{3}H, 12, 13-^{14}C]farnesyl pyrophospates (7.34b) and (7.34c) to trichodiene (7.38) by way of enzyme-bound (3R)-nerolidyl pyrophosphate (7.39).

from 1S-[1-^{3}H, 12,13-^{14}C]farnesyl pyrophosphate (7.34b), gave the allylic alcohol (7.44) which was devoid of tritium, the corresponding allylic alcohol derived from 1R-[1-^{3}H, 12,13-^{14}C] farnesyl pyrophosphate (7.34c) retained the full equivalent of tritium. These results establish that the cyclization of farnesyl pyrophosphate takes place with net *retention* of configuration at C-1 of the precursor, in contrast to the demonstrated *inversion* of configuration at this centre in the analogous intermolecular condensation reactions catalysed by prenyl transferase. This apparent contradiction is a consequence of the required isomerization of the *trans*-2,3-double bond of the farnesyl pyrophosphate substrate which must precede cyclization itself and is fully consistent with the proposed intermediacy of nerolidyl pyrophosphate.

In further experiments with trichodiene synthase, we have directly confirmed the intermediacy of nerolidyl pyrophosphate (7.39) in the enzymatic conversion of farnesyl pyrophosphate to trichodiene and established the full stereochemical course of this transformation (Cane and Ha, 1986, 1988). Incubation of (1Z)-[1-^{3}H, 12,13-^{14}C] nerolidyl pyrophosphate (7.39c) with trichodiene synthase gave trichodiene (7.38c) which was shown by the usual chemical degradation sequence to be labelled exclusively at

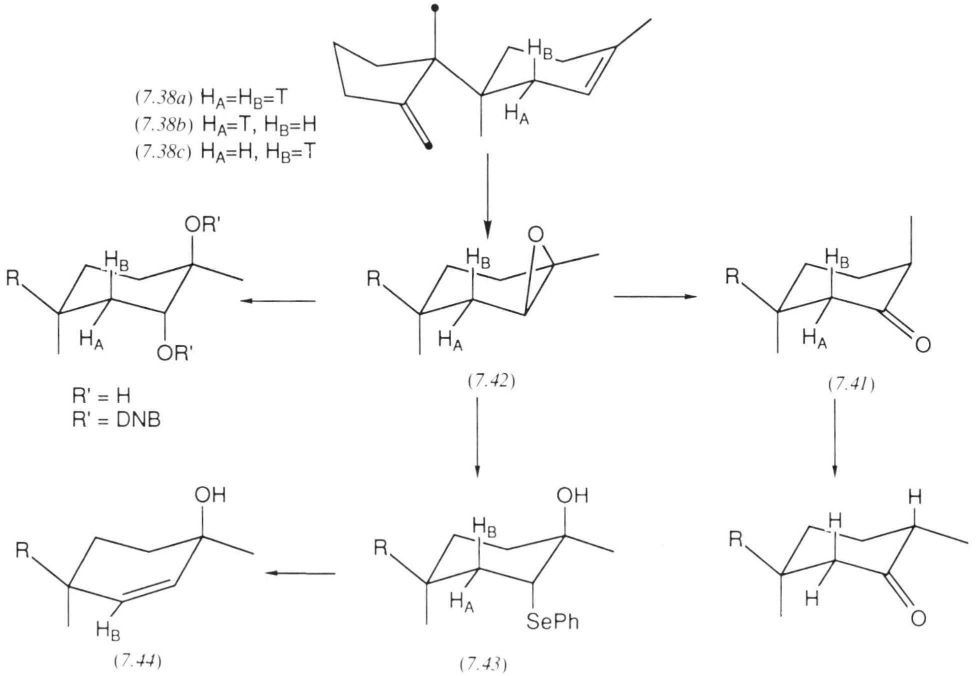

Fig. 7.20 Degradation of trichodiene (7.38) formed by trichodiene synthase-catalysed cyclization of stereospecifically tritiated samples of farnesyl pyrophosphate and nerolidyl pyrophosphate.

H-11β (Figs 7.19 and 7.20). By incubating a mixture of (3S)-[1-³H]nerolidyl pyrophosphate and (3RS)-[12,13-¹⁴C]nerolidyl pyrophosphate with trichodiene synthase, it was established that only the 3R enantiomer of the tertiary allylic pyrophosphate serves as a substrate for the cyclase, based on the absence of tritium in the derived [¹⁴C]trichodiene (Fig. 7.21) To demonstrate that nerolidyl pyrophosphate is a true intermediate in the cyclization, we carried out a competitive incubation of [1-³H]farnesyl pyrophosphate and [12,13-¹⁴C]nerolidyl pyrophosphate and examined the ³H/¹⁴C ratio of the resulting trichodiene and recovered farnesol and nerolidol as a function of time. The results of the latter experiment established that nerolidyl pyrophosphate is normally an enzyme-bound intermediate of the conversion of farnesyl pyrophosphate to trichodiene and indicated that the (3R)-enantiomer of (7.39) has a V_{max}/K_m approximately 1.5–2.0 times that of farnesyl pyrophosphate.

The above results are fully consistent with the mechanistic scheme illustrated in Fig. 7.19. Trichodiene synthase is therefore an isomerase–

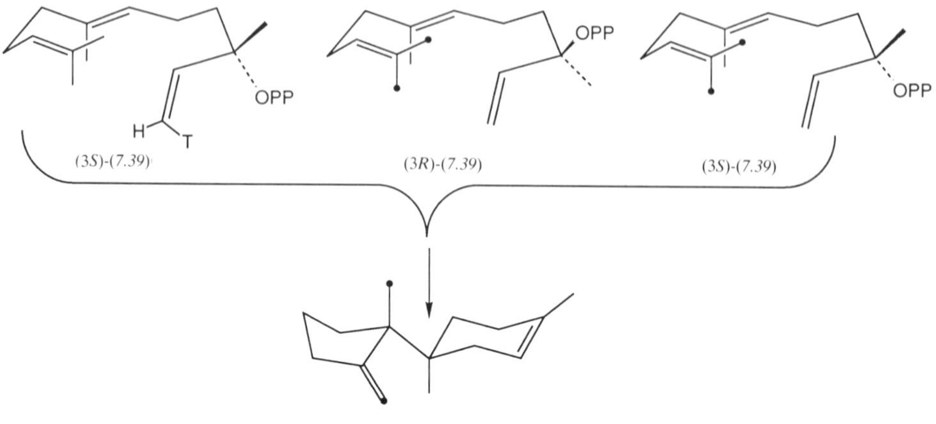

Fig. 7.21 Competitive incubation of (3S)-[1-³H]nerolidyl pyrophosphate and (3RS)-[12,13-¹⁴C]nerolidyl pyrophosphate (7.39) with trichodiene synthase with exclusive utilization of the 3R enantiomer.

cyclase, catalysing both the isomerization of the original *trans* double bond of its substrate as well as the subsequent conversion to cyclic product. The initial isomerization of farnesyl pyrophosphate to nerolidyl pyrophosphate takes place with net suprafacial stereochemistry, presumably by way of the corresponding transoid allylic cation–pyrophosphate anion pair. This conclusion is in accord with the results of earlier studies conducted in this laboratory on a farnesyl pyrophosphate–nerolidyl pyrophosphate isomerase isolated from *Gibberella fujikuroi* (Cane *et al.*, 1981c). Rotation about the 2,3-bond of the (3R)-nerolidyl pyrophosphate intermediate and subsequent ionization will generate the cisoid ion-pair which is immediately juxtaposed over the neighboring 6,7- double bond, allowing cyclization to the bisabolyl cation. Based on the demonstrated intermediacy of the 3R-enantiomer of (7.39) and the fact that cyclization takes place on the 1-*re* face of the vinyl double bond, generation of the bisabolyl cation must take place by way of the *anti–endo* conformation of nerolidyl pyrophosphate. These results are also consistent with a large body of data bearing on the enzymic cyclization of geranyl and linalyl pyrophosphate to monoterpenes (Croteau, 1987) as well as numerous studies of the solvolysis of geraniol, linalool, and nerol derivatives (Gotfredsen *et al.*, 1977; Poulter and King, 1982).

   The above studies of trichodiene biosynthesis were carried out with relatively crude preparations of trichodiene synthase. The cyclase itself has recently been purified to homogeneity from *Fusarium sporotrichioides* by Hohn and shown to be a homodimer of subunit molecular weight M_r 45 000. (Hohn and van Middlesworth, 1986) The observed K_m for farnesyl pyrophosphate is 25 nM and the only required cofactor is a divalent cation, with Mg^{2+} being preferred. With the availability of purified cyclase, in collaboration with Dr Hohn, we have been investigating the action of trichodiene synthase on a group of anomalous substrates in order to factor the normal enzymatic reaction sequence into its respective isomerase and cyclase components. Both (7R)- and (7S)-6,7-dihydrofarnesyl pyrophosphate (7.45) proved to be only modest competitive inhibitors of trichodiene synthase, with K_I values of 235 nM and 390 nM, respectively. These values can be compared with the K_I for inorganic pyrophosphate itself, 495 nM. While each of the dihydrofarnesyl pyrophosphate substrates is, in principle, capable of undergoing enzyme-catalysed isomerization to the corresponding enantiomer of 6,7-dihydronerolidyl pyrophosphate, further cyclization of the resulting intermediate is prevented by the absence of the 6,7- double bond. Indeed, incubation of either enantiomer of (7.45) with trichodiene synthase gave rise in each case to a mixture of acyclic products consisting of ca. 80–85% of the farnesene isomers, (7.46), (7.47), and (7.48), 10–15% of dihydronerolidol (7.49), and 5% trans-dihydrofarnesol (7.50). (Fig. 7.22) These products result from abortive ionization–isomerization–ionization of the anomalous dihydrofarnesyl pyrophosphate substrate and provide indirect evidence for the normally cryptic farnesyl to nerolidyl pyrophosphate isomerization. By contrast, 95% of the products of nonenzymatic solvolysis of (7.45) are allylic alcohols consisting largely of a 3:1 mixture of dihydronerolidol and trans-dihydrofarnesol. The increased proportion of olefinic products resulting from the enzymatic reaction is presumably due to the sequestering of the substrate in an active site which usually must rigorously exclude water to avoid premature quenching of the highly electrophilic intermediates of the normal isomerization–cyclization process. Completely consistent results have been reported by Croteau for analogous incubations of 6,7-dihydrogeranyl pyrophosphate with a mixture of monoterpene cyclases from sage (Salvia officinalis) (Wheeler and Croteau, 1986). The processing of the anomalous dihydrofarnesyl pyrophosphate substrates by trichodiene synthase takes place with a V_{max} ca.10% that for farnesyl pyrophosphate. Interestingly, examination of the polar products of the abortive cyclization led to recovery of a mixture of pyrophosphate esters containing as much as 25% of the cis-6,7-dihydrofarnesyl pyrophosphate isomer but none of the corresponding dihydronerolidyl pyrophosphate derivative.

In the meantime, Hohn has isolated the gene for trichodiene synthase by screening a λgt11 library of F. sporotrichioides DNA with antibodies to trichodiene synthase (Hohn and Beremand, 1989). The complete amino acid sequence of trichodiene synthase has been deduced from the corresponding nucleotide sequence which revealed the gene to consist of an 1182 nucleotide ORF containing an unusual, in frame, 60 nucleotide intron. Although plasmids containing the intron sequence did not give rise to

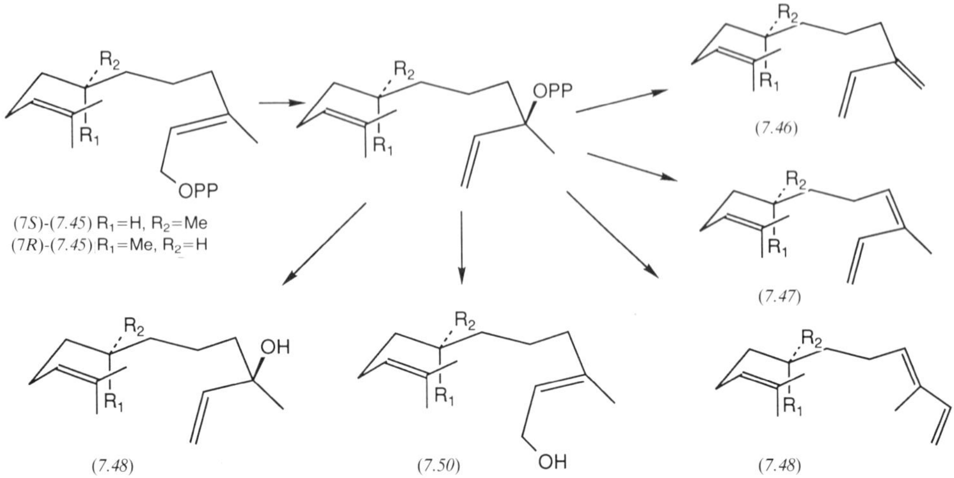

Fig. 7.22 Incubation of (7*R*)- and (7*S*)-6,7-dihydrofarnesyl pyrophosphates (*7.45*) with trichodiene synthase and formation of abortive products of the normal isomerization–cyclization sequence.

measurable cyclase activity, excision of the intron and subcloning has now resulted in low level expression of trichodiene synthase activity in *E. coli* (Hohn and Plattner, 1988). The cloning, sequencing, and expression of the trichodiene synthase gene represent a major milestone in the study of terpenoid biosynthesis. Although, not unexpectedly, the protein has no apparent homology to any other known sequence, the availability of recombinant enzyme opens the way for a host of studies, including the identification of the active site and the eventual determination of the three-dimensional structure of the synthase. Furthermore, since there are presumably some 200 distinct sesquiterpene synthases, each mediating the cyclization of the common substrate farnesyl pyrophosphate to a distinct product, it should soon be possible to make direct comparisons at the genetic level among this fascinating family of enzymes.

7.3.2 Pentalenene Synthase

In addition to cyclizations proceeding through a bisabolyl cation or its equivalent, a large and varied group of sesquiterpenes are formed by initial cyclization of C-1 of farnesyl pyrophosphate to the distal double bond with generation of 10- or 11-membered ring (pathways c and d, Fig. 7.17) (Cane, 1981). We have been interested in the biosynthesis of several of these compounds containing a dimethylcyclopentane ring and formally derivable from humulene. (Cane and Nachbar, 1978; Cane *et al.*, 1981b). In connec-

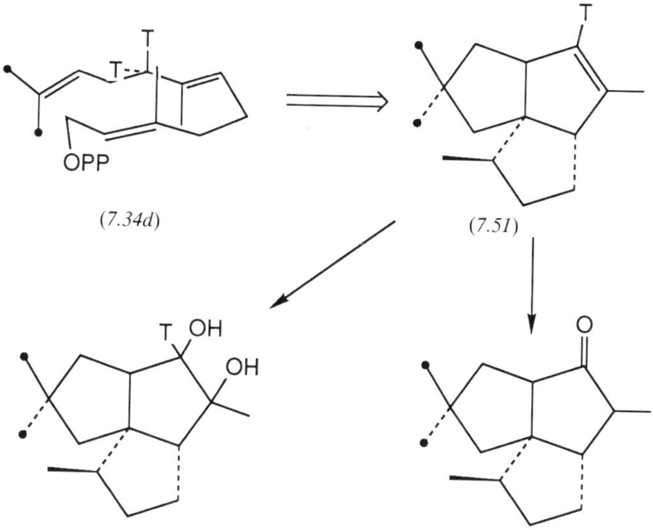

(7.34d) (7.51)

(7.52) α-*cis* diol
(7.53) β-*cis* diol
(7.54)

Fig. 7.23 Conversion of [8-^{3}H, 12,13-^{14}C]farnesyl pyrophosphate (*7.34d*) to pentalenene by pentalenene synthase and degradation of (*7.51*) to locate the tritium label.

tion with these interests, we have been examining the mechanism of the enzymatic conversion of farnesyl pyrophosphate to pentalenene (*7.51*), (Seto and Yonehara, 1980) the parent hydrocarbon of the pentalenolactone family of sesquiterpene antibiotics (Cane and Tillman, 1983). Incubation of [8-^{3}H, 12,13-^{14}C]farnesyl pyrophosphate (*7.34d*) with crude pentalenene synthase obtained from *Streptomyces* UC5319 gave pentalenene which retained 50% of the original tritium activity, as inferred from the ^{3}H/^{14}C ratios of the derived crystalline diols (*7.52*) and (*7.53*) (Fig. 7.23) Hydroboration–oxidation of labelled pentalenene gave the ketone (*7.54*), which was devoid of tritium, thereby locating the tritium in (*7.51*) at C-7.

The observed labelling results can be explained by the pathway illustrated in Fig. 7.24 in which farnesyl pyrophosphate is first converted to humulene (*7.55*). Reprotonation at C-10 initiates cyclization of humulene to the protoilludyl cation (*7.56*), which can undergo a hydride shift and further cyclization to pentalenene with concomitant loss of one of the protons originally at C-8 of farnesyl pyrophosphate. In this scheme, the conformation of the presumed humulene intermediate is deduced from the known relative and absolute configuration of the derived pentalenene, while the absolute sense of folding of the farnesyl pyrophosphate precursor has been

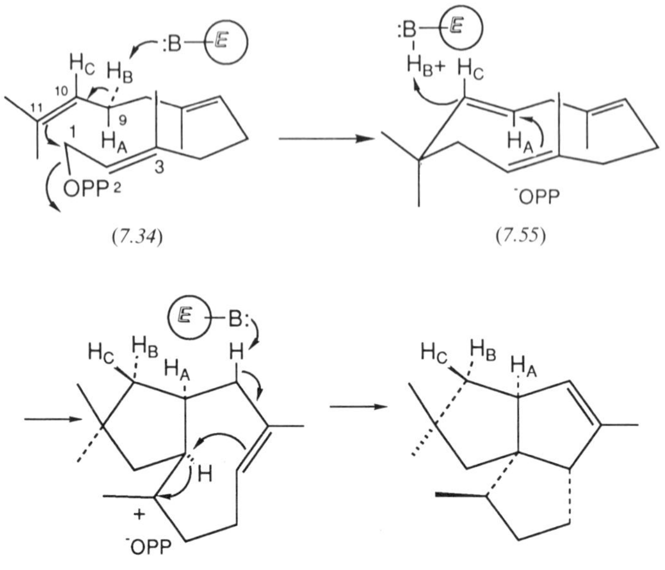

Fig. 7.24 Mechanism and stereochemistry of the cyclization of farnesyl pyrophosphate (7.34) to pentalenene (7.51).

inferred from the results of whole cell feeding experiments using ^{13}C-labelled precursors (Cane et al., 1981b).

If both the formation and further cyclization of humulene were to take place at the same active site, there is a possiblity that the proton abstracted from C-9 of farnesyl pyrophosphate in the course of the initial cyclization step could be redonated to C-10 of the transiently generated humulene provided that exchange with the external medium be slow with respect to reprotonation. The postulated proton return was in fact demonstrated by incubation of [9-^{3}H, 12,13-^{14}C]farnesyl pyrophosphate (7.34e) with penta-lenene synthase (Cane et al., 1984) (Figs 7.24 and 7.25). The resulting pentalenene retained the bulk of the original tritium, based on the ^{3}H/^{14}C ratio of the derived diols, (7.52) and (7.53). Base-catalysed exchange of the corresponding pentalen-7-one (7.54) resulted in loss of half the tritium label, establishing the presence of one equivalent of tritium at the bridgehead, C-8, as expected. The remainder of the tritium was located at C-1 by feeding enzymatically labelled pentalenene to a culture of Streptomyces UC5319. Whereas the derived epipentalenolactone F methyl ester (7.57) had the same ^{3}H/^{14}C ratio as the pentalenene precursor, half the tritium label was lost from the corresponding sample of pentalenic acid methyl ester (7.58). The demonstrated internal return of one of the original H-9 protons of

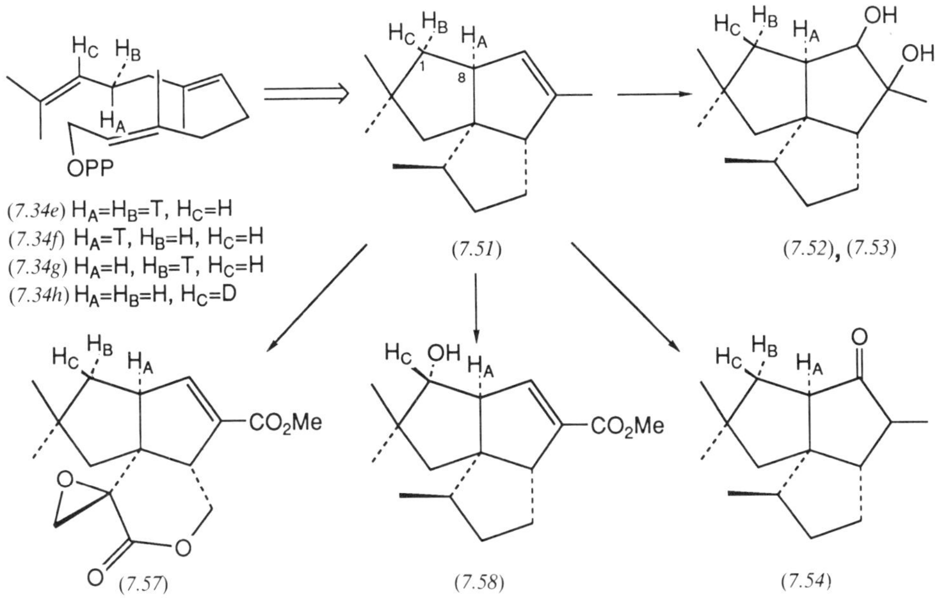
OPP
(7.34e) $H_A=H_B=T$, $H_C=H$
(7.34f) $H_A=T$, $H_B=H$, $H_C=H$
(7.34g) $H_A=H$, $H_B=T$, $H_C=H$
(7.34h) $H_A=H_B=H$, $H_C=D$
(7.51)
(7.52), (7.53)
OH
OH
CO2Me
CO2Me
(7.57)
(7.58)
(7.54)

Fig. 7.25 Chemical and microbial degradation of pentalenene (*7.51*) derived from incubation of stereospecifically tritiated or deuteriated farnesyl pyrophosphates (*7.34e–h*) with pentalenene synthase.

farnesyl pyrophosphate establishes that the formation of pentalenene is catalysed by a single enzyme. Moreover, the efficiency of proton return suggests that whatever base is responsible for the abstraction of the original H-9 proton must carry no additional hydrogen atoms.

The stereochemical course of the inital ring-forming reaction was established by incubating (9*R*)- and (9*S*)-[9-^{3}H, 4,8-^{14}C]farnesyl pyrophosphate (*7.34f*) and (*7.34g*) with pentalenene synthase in separate experiments and carrying out the analogous set of chemical and microbial degradations (Cane *et al.*, 1988b) (Figs 7.24 and 7.25). In this manner it was shown that H-9*re* of farnesyl pyrophosphate corresponds to H-8 of pentalenene, while H-9*si* undergoes an intramolecular transfer to C-1 of (*7.51*). Since the formation of the postulated humulene intermediate must involve electrophilic attack on the *si* face of the 10, 11- double bond of farnesyl pyrophosphate (Cane *et al.*, 1981b), the removal of the H-9*si* proton implies that the formal S_E' reaction takes place with net *anti* stereochemistry, in contrast to the previously demonstrated *syn* stereochemistry of the corresponding intermolecular allylic addition–elimination catalysed by prenyl transferase. These results indicate that there is no stereochemical imperative for enzymatic S_E' transformations and support the proposed RSR-CT conformation (Cane *et al.*, 1981b; Cane *et al.*, 1984; Cane, 1985) of the farnesyl pyrophosphate substrate which prevents access of the enzymatic base to H-9*re*.

Further confirmation of the stereochemical model has come from incubation of [10-^{2}H]farnesyl pyrophosphate (*7.34h*) with pentalenene synthase. Analysis of the resulting pentalenene by ^{2}H NMR established that the deuterium label resided exclusively in the H-1β position, implying that reprotonation of the intermediate humulene must occur on the 10-*re* face of the 9,10- double bond, as predicted. (Fig. 7.24). Taken together with the above tritium labelling studies, the latter result also establishes that introduction of the hydroxyl oxygen atom of pentalenic acid occurs with the usually observed retention of configuration (Hayaishi, 1962; Gautier, 1980), a conclusion confirmed by incubation of [10-^{3}H, 12,13-^{14}C]farnesyl pyrophosphate with pentalenene synthase and conversion of the resulting labelled pentalenene to [1-^{3}H, 12,13-^{14}C]pentalenic acid (Fig. 7.25).

Using ^{2}H NMR we have also established the stereochemical course of the displacement of pyrophosphate from C-1 of farnesyl pyrophosphate catalysed by pentalenene synthase (Harrison *et al.*, 1988). Cyclization of (1*R*)-[1-^{2}H]farnesyl pyrophosphate (*7.34i*) resulted in the formation of pentalenene labelled exclusively at H-3*re*, whereas incubation of (1*S*)-[1-^{2}H]farnesyl pyrophosphate (*7.34j*) gave pentalenene with deuterium at H-3*si*, implying that the condensation reaction has taken place with net *inversion* of configuration at C-1 of farnesyl pyrophosphate. (Fig. 7.26). In contrast to the reaction catalysed by trichodiene synthase, which results in net *retention* at this centre, the pentalenene synthase reaction does not require isomerization of the 2,3- double bond of the substrate and is therefore expected to involve inversion at C-1, analogous to the prototypical condensation catalysed by prenyl transferase.

Taken together, the experiments described above have provided a detailed picture of the stereochemical course of the enzymatic conversion of farnesyl pyrophosphate to pentalenene. We have previously speculated that the same base which mediates the transfer of the original H-9*si* proton of

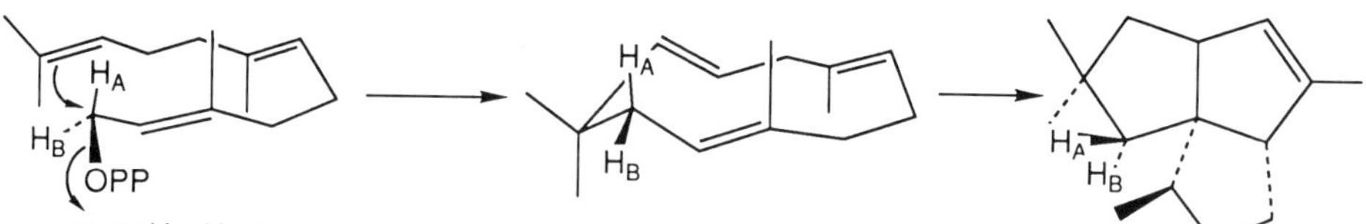

(*7.34i*) H$_A$=H, H$_B$=D
(*7.34j*) H$_A$=D, H$_B$=H   (*7.55*)   (*7.51*)

Fig. 7.26 Inversion of configuration at C-1 of farnesyl pyrophosphate (*7.34*) in the formation of pentalenene (*7.51*).

farnesyl pyrophosphate may also be responsible for eventual deprotonation from C-7 which generates the ultimate product of the cyclization, pentalenene (Cane *et al.*, 1981b; Cane *et al.*, 1984). During the course of the cyclization, the pyrophosphate ion can in principle stabilize positive charge at sites corresponding to C-1, C-2, or C-3 of (*7.34*). Little is known concerning the role of the pyrophosphate ion itself once the original farnesyl pyrophosphate substrate has undergone ionization. In collaboration with Croteau, we have shown that bornyl pyrophophate synthases from both sage and tansy (*Tanacetum vulgare*) exercise a remarkably tight restriction on the internal motion of the pyrophosphate moiety during the course of the cyclization of geranyl pyrophosphate (Croteau *et al.*, 1985). It is appealing to speculate that terpene synthases may mediate their characteristic cyclization reactions by taking advantage of ion-pairing between inorganic pyrophosphate and transiently generated cationic intermediates to steer the course of the various rearrangement reactions.

In parallel with the above mechanistic studies, we have also undertaken a programme to purify and characterize pentalenene synthase itself (Cane and Pargellis, 1987). The molecular weight of the partially purified synthase was estimated by gel filtration to be 57 000, while a slightly lower estimate of 45 000 was obtained from SDS polyacrylamide gel electrophoresis (PAGE), suggesting that the native enzyme is a monomer. The protein itself is notably lipophilic, which may account for the divergent estimates of the molecular weight. The K_m for farnesyl pyrophosphate is $0.8 \mu M$ and the enzyme has been shown to prefer Mg^{2+} as the only required cofactor. Whereas inorganic pyrophosphate alone is only a modest competitive inhibitor ($K_I > 30 \mu M$) and $10 \mu M$ pentalenene itself has little effect on the rate of cyclization, the combination of the two products at $10 \mu M$ each resulted in a seven-fold increase in the apparent K_M for farnesyl pyrophosphate, suggesting that both products can bind co-operatively at the active site to inhibit pentalenene synthase˙ competitively. We have recently purified pentalenene synthase to homogeneity on a small scale and have initiated experiments directed at the isolation and cloning of the relevant structural gene by screening of an EMBL4 genomic library of *Streptomyces* UC5319 DNA with oligonucleotide probes based on the amino acid sequence of tryptic peptides obtained from the purified synthase.

7.3.3 Additional Sesquiterpene Synthases

Although trichodiene synthase and pentalenene synthase are by far the best studied sesquiterpene cyclases, several other members of this class of enzymes have recently been described. We have reported the isolation of β-*trans*-bergamotene synthase from *Pseudeurotium ovalis* (Cane *et al.*, 1989a) and established that cyclization of farnesyl pyrophosphate to (*7.59*)

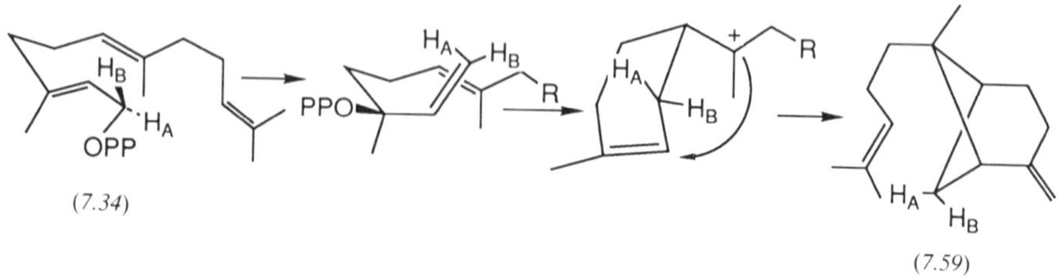

(7.34)

(7.59)

Fig. 7.27 Cyclization of farnesyl pyrophosphate catalysed by β-*trans*-bergamotene (7.59) synthase with retention of configuration at C-1 of (7.34).

takes place with retention of configuration at C-1 (Fig. 7.27). Aristolochene (7.60) synthase (Fig. 7.28) has been isolated from *Aspergillus terreus* (Cane *et al.*, 1989b), and Hohn has purified this enzyme to homogeneity from *Penicillium roqueforti* and shown that the protein is a monomer of M_r 37 000 based on SDS-PAGE (Hohn and Plattner, 1989b). The K_m for farnesyl pyrophosphate was 0.5 μM and the calculated turnover number was ca. $0.04\,s^{-1}$. The latter rate may be compared to turnover values of $0.14\,s^{-1}$ and $0.1\,s^{-1}$ for trichodiene synthase (Hohn and van Middlesworth, 1986) and prenyl transferase (Laskovics and Poulter, 1981), respectively. A synthase mediating the cyclization of farnesyl pyrophosphate to 4-*epi*-aristolochene (7.61), the precursor of the phytoalexin capsidiol, has recently been isolated from tissue cultures of *Nicotiana tabacum* treated with elicitor (Whitehead *et al.*, 1988, 1989; Threllfall and Whitehead, 1988; Chappell and Nable, 1987). Finally, Croteau has reported the isolation and separation of a pair of sesquiterpene cyclases, humulene (7.55) synthase and caryophyllene (7.62) synthase (Fig. 7.29). (Croteau and Gundy, 1984; Dehal and Croteau, 1988).

7.4 PROBLEMS AND PROSPECTS

Within the last decade, the study of biosynthetic transformations at the cell-free level has transformed the field of natural products biosynthesis

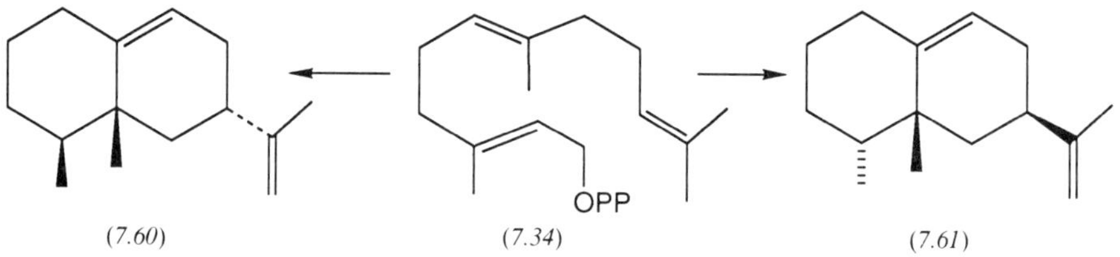

(7.60)   (7.34)   (7.61)

Fig. 7.28 Cyclization of farnesyl pyrophosphate (7.34) to aristolochene (7.60) and 5-*epi*-aristolochene (7.61).

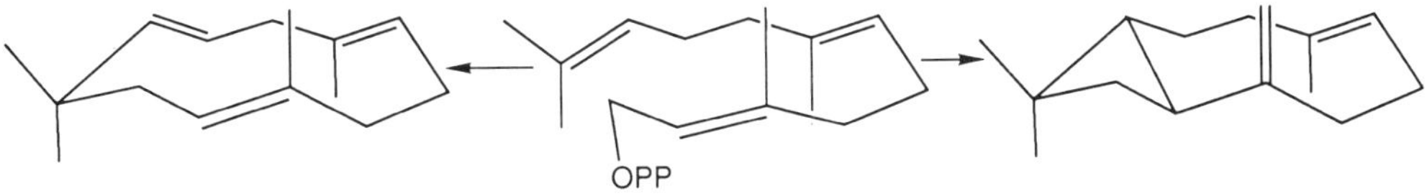

Fig. 7.29 Cyclization of farnesyl pyrophosphate (*7.34*) to humulene (*7.55*) and caryophyllene (*7.62*).

from a largely indirect, highly inferential science, to a vigorous branch of modern mechanistic enzymology. The synthetic virtuosity of biosynthetic enzymes has long been appreciated by natural products chemists; only recently has it been possible to translate that appreciation into understanding of the catalysts themselves. The combination of powerful enzymological and molecular genetic tools with high field NMR spectroscopy and organic synthesis has begun to unlock the secrets of some of Nature's most complex metabolic pathways. Nevertheless, several, largely technical, obstacles still must be overcome if progress in this emerging area is to be maintained. In spite of considerable effort, the detailed steps of many biosynthetic pathways are still largely undefined. As a consequence, studies of the enzymology of new biosynthetic transformations may often be hindered by the fact that neither the substrate nor the product of a given enzyme-catalysed reaction is known for certain. In some cases crude cell-free preparations may support the entire sequence of transformations linking the primary precursors to a fully mature natural product. In such instances it may be possible to study shorter and shorter segments of a pathway, leading to eventual recognition of the individual enzyme-catalysed steps. When the substrate, but not the product, of an enzyme-catalysed reaction is known, it may be necessary to obtain authentic samples of potential products, from either natural or synthetic sources, in order to allow identification and structure determination of the actual enzymatic product. The extremely low titres of many biosynthetic enzymes make such structure determinations extremely labour intensive, while limiting the quantities of enzyme available for kinetic studies or purification experiments. Although recombinant DNA methods can ultimately overcome problems of enzyme supply, in most cases the protein of interest must be purified at least once in order to allow isolation of the gene by oligonucleotide hybridization or antibody screening methods.

As techniques for enzyme isolation and purification continue to be improved, many of the present obstacles to further study of biosynthetic enzymes will be overcome. The use of molecular genetics for the study of biosynthesis of natural products is still in its infancy. As transformation systems for a wider range of higher plants and fungi become available, the isolation and expression of biosynthetic genes will become more routine. By using site-directed mutagenesis, it should be possible not only to identify the key catalytic groups at the active site of biosynthetic enzymes, but to modify the catalysts themselves to change their substrate specificity or the actual product produced. Studies of the regulation of biosynthetic genes should lead not only to a better understanding of the molecular biology of differentiation, but is expected to have important practical consequences in the control and enchancement of the production of antibiotics and other commercially important fermentation products. Finally, comparisons of DNA sequence information for families of biosynthetic genes should lead to a better understanding of the relationship between protein structure and function while giving new insights into the evolution of biosynthetic enzymes themselves.

REFERENCES

Aberhart, J. (1977) *Tetrahedron*, **33**, 1545.
Aberhart, D. J. and Lin, L. J. (1974) *J. Chem. Soc., Perkin Trans. I*, 2320.
Alefounder, P. R., Abell, C. and Battersby, A. R. (1988) *Nucl. Acids Res.*, **20**, 9871.
Alefounder, P. R., Hart, G. J., Miller, A. D., Beifuss, E., Abell, C., Leeper, F. J. and Battersby, A. R. (1989) *Bioorganic Chem.*, **17**, 121.
Arnstadt, K.-I., Schindbleck, G., and Lynen, F. (1975) *Eur. J. Biochem.*, **55**, 561.
Arnstein, H. R. V. and Morris, D. (1960) *Biochem. J.*, **76**, 357.
Baldwin, J. E. and Abraham, E. P. (1988) *Natural Prod. Rep.*, **5**, 129.
Baldwin, J. E., Abraham, E. P., Adlington, R. M., Chakravarti, B., Derome, A. E., Murphy, J. A., Field, L. E., Green, N. B., Ting, H.-H., and Usher, J. J. (1983a) *J. Chem. Soc., Chem. Commun.*, 1317.
Baldwin, J. E., Abraham, E. P., Adlington, R. M., Murphy, J. A., Green, N. B., Ting, H.-H., and Usher, J. J. (1983b) *J. Chem. Soc., Chem. Commun.*, 1319.
Baldwin, J. E., Adlington, R. M., Moroney, S. E., Field, L. D., and Ting, H.-H. (1984) *J. Chem. Soc., Chem. Commun.*, 984.
Baldwin, J. E., Adlington, R. M., Domayne-Hayman, B. P., Ting, H.-H., and Turner, N. J. (1986) *J. Chem. Soc., Chem. Commun.*, 110.
Baldwin, J. E., Killin, S. J., Pratt, A. J., Sutherland, J. D., Turner, N. J., Crabbe, J. C., Abraham, E. P., and Willis, A. C. (1987) *J. Antibiot.*, **40**, 652.
Baltz, R. H. and Seno, E. T. (1981) *Antimicrob. Agents Chemother.*, **20**, 214.
Baltz, R. H., Seno, E. T., Stonesifer, J., Matsushima, P. and Wild, G. M. (1981) *Microbiol.*, 371.
Barnard, G. F. (1984) in *Methods in Enzymology (Steroids and Isoprenoids)*, Vol. 110, (eds J. H. Law and H. C. Rilling), Academic Press, New York, pp. 155–167.
Battersby, A. R. and McDonald, E. (1979) *Acc. Chem. Res.*, **12**, 14.
Battersby, A. R., Fookes, C. J. R., McDonald, E. and Meegan, M. J. (1978) *J. Chem. Soc., Chem. Commun.*, 185.
Battersby, A. R., Brereton, R. G., Fookes, C. J. R., McDonald, E. and Matcham, G. W. J. (1980a) *J. Chem. Soc., Chem. Commun.*, 1124.

Battersby, A. R., Fookes, C. J. R., Matcham, G. W. J. and McDonald, E. (1980b) *Nature*, **285**, 17.

Battersby, A. R., Fookes, C. J. R., Gustafson-Potter, K. E., McDonald, E., and Matcham, G. W. J. (1982) *J. Chem. Soc., Perkin Trans. I*, 2428.

Battersby, A. R., Fookes, C. J. R., Matcham, G. W. J., McDonald, E., and Hollenstein, R. (1983) *J. Chem. Soc., Perkin Trans. I*, 3031.

Bloch, K. (1952) *Harvey Lectures*, **48**, 68.

Bloch, K. and Vance, D. (1977) *Ann. Rev. Biochem.*, **46**, 263.

Brown, S. A. (1972) *Biosynthesis*, **1**, 1.

Bowman, E., McQueney, M., Barry, R. J. and Dunaway-Mariano, D. (1988) *J. Am. Chem. Soc.*, **110**, 5575.

Cane, D. E. and Nachbar, R. B. (1978) *J. Am. Chem. Soc.*, **100**, 3208. (corrn **101**, 1980).

Cane, D. E. (1980) *Tetrahedron*, **36**, 1109.

Cane, D. E. (1981) in *Biosynthesis of Isoprenoid Compounds*, Vol. 1, (eds, J. W. Porter and S. L. Spurgeon), John Wiley, New York, pp. 283–374.

Cane, D. E. (1985) *Acc. Chem. Res.*, **18**, 220.

Cane, D. E. and Ha, H.-J. (1986) *J. Am. Chem. Soc.*, **108**, 3097.

Cane, D. E. and Ha, H.-J. (1988) *J. Am. Chem. Soc.*, **110**, 6865.

Cane, D. E. and Ott, W. R. (1988) *J. Am. Chem. Soc.*, **110**, 4840.

Cane, D. E. and Pargellis, C. (1987) *Arch. Biochem. Biophys.*, **254**, 421.

Cane, D. E. and Tillman, A. M. (1983) *J. Am. Chem. Soc.*, **105**, 122.

Cane, D. E. and Yang, C.-C. (1987) *J. Am. Chem. Soc.*, **109**, 1255.

Cane, D. E., Swanson, S. and Murthy, P. P. N. (1981a) *J. Am. Chem. Soc.*, **103**, 2136.

Cane, D. E., Rossi, T., Tillman, A. M., and Pachlatko, J. P. (1981b) *J. Am. Chem. Soc.*, **103**, 1838.

Cane, D. E., Iyengar, R. and Shiao, M.-S. (1981c) *J. Am. Chem. Soc.*, **103**, 914.

Cane, D. E., Liang, T.-C., and Hasler, H. (1982) *J. Am. Chem. Soc.*, **104**, 7274.

Cane, D. E., Hasler, H., Taylor, P. B., and Liang, T.-C. (1983) *Tetrahedron*, **39**, 3449.

Cane, D. E., Abell, C. and Tillman, A. M. (1984) *Bioorganic Chem.*, **12**, 312.

Cane, D. E., Ha, H.-J., Pargellis, C., Waldmeier, F., Swanson, S. and Murthy, P. P. N. (1985) *Bioorganic Chem.*, **13**, 246.

Cane, D. E., Liang, T.-C., Taylor, P. B., Chang, C. and Yang, C.-C. (1986) *J. Am. Chem. Soc.*, **108**, 4957.

Cane, D. E., Yang, C.-C. and Ott, W. R. (1988a) in *Biology of Actinomycetes '88* (eds Y. Okami, T. Beppu and H. Ogawara), Japan Scientific Societies, Tokyo, pp. 395–400.

Cane, D. E., Abell, C., Lattman, R., Kane, C. T., Hubbard, B. R. and Harrison, P. H. R. (1988b) *J. Am. Chem. Soc.*, **110**, 4081.

Cane, D. E., McIlwaine, D. B. and Harrison, P. H. M. (1989a) *J. Am. Chem. Soc.*, **111**, 1152.

Cane, D. E., Rawlings, B. J., Noguchi, H., Salaski, E. J. and Prabhakaran, P. C. (1989b) *J. Am. Chem. Soc.*, **111**, 8914.

Carr, L. G., Skatrud, P. L., Scheetz, M. E., Queener, S. W. and Ingolia, T. D. (1986) *Gene*, **48**, 257.

Chappell, J. and Nable, R. (1987) *Plant Physiol.*, **85**, 469.

Chayken, S., Law, J., Phillips, A. H., Tchen, T. T., and Bloch, K. (1958) *Proc. Nat. Acad. Sci., USA*, **44**, 998.

Chen, C. W., Lin, H.-F., Kuo, C. L., Tsai, H.-L. and Tsai, J. F.-Y. (1988) *Bio/Technology*, **6**, 1222.

Clayton, R. B. (1965) *Quart. Rev.*, **19**, 168.

Corcoran, J. W. (1981) in *Antibiotics IV. Biosynthesis* (ed. J. W. Corcoran), Springer-Verlag, Berlin, pp. 132–174.

Corcoran, J. W., Chick, M. and Darby, F. J. (1967) in *Proc. 5th Int. Congr. Chemother., Abstracts of Communications*, Wien, p. 35.

Cornforth, J. W. (1973) *Chem. Soc. Rev.*, **2**, 1.

Cornforth, J. W., Cornforth, R. H., Donninger, C. and Popjak, G. (1966a) *Proc. Roy. Soc. London, Ser. B.*, **163**, 492.

Cornforth, J. W., Cornforth, R. H., Popjak, G. and Yengoyan, L. (1966b) *J. Biol. Chem.*, **241**, 3970.

Croteau, R. (1981) in *Biosynthesis of Isoprenoid Compounds*, Vol. 1, (eds J. W. Porter and S. L. Spurgeon), John Wiley, New York, pp. 225–282.

Croteau, R. (1987) *Chem. Rev.*, **87**, 929.

Croteau, R. and Gundy, A. (1984) *Arch. Biochem. Biophys.*, **233**, 838.

Croteau, R. B., Shaskus, J. J., Renstrøm, B., Felton, N. M., Cane, D. E., Saito, A. and Chang, C. (1985) *Biochemistry*, **24**, 7077.

Davisson, V. J., Neal, T. R. and Poulter, C. D. (1985) *J. Am. Chem. Soc.*, **107**, 5277.

Dehal, S. S. and Croteau, R. (1988) *Arch. Biochem. Biophys.*, **261**, 346.

Dimroth, P., Ringlemann, E. and Lynen, F. (1976) *Eur. J. Biochem.*, **68**, 581.

Donadio, S., Tuan, J. S., Staver, M. J., Weber, J. M., Paulus, T. J. Maine, G. T., Leung, J. O., DeWitt, J. P., Vara, J. A, Wang, Y.-G., Hutchinson, C. R., and Katz, L. (1988) in *Genetics and Molecular Biology of Industrial Microorganisms*, American Society Microbiology, Washington, D.C., pp. 53–60.

Donninger, C. and Popjak, G. (1966) *Proc. Roy. Soc. London, Ser. B.*, **163**, 465.

Elson, S. W., Baggaley, K. H., Gillet, J., Holland, S., Nicholson, N. H., Sime, J. T., and Woroniecki, S. R. (1987) *J. Chem. Soc., Chem. Commun.*, 1736.

Evans, R., and Hanson, J. R. (1976) *J. Chem. Soc., Perkin Trans. 1*, 326.

Evans, J. N. S., Fagerness, P. E., Mackenzie, N. E. and Scott, I. A. (1984) *J. Am. Chem. Soc.*, **106**, 5738.

Fawcett, P. A., Usher, J. J., Huddleston, J. J. and Abraham, E. P. (1976) *Biochem. J.*, **157**, 651.

Fishman, S. E., Cox, K., Larson, J. L., Reynolds, P. A., Seno, E. T., Yeh, W.-K., Van Frank, R., and Hershberger, C. L. (1987) *Proc. Natl. Acad. Sci. USA*, **84**, 8248.

Gautier, A. E. (1980) *Sterischer Vearlauf Einiger Chemischer and Biochemischer Reactionen an Substraten mit Chiraler Methylgruppe*, Dissertation ETH (Zurich), No. 6583.

Gotfredsen, S., Obrecht, J. P. and Arigoni, D. (1977) *Chimia*, **31**, (2), 62.

Hale, R. S., Jordan, K. N. and Leadlay, P. F. (1987) *FEBS Lett.*, **224**, 133.

Hara, O., Anzai, H., Imai, S., Kumada, Y., Murakami, T., Itoh, R., Takano, E., Satoh, A. and Nagoka, K. (1988) *J. Antibiot.*, **41**, 538.

Harrison, P. H. M., Oliver, J. S. and Cane, D. E. (1988) *J. Am. Chem. Soc.*, **110**, 5922.

Hart, G. J., Miller, A. D., Leeper, F. J. and Battersby, A. R. (1987) *J. Chem. Soc., Chem. Commun.*, 1762.

Hayaishi, O. (1962) *Oxygenases*, Academic Press, New York.

Hendrickson, J. B. (1959) *Tetrahedron*, **7**, 82.

Herbert, R. B. (1981) *The Biosynthesis of Secondary Metabolites*, Chapman and Hall, London, pp. 28–49.

Hohn, T. M. (1988) personal communcation.

Hohn, T. M. and Beremand, P. D. (1989) *Gene*, **79**, 131.

Hohn, T. M. and Plattner, R. D. (1989a) *Arch. Biochem. Biophys.*, **275**, 92.

Hohn, T. M. and Plattner, R. D. (1986b) *Arch. Biochem. Biophys.*, **272**, 137.

Hohn, T. M. and van Middlesworth, F. (1986) *Arch. Biochem. Biophys.*, **251**, 756.

Horiguchi, M. (1972) *Biochim. Biophys. Acta*, **261**, 102.

Hunaiati, A. A. and Kolattukudy, P. E. (1984) *Antimicrob. Agents Chemother.*, **25**, 173.

Hutchinson, C. R. (1986) *Natural Prod. Rep.*, **3**, 133.

Hutchinson, C. R., Sherman, M. M., Vederas, J. C. and Nakashima, T. T. (1981) *J. Am. Chem. Soc.*, **103**, 5953.

Hutchinson, C. R., Bibb, M. J., Biro, S., Collins, J. F., Motamedi, H., Shafiee, A. and Punekar, N. (1988) in *Biology of Actinomycetes '88* (eds Y. Okami, T. Beppu and H. Ogawara), Japan Scientific Societies, Tokyo, pp. 76–81.

Imai, S., Seto, H., Sasaki, T., Tsuruoka, T., Ogawa, H., Satoh, A., Inouye, S., Niida, T. and Otake, N. (1984) *J. Antibiot.*, **37**, 1505.

Imai, S., Seto, H., Sasaki, T., Tsuruoka, T., Ogawa, H., Satoh, A., Inouye, S., Niida, T. and Otake, N. (1985) *J. Antibiot.*, **38**, 687.

Jensen, S. E., Westlake, D. W. S. and Wolfe, S. (1985) *J. Antibiot.*, **38**, 263.

Jordan, P. M., Mgbeje, B. I. A., Alwan, A. F. and Thomas, S. D. (1987) *Nucl. Acids Res.*, **15**, 10583.

Jordan, P. M., Mgbeje, B. I. A., Thomas, S. D. and Alwan, A. F. (1988b) *Biochem. J.*, **249**, 613.

Jordan, P. M., Warren, M. J., Williams, H. J., Stolowich, N. J., Roessner, C. A., Grant, S. K. and Scott, A. I. (1988a) *FEBS Lett.*, **235**, 189.

Kinoshita, K., Takenaka, S. and Hayashi, M. (1988) *J. Chem. Soc., Chem. Commun.*, 943.

Kluender, H., Bradley, C. H., Sih, C. J., Fawcett, P. and Abraham, E. P. (1973) *J. Am. Chem. Soc.*, **95**, 6149.

Laskovics, F. M. and Poulter, C. D. (1981) *Biochemistry*, **20**, 1893.

Leeper, F. J. (1985) *Natural Prod. Rep.*, **2**, 19.

Leskiw, B. K., Aharonowitz, Y., Mevarech, M., Wolfe, S., Vining, L. C., Westlake, D. W. S. and Jensen, S. E. (1988) *Gene*, **62**, 187.

Lynen, F., Eggerer, H., Henning, U. and Kessel, I. (1958) *Angew. Chem.*, **70**, 738.

Lynen, F., Agranoff, B. W., Eggerer, H., Henning, U. and Moeslein, E. M. (1959) *Angew. Chem.*, **71**, 657.

Malpartida, F. and Hopwood, D. A. (1984) *Nature (London)*, **309**, 462.

Malpartida, F. and Hopwood, D. A. (1986) *Mol. Gen. Genet.*, **205**, 66.

Malpartida, F., Hallam, S. E., Kieser, H. M., Motamedi, H., Hutchinson, C. R., Butler, M. J., Sugden, D. A., Warren, M., McKillop, C., Bailey, C. R., Humphreys, G. O. and Hopwood, D. A. (1987) *Nature (London)*, **325**, 818.

Mann, J. (1978) *Secondary Metabolism*, Oxford University Press, Oxford, pp. 21–77.

Mash, E. A., Gurria, G. M. and Poulter, C. D. (1981) *J. Am. Chem. Soc.*, **103**, 3927.

Miller, A. D., Hart, G. J., Packman, L. C. and Battersley, A. R. (1988) *Biochem J.*, **254**, 915.

Monroney, S. E. (1985) D. Phil. Thesis, Oxford University.

Murakami, T., Anzai, H., Imai, S., Satoh, A., Nagaoka, K. and Thompson, C. J. (1986) *Mol. Gen. Genet.*, **205**, 42.

Neuss, N , Nash, C. H., Baldwin, J. E., Lemke, P. A. and Grutzner, J. B. (1973) *J. Am. Chem. Soc.*, **95**, 3797.

Nozoe, S. and Machida, Y. (1972) *Tetrahedron*, **28**, 5105.

Ogawa, H., Imai, S., Shimizu, T., Satoh, A. and Kojima, M. (1983) *J. Antibiot.*, **36**, 1040.

Ogura, K., Nishino, T., Shinka, T. and Seto, S. (1984) in *Methods in Enzymology (Steroids and Isoprenoids)*, Vol. 110, (eds J. H. Law and H. C. Rilling), Academic Press, New York, pp.167–171.

O'Hagan, D., Robinson, J. A. and Turner, D. L. (1983) *J. Chem. Soc., Chem. Commun.*, 1337.

Omura, S. Sadakane, N., Kitao, C., Matsubara, H. and Nakagawa, A. (1980) *J. Antibiot.*, **33**, 913.

Pang, C.-P., Chakravarti, B., Adlington, R. M., Ting, H.-H., White, R. L., Jayatilake, G. S., Baldwin, J. E. and Abraham, E. P. (1984) *Biochem. J.*, **222**, 789.

Parker, W., Roberts, J. S. and Ramage, R. (1967) *Quart. Rev.*, **21**, 331.

Popjak, G. and Cornforth, J. W. (1960) *Adv. Enzymol. Relat. Subj. Biochem.*, **22**, 281.

Popjak, G. and Cornforth, J. W. (1966) *Biochem. J.*, **101**, 553.

Poulter, C. D. and Rilling, H. C. (1978) *Acc. Chem. Res.*, **11**, 307.

Poulter, C. D. and Rilling, H. C. (1981) in *Biosynthesis of Isoprenoid Compounds*, Vol. 1, (eds J. W. Porter and S. L. Spurgeon), John Wiley, New York, pp. 161–224.

Poulter, C. D. and King, C.-H., R. (1982) *J. Am. Chem. Soc.*, **104**, 1420, 1422.

Poulter, C. D., Argyle, J. C. and Mash, E. A. (1978) *J. Biol. Chem.*, **253**, 7227.

Poulter, C. D., Wiggins, P. L. and Le, A. T. (1981) *J. Am. Chem. Soc.*, **103**, 3926.

Ramos, F. R., López-Nieto and Martin, J. F. (1985) *Antimicrob. Agents Chemother.*, **27**, 380.

Ray, W. J., Jr. (1983) *Biochemistry*, **22**, 4625.

Rilling, H. C. (1984) in *Methods in Enzmology (Steroids and Isoprenoids)*, Vol. 110, (eds J. H. Law and H. C. Rilling), Academic Press, New York, pp. 145–152.

Roberts, G. and Leadlay, P. F. (1983) *FEBS Lett.*, **159**, 13.

Roberts, G. and Leadlay, P. F. (1984) *Biochem. Soc. Trans.*, **12**, 642.

Rogers, T. O. and Birnbaum, J. (1974) *Antimicrob. Agents Chemother.*, **5**, 121.

Ruzicka, L. (1959) *Proc. Chem. Soc. (Lond.)*, 341.

Ruzicka, L. (1963) *Pure Appl. Chem.*, **6**, 482.

Ruzicka, L. and Stoll, M. (1922) *Helv. Chim. Acta*, **5**, 923.

Saito, A. and Rilling, H. C. (1981) *Arch. Biochem. Biophys.*, **208**, 508.

Samson, S. M., Belagaje, R., Blakenship, D. T., Chapman, J. L., Perry, D., Skatrud, P. L., Van Frank, R. M., Abraham, E. P., Baldwin, J. E., Queener, S. W. and Ingolia, T. D. (1985) *Nature (London)*, **318**, 191.

Samson, S. M., Chapman, J. L., Belagje, R., Queener, S. W. and Ingolia, T. D. (1987a) *Proc. Natl. Acad. Sci. USA*, **84**, 5705.

Samson, S. M., Dotzlaf, J. E., Slisz, M. L., Becker, G. W., Van Frank, R. M., Veal, L. E., Yeh, W.-K., Miller, J. R., Queener, S. W. and Ingolia, T. D. (1987b) *Bio/Technology*, **5**, 1207.

Sasarman, A., Nepveu, A., Echelard, Y., Dymetryszyn, J., Drolet, M. and Goyer, C. (1987) *J. Bacteriol.*, **169**, 4257.

Scheidegger, A., Küenzi, M. T. and Nüesch, J. (1984) *J. Antibiot.*, **37**, 522.

Scott, A. I., Beadling, L. C., Georgopapadakou, N. M. and Subbarayan, C. R. (1974) *Bioorganic Chem.*, **3**, 238.

Scott, A. I., Burton, G., Jordan, P. M., Matsumoto, H., Fagerness, P. E. and Pryde, L. M. (1980) *J. Chem. Soc., Chem. Commun.*, 384.

Scott, A. I., Stolowich, N. J., Williams, H. J., Gonzalez, M. D., Roessner, C. A., Grant, S. K. and Pichon, C. (1988a) *J. Am. Chem. Soc.*, **110**, 5898.

Scott, A. I., Roessner, C. A., Stolowich, N. J., Karuso, P., Williams, H. J., Grant, S. K., Gonzalez and Hoshino, T. (1988b) *Biochemistry*, **27**, 7984.

Sedgwick, B. and Cornforth, J. W. (1977) *Eur. J. Biochem.*, **75**, 465.

Sedgwick, B., Cornforth, J. W., French, S. J., Gray, R. T., Kelstrup, E. and Willadsen, P. (1977) *Eur. J. Biochem.*, **75**, 481.

Sedgwick, B., Morris, C. and French, S. J. (1978) *J. Chem. Soc., Chem. Commun.*, 93.

Seno, E. T. and Baltz, R. H. (1981) *Antimicrob. Agents Chemother.*, **20**, 370.

Seno, E. T. and Baltz, R. H. (1982) *Antimicrob. Agents Chemother.*, **21**, 758.

Seto, H. and Yonehara (1980) *J. Antibiot.*, **33**, 92.

Seto, H., Imai, S., Tsuruoka, T., Satoh, A. and Kojima, M. (1982) *J. Antibiot.*, **35**, 1719.

Seto, H., Sasaki, T., Imai, S., Tsuruoka, T., Ogawa, H., Satoh, A., Inouye, S., Niida, T. and Otake, N. (1983a) *J. Antibiot.*, **36**, 96.

Seto, H., Imai, S., Tsuruoka, T., Ogawa, H., Satoh, A., Sasaki, T. and Otake, N. (1983b) *Biochem. Biophys. Res. Commun.*, **111**, 1008.

Seto, H., Imai, S., Sasaki, T., Shiotohno, K., Tsuruoka, T., Ogawa, H., Satoh, A., Inouye, S., Niida, T. and Otake, N. (1984) *J. Antibiot.*, 37, 1509.

Shafiee, A. and Hutchinson, C. R. (1987) *Biochemistry*, **26**, 6204.

Sherman, M. M. and Hutchinson, C. R. (1987) *Biochemistry*, **26**, 438.

Shimotohno, K. W., Seto, H., Otake, N., Imai, S. and Murakami, T. (1988) *J. Antibiot.*, **41**, 1057.

Sood, G. R., Robinson, J. A. and Ajaz, A. A. (1984) *J. Chem. Soc., Chem. Commun.*, 1421.

Stanzak, R., Matsushima, P., Baltz, R. H. and Rao, R. N. (1986) *Bio/Technology*, **4**, 229.

Stark, W. M., Hart, G. J. and Battersby, A. R. (1986) *J. Chem. Soc., Chem. Commun.*, 465.

Tamm, Ch. and Breitensein, W. (1980) in *The Biosynthesis of Mycotoxins*, (ed. P. S. Steyn), Academic Press, New York, pp. 69–104.

Thomas, S. D. and Jordan, P. M. (1986) *Nucl. Acids Rec.*, **14**, 6215.

Threllfall, D. R. and Whitehead, I. M. (1988) *Phytochemistry*, **27**, 2567.

Trebst, A. and Gseike, F. (1967) *Z. Naturforsch., B: Anorg. Chem., Org. Chem., Biochem., Biophys., Biol.*, **22**, 989.

Vederas, J. (1987) *Natural Prod. Rep.*, **4**, 277.

Volpe, J. J. and Vagelos, P. R. (1973) *Ann. Rev. Biochem.*, **42**, 21.

Wakil, S. J. and Stoops, J. K. (1983) in *The Enzymes*, Vol. XVI, (ed. P. D. Boyer), Academic Press, New York, pp. 3–61.

Wallach, O. (1887) *Justus Liebig's Ann. Chem.*, **239**, 1.

Warren, S. and Williams, M. R. (1971) *J. Chem. Soc. (B)*, 618.

Wawszkiewicz, E. J. and Lynen, F. (1964) *Biochem. Z.*, **340**, 213.

Wheeler, C. J. and Croteau, R. (1986) *Arch. Biochem. Biophys.*, **246**, 733.

Whitehead, I. M., Threllfall, D. R. and Ewing, D. F. (1988) *Phytochemistry*, **27**, 1365.

Whitehead, I. M., Ewing, D. F. and Threllfall, D. R. (1989) *Phytochemistry*, **27**, 775.

Wilson, B. A., Bantia, S., Salituro, G. M., Holbrooks, A. McE. and Townsend, C. A. (1988) *J. Am. Chem. Soc.*, **110**, 8238.

Wolfe, S., Westlake, D. W. S. and Jensen, S. E. (1985) U. S. Pat. 4,510,246.

Yue, S., Duncan, J. S., Yamamoto, Y. and Hutchinson, C. R. (1987) *J. Am. Chem. Soc.*, **109**, 1253.

8 | Enzymes in the food industry

David R. Berry and Alistair Paterson

8.1 INTRODUCTION

The use of enzymes such as rennet in cheese making and barley amylases in brewing is as old as the food and beverage industry itself. However, the production of the amylase represents the first example of the industrial production of an enzyme for use in the food industry. The quantity and variety of enzymes used in the food and beverage industry has increased dramatically in the past decade. (Godfrey and Reichelt, 1983; Peppler and Reed, 1987). More than 20 different enzymes are regularly used in the food and beverage industry at the present time. Some of those are shown in Table 8.1. The conversion of starch to different commercial products, such as glucose and maltose syrups and dextrin preparations, represents the major application of food enzymes (Fig. 8.1) having a market value in excess of $100 million in 1988 (West, 1988). The use of enzymes in the dairy industry for cheese making and for the production of low lactose dairy products is the second largest market at $65 million.

Since it would not be feasible to describe all of these enzymes and their applications in one chapter, a number of enzymes are described that are particularly important in the processing of plant material, including cereals; the applications of these enzymes are illustrated in some of these processes, and in the cheese industries. Where possible an attempt has been made to indicate the relationship between the properties of the enzymes and their substrates and the mechanisms by which the substrates are attacked.

Table 8.1 Selected enzymes in use in the food industry

Enzyme	Major source	Enzyme classification	Applications
Amylase	*Bacillus subtilis* *Aspergillus* spp. Barley	3.2.1.1	Glucose syrup production brewing, fruit processing baking
Cellulase	*Aspergillus niger* *Trichoderma reesei*	3.2.1.4	Fruit processing, flavour production
Dextranase	*Penicillium* spp. *Chaetomium* spp.	3.2.11	Sugar processing
Glucanase	*Bacillus subtilis* *Penicillium emersonii*	3.2.1.6	Brewing, wheat processing
Glucose isomerase	*Bacillus coagulans* *Streptomyces albus* *Arthrobacter*	5.3.1.5	Production of high fructose syrups from starch hydrolysates
Glucose oxidase	*Aspergillus niger*	1.1.3.4	Removal of oxygen from food products and beverages
Lactase (β-galactosidase)	*Aspergillus* spp. *Bacillus* spp.	3.1.1.12	Milk processing
Lipase	*Aspergillus niger* *Candida* spp. *Mucor* spp. *Rhizopus* spp.	3.3.3	Flavour production, enzyme modified cheese, fat modification, emulsifier synthesis
Pectinase	*Aspergillus* spp.	3.2.1.15	Fruit processing
Protease	Animal e.g. trypsin *Aspergillus* spp. *Bacillus* spp. Plant e.g. papain	3.4.21.4 3.4.23.6 3.4.21.14 3.4.22.2	Baking, brewing, protein hydrolysates, flavour production
Xylanase	*Aspergillus* spp. *Trichoderma* spp.	3.2.1.32	Wheat processing baking

However, because of the complexity of many of the food substrates and the lack of purity of many commercial enzyme preparations, a detailed understanding of the mechanism of the enzymic reaction is frequently not available. In view of the economic importance of this group of enzymes, this is surprising and regrettable. However, for the food industry, the prime goal is to obtain an enzyme which offers a high performance for the applications in question at an economic price. In the first part of this chapter (section 8.2 to 8.4), the characteristics of several important classes of enzyme are discussed and the second part (sections 8.5 to 8.7) describes their application to the processing of cereals and to cheese making.

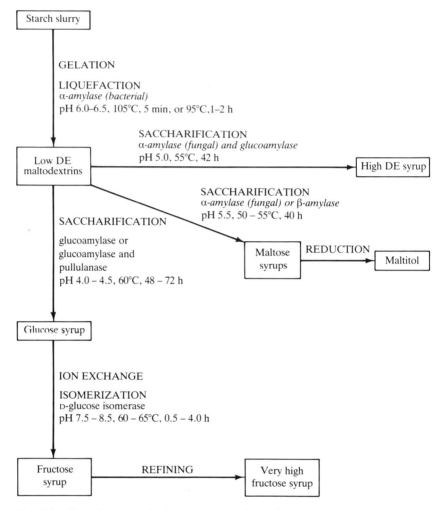

Fig. 8.1 Use of enzymes in the processing of starch to syrups.

8.1.1 The human diet and enzymes

Much of the current human diet consists of plant material, either in the form of processed seeds or raw or processed vegetal tissue. Furthermore, the meat we eat is largely taken from herbivores that also extract both energy and essential nutrients from ingested plant material. Consequently a major area of importance to the food enzymologist is the breakdown of plant cell walls and the hydrolysis of plant storage polymers with conversion to low molecular weight products that can be taken up through the gut wall or used as a food ingredient.

Traditionally, food processing has been carried out by a mixture of mechanical and thermal treatments supplemented with crude natural enzymes. These include the enzymes of malt, in which the production of cereal enzymes necessary for hydrolysis of grain storage polymers is induced to release energy for germination and synthesis of the tissues of the seedling. Other enzymes include chymosin, the active part of rennin, which cleaves milk caseins in the stomach of the new-born calf to convert a liquid to a semi-solid food, and proteases and lipases secreted by fungi which assist in the development of texture and flavour in cheeses.

Enzymes are today of greater importance in the food and feed industries than they have been previously because the consumer is concerned with food quality and with the option to choose foods perceived as being 'natural', since supply exceeds demand in Western Europe and North America. Manufacturers are now also sufficiently technically aware and educated to consider new processing options. Consequently, process steps that were previously thermal and mechanical, with degradations of flavour and food properties, are now being replaced with enzymatic treatments. A further area of application is in the treatment of animal feeds, including the direct feeding of enzymes to animals as probiotics, which is likely to be a developing market as the agricultural market learns to appreciate the potential benefits.

8.2 AMYLASES AND STARCH HYDROLYSIS

Starch is the most abundant form of carbohydrate in use in the food industry (Kennedy *et al.*, 1988). Although it can be consumed and digested without prior processing, in many food processes, including the traditional brewing processes, it is hydrolysed to syrups which contain glucose, maltose and dextrins in different quantities. These may be subsequently modified using glucose isomerase to produce, for example, high fructose corn syrups which are sweeter. The use of enzymes to produce starch hydrolysates is largely replacing acid hydrolysis because the degree of hydrolysis can be more precisely controlled. Also, by using different enzymes the composition of the product can be manipulated (Fig. 8.1). Since the conditions of hydrolysis are also milder, a cleaner product is obtained which does not require further treatment to remove colour and ash (Norman, 1981).

The study of enzymes which break down starch, the amylases, has been restricted to some extent by the variability of the substrate starch. Starch is described as 'a heterogeneous polysaccharide composed of two high molecular weight entities called amylose and amylopectin' (Fogarty and Kelly, 1979). Amylose is predominantly composed of α-1,4-linked D-glucose; however the action of debranching enzymes indicates that a few α-1,6-linked branch points do occur (Kjolberg and Manners, 1963). In amylopectin the frequency of α-1,6 linkages is much higher, in the proportion of 1 to 15 α-1,4 linkages. Typically starch contains 75% amylopectin and 25% amylose; however the amylose can vary from less than 5% in waxy maize to 28% in sorghum and sago (Fogarty and Kelly, 1979). It is not surprising therefore, that the mode of action of amylases is dependent upon the origin of the starch and the degree to which it has been gelatinized. In its native state in plants starch exists in the form of insoluble granules but if an aqueous suspension of starch is heated above 60°C then the hydrogen bonds which are involved in maintaining the starch structure are weakened, the granules swell, and the starch is dispersed into solution. This gelatinization process is a prerequisite of most enzymic amylolytic processes.

A great diversity of amylolytic enzymes has evolved in different groups of micro-organisms. These can be divided into three classes; *endo*-amylases which hydrolyse α-1,4 bonds throughout the starch molecule, e.g. α-amylases; exoamylases which remove glucose or maltose from the non-reducing ends of the starch molecule by cleaving 1,4-glycosidic bonds, e.g. glucoamylases and β-amylases; and debranching enzymes which cleave α-1,6-bonds, e.g. amylo-1,6-glucosidase (Fig. 8.2).

8.2.1 α-Amylases

α-Amylases (EC 3.2.1.1) are extensively used in industrial processes because in addition to breaking down the starch into a mixture of glucose, maltose, oligosaccharides and dextrin their mode of action results in a rapid reduction in the viscosity of the gelatinized starch, a process referred to as liquefaction. Since the gelatinization process involves heating the starch suspension to temperatures usually *c.* 105–110°C, then it is advantageous if thermostable amylases are available so that the process of liquefaction can be initiated before the gelatinized starch cools down and solidifies. The traditional source of α-amylase in the food and beverage industry was barley; however barley α-amylase is not stable above 70°C. Several industrial processes have been described in the past decade for producing corn syrups from maize (*Zea mays*) starch using bacterial α-amylases which are stable up to 100°C.

The properties of the most extensively used amylases, those from *Bacillus amyloliquefaciens* and from *Bacillus licheniformis*, have been discussed by Norman (1981). Whereas the *B. amyloliquefaciens* enzyme has a temperature optimum of 70°C, that of *B. licheniformis* is 92°C. Under industrial conditions of high levels of starch (30–40%) and optimum pH and calcium concentrations, operating temperatures of up to 110°C can be used for short periods (of the order of one hour) before the enzyme becomes inactivated. *B. licheniformis* α-amylase has a broader pH optimum (pH 5.0–8.0) than the

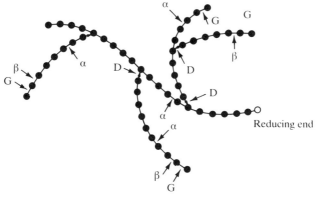

Fig. 8.2 Site of action of different amylase enzymes on starch; α = α-amylase, β = β-amylase, G = glucoamylase, D = α-1,6-amyloglucosidase or dextrinase.

B. amyloliquefaciens enzyme optimum (pH 5.9). Both enzymes are dependent upon Ca^{2+} for stability but the level which gives maximum activity with the *B. licheniformis* enzyme (3.4 ppm) is much lower than that required by the *B. amyloliquefaciens* enzyme (c. 150 ppm).

The mode of action of α-amylases is dependent upon the nature of the substrate as well as on the individual enzyme being used. Some 30 microbial α-amylases have been described, many of them with different structures and properties. The nature of the product not only varies with the enzyme used but with the length of incubation. However, the spectrum of oligomers shown in Fig. 8.3 is typical. The enzyme is always *endo*-acting and can cleave both amylose and amylopectin. It is not limited by branch points although the rate of cleavage of 1,4 bonds close to branch points may be reduced. Different models have been presented for the mode of action of α-amylases (Thoma, 1976) which include (1) random attack, e.g. any α-1,4 bond, (2) preferred attack, e.g. close to a branch, (3) multiple attack and at several points on the amylase chain, (4) multiple site attack at different distances from branch points. A variety of different amylases have been reported in the microbial kingdom. One from *Streptomyces hygroscopicus* has been reported to hydrolyse maltotriose by a remarkably circuitous mechanism which involves either the condensation of two triose molecules to form a maltohexose which is then cleaved to form three maltose units, or a transglycosylation reaction to produce maltose and maltotetraose which is subsequently hydrolysed (Hideka and Adachi, 1980).

Most α-amylases have a molecular weight of *ca* 50 000. Several have been studied in detail and some information on the structure of the enzyme is becoming available (MacGregor, 1988). The amino acid sequences have been determined for α-amylases from several organisms (*B. amylolique-*

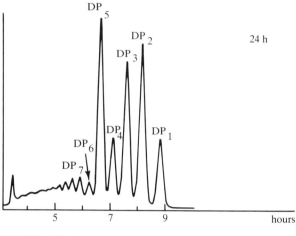

(a) *B.licheniformis*

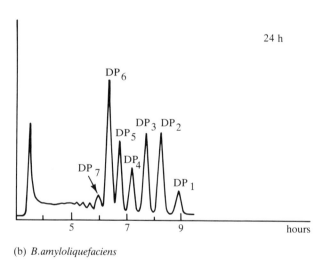

(b) *B.amyloliquefaciens*

Fig. 8.3   Size distribution of products formed by the action of different α-amylases on starch, obtained using liquid chromatography.

faciens, B. subtilis, Aspergillus oryzae, Saccharomycopsis fibuligera, Strep. hygroscopicus, and barley) and secondary structures which account for the differences in specificity of different α-amylases have been proposed. It is postulated that in addition to the main catalytic site, α-amylases contain several subsites which are important in binding the substrate to the enzyme. Although all α-amylases appear to have the same subsites close to the active

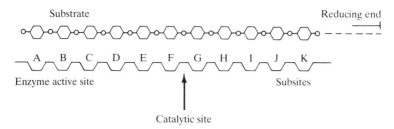

Substrate   Reducing end

A   B   C   D   E   F   G   H   I   J   K

Enzyme active site   Subsites

Catalytic site

Fig. 8.4 Diagrammatic representation of interaction between α-amylase and starch. A–K represent different subsites of the enzyme each of which binds to one glucose unit. The catalytic site is considered to be located between adjacent subsites and to cleave the glycosidic bond above it (from MacGregor, 1987).

site, (D–H in Fig. 8.4) α-amylases with different substrate affinities appear to have different peripheral subsites, (A–C and I–K in Fig. 8.4). In terms of protein structure, α-amylases are considered to have eight parallel strands and eight helical regions joined by loops which are considered to be important in forming the subsites of the active centre (Fig. 8.5).

The effect of pH on activity suggests that carboxyl and imidazole residues are involved in the active site of α-amylase but that sulphydryl groups are not involved (Fogarty and Kelly, 1979).

The genes for several *Bacillus* α-amylases have been cloned (Meade, *et al.*, 1987) and the complete nucleotide sequences of the α-amylase genes of *B. stearothermophilus*, *B. amyloliquefaciens* and *B. subtilis* have been determined. The genes of *B. amyloliquefaciens* and *B. stearothermophilus* show some degree of homology (67%). However the homology between these and the *B. subtilis* gene is only 30%, indicating a much reduced homology between these enzymes. Genes for barley and wheat α-amylase have also been cloned. Using a cDNA probe, two different α-amylase genes were identified in barley which had only a 77% homology (Rogers and Milliman, 1983; Chandler *et al.*, 1984) so clearly do not code for identical enzymes. These may correspond to the genes for two of the four isozymes for α-amylase which have been reported in barley.

8.2.2 β-Amylases

These are enzymes which hydrolyse α-1,4 glycosidic bonds but cannot bypass α-1,6 linkages. The most well known of these is the barley β-amylase which is active during the malting and mashing stages of the brewing process. The enzyme cleaves alternate glycosidic bonds starting at the non-reducing end and releasing maltose. This enzyme is readily inactivated at temperatures above 65°C and has a pH optimum of 5.2.

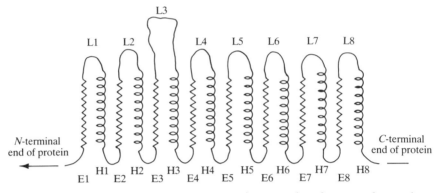

Fig. 8.5 Schematic representation of the major secondary features of α-amylase. E1–E8 represent parallel β strands; H1–H8 represent helices of the βα-structure; L1–L8 represent loops which join the C-terminal end of the β-strand to the N-terminal end of the adjacent helix which are believed to be important in forming the subsites of the active site (from MacGregor, 1987).

The demand for high maltose syrups for the confectionery, soft drinks and brewing industries (Kennedy *et al.*, 1988) as well as a source of sugar for diabetics (Yuen, 1974) has increased in recent years. It is produced by treating liquefied starch with β-amylases of either plant or microbial origin. At the present time Japan is the major producer of high maltose syrups and powder, manufacturing over 20 000 tons per year using a process based on a *Pseudomonas* isoamylase and a β-amylase from soybean. Crystalline maltose which can be obtained from high maltose syrups can be hydrogenated to give maltitol and isomaltitol or isomerized to give maltilose. Maltitol has a sweetness equivalent to 90% that of sucrose but is not absorbed from the gut so it makes an ideal diabetic sweetener (Saha and Zeikus, 1987).

The kinetics of maltose production from soluble starch using a β-amylase and a debranching enzyme have been investigated by Siraishi *et al.* (1987). The studies indicated that the rate of maltose formation was related to the rate at which linear portions of substrate were produced by the action of the debranching enzymes. β-Amylases have been isolated from several microorganisms and the enzymes from *Bacillus cereus* (Takasaki and Yamanobe, 1981) and *Bacillus polymyxa* (Griffin and Fogarty, 1973) have been investigated. Microbial β-amylases appear to have a similar mechanism of action to that of plant β-amylases. They have a similar *exo*-mechanism of attack on periodate treated amylose, in that they remove maltose units from the non-reducing end of the amylose, and are inhibited by reagents such as *p*-chloromercuribenzoate which inhibit sulphydryl reactions and by cyclodextrins, (Schardinger dextrins). The effect of sulphydryl reagents indicates that there are cysteine residues at the active site of the molecule which are involved in maintaining the correct conformation of the enzyme. The molecular weights of the β-amylases of *B. cereus* and *B. polymyxa* (35 000 (Takasaki and Yamanobe, 1981) and 59 000 (Fogarty and Griffin, 1979) respectively) are much smaller than those of plant enzymes which can be as high as 152 000. The *B. cereus* enzyme has a pH optimum of 7.0 and a temperature optimum of 50°C, which makes it more useful for food processing than the plant enzymes which have a pH optimum *c.* 5.0.

In addition to maltohydrolases, exoamylases have been described which produce maltotetraose (Fogarty and Kelly, 1979) and maltohexaose (Kainuma *et al.*, 1975) as the major products of hydrolysis. These enzymes may find application in the food industry for producing pure oligosaccharides.

8.2.3 Glucoamylases

Glucoamylases or amyloglucosidases (EC 3.2.1.3) are *exo*-acting enzymes which hydrolyse primarily α-1,4 linkages from the non-reducing end of amylose and related compounds, releasing glucose. However, glucoamylases are also capable of hydrolysing α-1,6 bonds at a slow rate so it is possible to achieve almost 100% conversion of starch to glucose using a combination of α-amylase and glucoamylase. This combination of enzymes is widely employed for the production of glucose syrups or liquid dextrose which is used in soft drinks, caramel, baking and brewing. A process has been developed in which immobilized glucoamylase was used to convert liquefied starch into high glucose syrup. This process gave shorter reaction times but the yield of glucose was 2–3% lower than that of the non-immobilized system, sufficient to make it uneconomic (Norman, 1981).

Glucoamylases have been isolated from over 20 species of fungi and those from *Aspergillus* spp. and *Rhizopus* have been studied in some detail. The rate of hydrolysis of α-1,4 bonds varies with both the molecular weight and structure of the substrate, increasing linearly with increasing molecular weight up to maltopentaose. The *A. awamori* enzyme has been reported to hydrolyse soluble starch, amylose, amylopectin and maltose at relative rates of 100, 90, 79 and 17, respectively (Yamasaki *et al.*, 1977). Studies on purified preparations of glucoamylases from *Rhizopus* and *Aspergillus* have indicated that they are capable of hydrolysing α-1,6-glucosidic bonds; however in the case of the *A. niger* enzyme the activity against isomaltose α-1,6-linked maltose was 100 times lower than against normal α-1,4-linked maltose (Fogarty, 1983). Branched substrates are not completely degraded by glycoamylases; however, these restrictions are readily overcome when they are used in conjunction with α-amylase. Amyloglucosidases can attack raw starch, but the capacity to degrade raw starch varies between different enzymes and appears to be correlated with a good debranching activity.

Most glucoamylases have pH optima between 4.5 and 5.0 and temperature optima between 40 and 60°C. Molecular weight values have been determined varying between 28 000 D in *Cephalosporium eichhorniae* and 112 000 in *A. niger* (Fogarty, 1983).

Glucoamylases are capable of catalysing a reverse of the normal hydrolysis reaction to produce mainly maltose and isomaltose. This reaction is important in industrial processes in which high concentrations of up to 40% w/v sugars can occur which favour some maltose formation. These biosynthetic reactions can also be catalysed by transglucosidases (EC 2.4.1.24) which are frequently present in crude preparations of glucoamylases. Transglucosidases have been purified and characterized by Fogarty (1983) and found to have a broad hydrolytic substrate specificity but to be most active against maltose.

The mode of action of glucoamylases has not been investigated in detail. However, investigations using *N*-bromosuccinimide to oxidize tryptophanyl residues combined with UV absorbance and fluorescence spectroscopy have indicated that two tryptophanyl residues are involved, one at the active site and the other at a site which is involved in binding of the substrate. Using the base sequence of the gene *ATA1* of *Saccharomyces diastaticus*, Yamashita *et al.* (1984) have presented a model of the structure of glucoamylase and the proteolytic cleavages which are involved in its biosynthesis (Fig. 8.6). They have identified two glycosylated subunits H and Y of molecular weight 53–68 K and 14 K. Each is glycosylated in the *N*-position and can be deglycosylated by *endo*glycosylase H. The *N*-terminal 33 amino acids are a signal sequence and are cleaved off during secretion. The remainder can be divided into two domains, one of which is rich in threonine and serine, which can be cleaved off without loss of activity, and the functional domain containing H and Y. A similar two-domain structure has also been described for *A. niger* glucoamylase (Boel *et al.*, 1984).

Interest in developing yeast strains with an improved amylolytic activity for the brewing and distilling industries has led to extensive biochemical and genetic investigation of glucoamylases from *A. awamori* (Cole *et al.*, 1988; Meade *et al.*, 1987), *S. diastaticus* (Erratt and Nasim, 1987) and other yeasts (Tubb, 1986). The gene for *A. awamori* glucoamylase was identified using a cDNA probe which was complementary to glucoamylase mRNA. The gene isolated was not initially expressed in yeast; however, subsequent investigations indicated that the initiation and termination DNA sequences which are necessary were not recognized by yeast during protein synthesis and that the gene also contained intron segments which were not properly processed in yeast. Expression and excretion of *A. awamori* glucoamylase in yeast was obtained by constructing an intron-free gene, and inserting it between a yeast promoter and terminator obtained from the enolase gene (*eno1*). Using the natural glucoamylase leader sequence from *A. awamori*, over

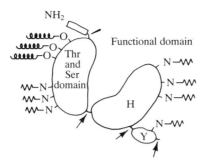

Fig. 8.6 Schematic representation of the structure and proteolytic processing of glucoamylase (from Yamashita, 1986).

95% of the glucoamylase was excreted by the yeast. This secreted enzyme was glycosylated at both *N*- and *O*-linked glycosylation sites (Innis *et al.*, 1985). Enzymes produced in this manner can be obtained in a relatively pure state since yeast does not secrete many other proteins.

The genes for glucoamylase have been sequenced in *A. awamori, A. niger, S. diastaticus, Rhizopus oryzae* and *Saccharomycopsis fibuligera*. In *A. niger* there are two forms of glucoamylase differing by about 10 000 in molecular weight. These were derived from a large mRNA and a smaller mRNA which was produced from the larger one by the excision of an intervening sequence (Boel *et al.*, 1984). The amino acid sequences of glucoamylases which have been published have been compared by van Arsdell (1987) who identified five highly homologous sequences which were considered to be conformationally similar to each other. A sequence Ala–Tyr–Thr–Gly, which has been reported to be associated with an essential tryptophan residue in the structure of *A. niger* glucoamylase (Clarke and Svensson, 1984), has also been identified in *S. fibuligera* glucoamylase and *A. oryzae* amylase.

8.2.4   Debranching enzymes

Debranching enzymes are characterized by their ability to degrade amylopectin and other polymers containing α-1,6 glucosidic linkages. These appear to fall into two groups, pullulanases (EC 3.2.1.41) which can hydrolyse pullulan, amylopectin and dextrin but cannot degrade native glycogen, and isoamylases (EC 3.2.1.68) which are capable of totally debranching glycogen but cannot degrade pullulan. Pullulan is a polymer of α-1,4 and α-1,6-linked glucose in which the α-1,6 bonds occur more or less after every third glucose unit. Pullulanases are variable in their properties and degrade amylopectin, dextrin and pullulan at different rates. It is considered that the features of the substrate which affect the activity of pullulanases include (i) outer chain length, (ii) inner chain length and (iii) whether or not the α-1,6 linkage constitutes a branch point (Lee and Whelan, 1971).

Critical work on the mechanism of debranching enzymes has been carried out on the amylo-1,6-glucosidase from rabbit skeletal muscle. Gillard *et al.* (1980) have proposed that the enzyme has a single overlapping or strongly interacting polymer binding site flanked on one side by the glucosidase site and on the other by the transferase site. A recent investigation of the binding constants of different substrates indicated that the binding of glycogen exhibited a lower dissociation constant than other oligomers and proposed that it may bind at several sites, each having a low affinity (Takrama and Kadsen, 1988).

The main industrial use of debranching enzymes which have no α-1,4 activity is in conjunction with α-amylase for the production of maltose syrups.

8.3 ENZYMES ACTING ON GLUCOSE AND OLIGOSACCHARIDES

8.3.1 Glucose isomerase

The production of high fructose syrups using glucose isomerase, is now a major industry with world wide production being over three million tonnes. Glucose isomerase has been isolated from several organisms although the enzymes from *Bacillus coagulans*, *Streptomyces albus*, *Actinoplanes missouriensis* and *A. olivocinereus* and *Arthrobacter* spp., are the ones which have been most used commercially (Bucke, 1983). The gene for *Escherichia coli* glucose isomerase has been cloned (Wovcha and Brooks, 1980). The enzyme is widely used in the immobilized form. This has been achieved using several different techniques such as that adopted by NOVO (Hemmingsen, 1979). Glutaraldehyde is used to cross-link the protein of the producer cell and the cross-linked preparation is formed into cylinders which give a good flow characteristic in column reactors. In a variation of this method Giste–Brocades have immobilized the enzyme by entrapment in gelatin and cross-linking with glutaraldehyde (Roels and Van Tilberg, 1979).

It is generally accepted that the enzyme is in fact a xylose isomerase (EC 5.3.1.5) which is induced by growth of the producer organism on xylose or xylan. The mechanism of reaction of aldose–ketose isomerases has been discussed by Rose (1975) who considered that the reaction proceeded through a *cis*-enediol intermediate. It has been demonstrated that D-glucose is the substrate for α-glucose isomerase activity in the *Bacillus* enzyme (McKay and Tavlarides, 1979). The isomerization reaction is inhibited by polyols such as ribitol, xylitol and glucitol which appear to be able to compete for the active site with glucose which binds in the α-pyranose form.

As has been observed for the other enzymes described above, a wide range of protein sizes are known. The *Actinoplanes missouriensis* enzyme has a molecular weight of 80 000 and contains two 40 000 subunits; however the *Streptomyces* enzyme has a molecular weight of 157 000 and contains 44 000 subunits (Rose, 1975). The *A. olivocinereus* enzyme has been reported to have a molecular weight of 160 000 and to be composed of four non-identical subunits (Bucke, 1983).

8.3.2 Glucose oxidase

Glucose oxidase (EC 1.1.3.4) is widely used as a laboratory reagent for assaying glucose, but its main use in the food industry is to remove small amounts of oxygen from food products or glucose from diabetic drinks. The possibility of incorporating it into membranes which could be used as wrapping materials is also being explored. Such membranes would help to reduce oxidative degradation of food products. It acts by the formation of gluconic acid by a two-step reaction which involves the uptake of molecular oxygen and the production of hydrogen peroxide (Fig. 8.7). The commercial enzyme is normally produced from *A. niger*. It has a molecular weight of 160 000 and contains 2 FAD cofactor subunits. The K_m for glucose is 30 mM (Bucke, 1983) and the enzyme has been reported to be composed of six identical subunits (Hayashi and Nakamura, 1981). β-glucose is a more active substrate than α-glucose.

Glucose oxidase

$$\text{Glucose} + 2H_2O + O_2 \rightarrow \text{Gluconic Acid} + H_2O_2$$

8.3.3 Lactases

Lactose is a disaccharide glucose, β-D-galactoside, which occurs in milk and milk products. It is not sweet and cannot be absorbed directly from the intestine of many adult mammals. It can be hydrolysed to glucose and galactose by the enzyme lactase or β-galactosidase (EC 3.2.1.23). Lactase is being used increasingly in the dairy industry for several purposes. For example the lactose from large quantities of surplus whey can be hydrolysed to a glucose/galactose mixture which is more soluble and has a sweetening power close to that of sucrose. Unlike lactose these sugars can also be metabolised by yeast so lactase-treated whey can be used as a substrate for ethanol production. The enzymic hydrolysis of lactose from milk and whey can alter the process characteristics of several dairy products. Sandiness in ice cream is reduced and its storage properties improved if β-galactosidase is

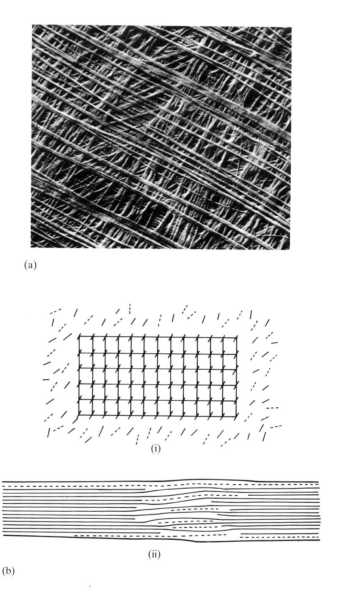

Fig. 8.7 Arrangement of cellulose chains in the plant cell wall. (a) Electron micrograph of lamellae from cell wall of *Cladophila prolifera*. Magnification *ca.* 12 000× (from Preston, 1986). (b) Diagrammatic representation of the structure of the cellulose microfibril (from Preston 1986); (i) transverse view; (ii) longitudinal view. Solid lines, glucan chains; broken lines, chains of other sugars. The rectangles in the crystalline centre of (a) are the basal planes of the Sponsler and Dore unit cells.

added. The setting time of yoghurts and cheeses can also be reduced. Whey syrups in which some of the lactose in the whey has been hydrolysed, can replace egg protein and sucrose in several bakery and confectionery products (Nijpels, 1981).

In parts of the world where milk does not form a part of the normal adult diet, many people are sensitive to lactose. This lactose intolerance which is defined clinically and is associated with abdominal pains, diarrhoea and flatulence, has led to a need to treat milk products with lactase to yield products which are suitable for lactose-intolerant individuals.

Most commercial lactase is produced from either the yeast *Kluyveromyces lactis* or from *A. niger*. The properties of both enzymes have been discussed by Nijpels (1981) and Godfrey (1983). The *Kluyveromyces* enzyme is most active at moderate temperatures (35–40°C) and pH 6.6–6.8, but it still retains some activity at 4°C which is valuable for treating some refrigerated dairy products. The *Aspergillus* enzyme is more active at higher temperatures (50°C) and a more acid pH (3.5–4.5) which is valuable for treating acid whey. This enzyme is frequently used in the immobilized form.

8.3.4 Invertase

Invertase (EC 3.2.1.26) is used in the confectionery industry for producing soft-centred chocolates.It hydrolyses terminal non-reducing fructofuranoside residues and usually acts on sucrose but is also active against raffinose (D-galactosyl-1,6-D-glucosyl-1,2-α-D-fructoside). The commercial enzyme preparation is obtained from *S. cerevisiae* which has been grown on molasses. It has a pH optimum of 4.5 and an optimum temperature of 50–60°C. The enzyme has been cloned in several laboratories and detailed molecular biological studies have been carried out on its structure and biosynthesis (Carlson *et al.*, 1983). Extracellular invertase is a glycoprotein, molecular weight 200 000 with approximately 50% by weight being mannan. A smaller cytoplasmic form has also been described. Both forms are derived from the same gene, but only the extracellular form is heavily glycosylated.

8.4 THE PLANT CELL WALL AND ITS BREAKDOWN

As indicated previously, the breakdown of the plant cell wall to release food components is an important part of food technology. The dominant material in this cell wall is regarded as being the heteropolymer lignocellulose. Lignocellulose is built up in stages, initially by plant growth with synthesis of the new primary cell walls, followed by secondary thickening and maturations that yield a material which provides structural strength, as in cereal stems, or resistance to desiccation, as in the desert shrub. In most lignocelluloses, the crystalline polymer cellulose, which forms 99% of cotton fibre, dominates. Celluloses are formed from linear chains of glucose units linked by glycoside bonds into β-1,4-glucan chains that can interlink by hydrogen bonding to produce an insoluble crystalline polymer. Behind the growing apical tip of the plant, thin primary cell walls are thickened by apposition of a sequence of lamellae formed of cellulose chains, grouped to form microfibrils, with each layer being of differing orientation to the preceding one, generating a latticework (Preston, 1986) (Fig. 8.7).These structural assemblages are embedded in non-cellulosic polysaccharides and non-sugar polymers such as the lignin in wood and phloem schlerenchyma (Fengel and Wegener, 1984), and cutin and suberin in the plant outer epidermis. Thus the final product is a laminate with a cellulosic matrix stabilized by a resin that varies in composition according to the plant and the individual tissue.

Chemists have divided non-cellulosic polysaccharides into the following classes (Preston, 1986)

1. unbranched homogeneous straight chains such as 1,4-linked xyloses in xylans or mannose yielding mannans
2. branched chains containing a variety of sugars either as sidechains or interspersed through the backbone, as in xyloglucans or galactogluco-mannans
3. chains of β-1,3-linked monosaccharides that are helical
4. polymers of galacturonic acid interspersed with D-galactosyl, L-arabinosyl, L-rhamnosyl and D-xylosyl units, with varying degrees of esterification of the acid carboxyl groups to methanol (pectins).

The first three of these polysaccharide classes are commonly termed 'hemicelluloses' and the latter 'pectin'. However, the term hemicellulose is misleading in that it implies that members of this class of heteropolymer are related to cellulose and are similar to each other whereas they differ substantially between monocotyledons (grasses and bamboos) and di-cotyledons ('broad-leaved' plants) and between the vegetative parts of the plants and the seeds. Major classes of hemicelluloses in monocotyledons, arabinoxylans, and in dicotyledons, xyloglucans are shown in Fig. 8.8. Current thinking is that cellulose is not directly linked to the non-sugar polymer components, but that the polyphenolic lignin is covalently bonded to hemicelluloses to form lignocarbohydrate complexes, yielding a hypo-thetical cell wall shown in Fig. 8.9 as proposed by Goring and his coworkers (Kerr and Goring, 1975; Goring, 1977).

It can be seen therefore that to degrade an intact plant cell wall to its component monosaccharides would require an extensive repertoire of enzymes. Often in processing, selective breakdown is introduced by thermal treatments at appropriate pH values, the effect of which can be shown by cooking vegetables in acidic and in alkaline water; the latter promotes

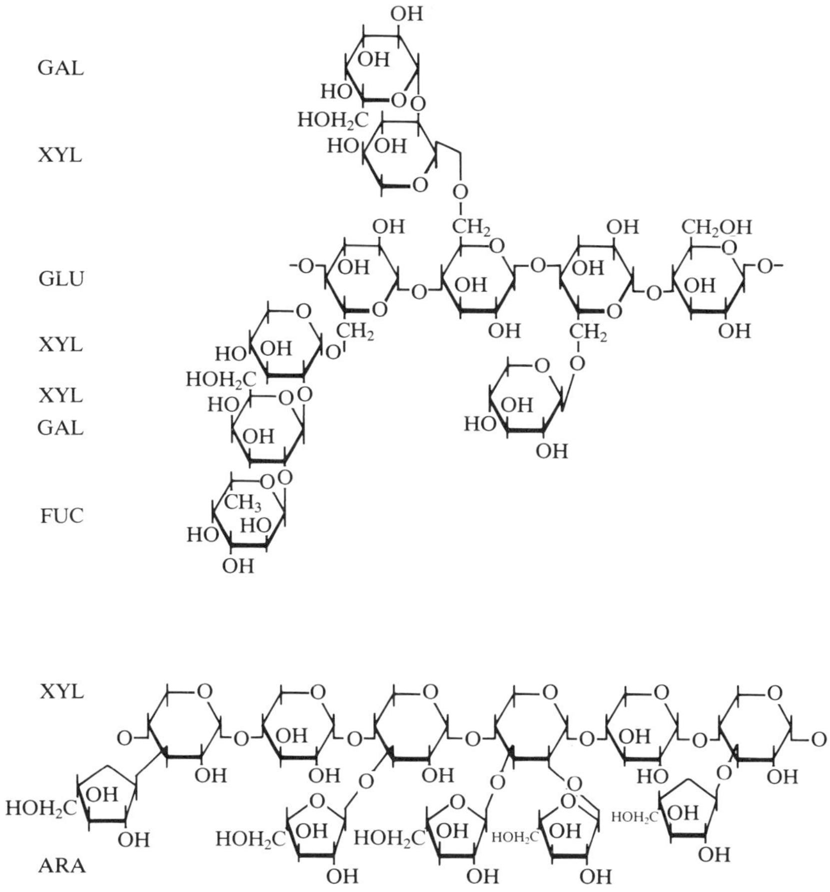

Fig. 8.8 Typical structures of two different hemicelluloses (from Harder and Hens, 1989). (a) Xyloglucans: common in fruits, vegetables, cereals and seeds. (b) Arabinoxylans: often found in cereals.

hemicellulose rather than pectin breakdown which leads to a mushy product (McGee, 1984). It has been calculated that a heteropolymer such as xyloglucan would require more than five differing enzymic activities whereas the homopolymer cellulose is known to require a minimum of three.

8.4.1 Pectinases

Pectins are structural carbohydrates which occur in the primary wall and middle lamellae of plant cells. They are complex polysaccharides which have

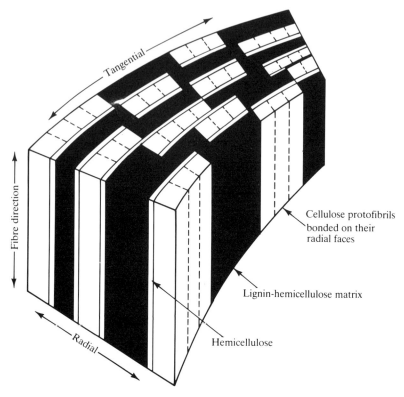

Fig. 8.9 Interrupted lamella model for the ultrastructural arrangement of carbohydrates and lignin in the cell wall according to Goring (Kerr and Goring, 1975; Goring, 1977).

a backbone of 1,4-linked D-polygalacturonic acid in which the carboxyl group at C6 is esterified with a methyl group in some residues.

Pectic enzymes are important in two contexts. They are involved in the changes in texture of fruit and vegetables during ripening and post-harvest storage, and commercial preparations can be used in the processing of fruit and vegetables as in the production of juices and the extraction of gelling agents (Rambouts and Pilnik, 1976). Pectic enzymes can be divided into four classes (Fig. 8.10) depending upon the substrate and site of action (Pilnik and Rambouts, 1982).

Pectin lyase (EC 4.2.2.10) acts on highly methylated polygalacturonic acid, cleaving glucosidic linkages adjacent to methyl ester groups by an elimination mechanism (Fig. 8.10). The formation of a double bond at the non-reducing end of each glycosidic linkage cleaved results in an increase in absorbance at 235 nm which can be used for assaying the enzyme. A highly

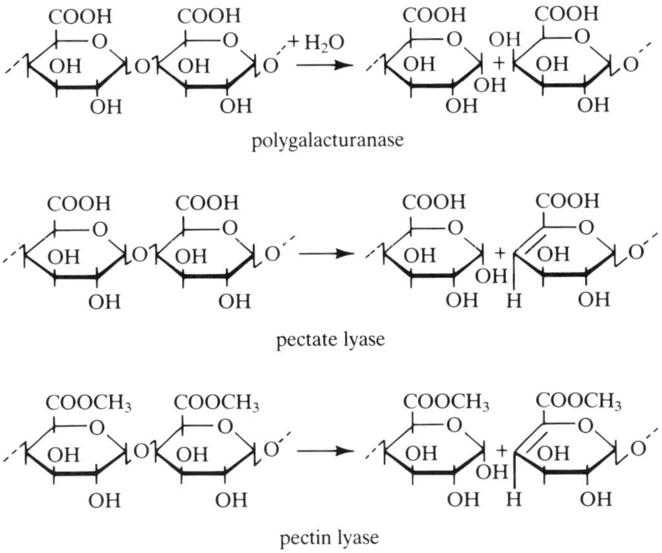

Fig. 8.10 Mode of action of different pectic enzymes. (i) Hydrolysis of glycosidic bond by polygalacturonase. (ii) Elimination cleavage of glycosidic bond by pectate lyase. (iii) Elimination cleavage of glycosidic bonds by pectin lyase.

esterified substrate is preferred by the enzyme and the affinity of the enzyme for the substrate increases with the degree of esterification. The activity of the enzyme is also dependent upon the pH and the presence of divalent cations such as Ca^{2+}.

Pectin lyase is an endoenzyme which cleaves the polygalacturonate backbone at random, resulting in a rapid reduction in viscosity. This property is also used for assaying the enzyme. The smallest substrate which can be degraded varies with the origin of the enzyme but tetra- and trimethyltrigalacturonates can act as substrate for *Aspergillus* enzymes.

Most pectin lyases are of fungal origin from, for example, *A. niger* and *Penicillium* spp. although some are of bacterial origin e.g. *Bacillus erwinia*. They do not occur in higher plants.

Pectate lyase enzymes differ from pectin lyases in that they act on pectins which are either partially or completely de-esterified. There are two classes of pectate lyase, endolyases (EC 4.2.2.2) which cleave glycosidic linkages at random and exolyases (EC 4.2.2.) which remove two dimers from the reducing end of the chain. These enzymes are only active in cleaving linkages which are adjacent to free carboxyl groups. Most pectate lyases are of bacterial origin. They are small proteins of molecular weight between 30 000 and 40 000, with an optimum pH of between 8 and 9 and they have a requirement for calcium (Pilnik and Rambouts, 1981).

Polygalacturonic acid which has not been esterified at all can be depolymerized by the action of polygalacturonases. Endopolygalacturonases (EC 1.2.1.15) occur in plants, fungi and bacteria and attack internal linkages resulting in a rapid reduction in viscosity. Moderately esterified pectin can act as a substrate but the rate of reaction decreases with increasing levels of esterification and only linkages adjacent to free carboxyl groups are attacked. An enzyme preparation from *A. niger*, frequently used in fruit processing, has a pH optimum of 5.0 and a molecular weight of 35 000. *Exo*-acting pectate lyases (EC 3.2.1.67) have been reported in plants, fungi and bacteria. The plant enzymes have attracted some interest and have been reported to remove monomers from the non-reducing end of polygalacturonate (Pressey and Avants, 1973).

Pectins can be de-esterified by the action of pectinesterases (EC 3.1.1.11). Pectinesterases are highly specific for the methyl ester of polygalacturonic acid and esters of polymannuronic acid are not attacked. Ethyl, propyl and allyl esters of polygalacturonic acid are attacked by some pectinesterases but usually at a lower rate. The reaction involves the release of methanol, which can be monitored by GLC, and the formation of free carboxyl groups which results in a pH change. Plant pectinesterases are considered to act either at the free reducing end of the chain or adjacent to free carboxyl groups and to proceed along the molecule producing regions of free polygalacturonic acid (Pilnik and Rambouts, 1981). Multiple forms of pectinesterases have been found in citrus fruits. These have been studied extensively since their action can lead to the coagulation of calcium pectate and self-clarification in lemon and lime juices. They are all inhibited by pectic acid, the end product, so continued de-esterification requires the synergistic action of polygalacturonase or pectate lyase to remove regions of free polygalacturonate (Pilnik and Rambouts, 1981).

8.4.2 Cellulose hydrolysis

Cellulose hydrolysis has been extensively studied in many laboratories in the past three decades. This is partly because of its economic potential in the transformation of the renewable resource into bulk chemicals and the nutritional importance in plant cell breakdown in the ruminant, but also because the nature of an attack by soluble enzymes on an insoluble crystalline polymer is an interesting problem. Almost all our understanding of cellulose breakdown is through study of microbial attack, since this is the source of depolymerization both in the herbivore rumen and in the external environment. Both bacteria and fungi can degrade crystalline cellulose although it is clear that they have evolved different strategies for breaking down the hydrogen bonding of the crystal. Fungal enzymes have been more extensively studied because they are secreted in large quantities (up to 2%

by weight) during growth in submerged batch liquid fermenters which has assisted purification and study. In contrast many of the bacterial enzymes exist as tight multienzyme complexes as in the thermophile *Clostridium thermocellum*(cellulosomes) from which it is difficult to recover individual active enzyme species.

8.4.3   The extracellular cellulases of *Trichoderma reesei*

The extracellular hydrolytic cellulases of the imperfect fungi *Trichoderma reesei* or *T. viride* have been studied extensively as a model system for fungal cellulases. It has proved difficult to elucidate the mode of attack on crystalline cellulose. A hypothesis used extensively suggests that there are three classes of hydrolytic enzyme (Fig. 8.11) (Montenecourt, 1983)

1. *endo*-acting β-1,4-glucan glucanohydrolases (EC 3.2.1.4) referred to as *endo*-glucanases which can cleave only chemically modified non-crystalline carboxymethyl and acid-swollen celluloses yielding oligosaccharides. These are unable to hydrolyse within three sugar residues of the non-reducing end of the polysaccharide chain.
2. *exo*-acting β-1,4-glucan cellobiohydrolases (EC 3.2.1.91), (cellobiohydrolases) which can hydrolyse native crystalline celluloses releasing

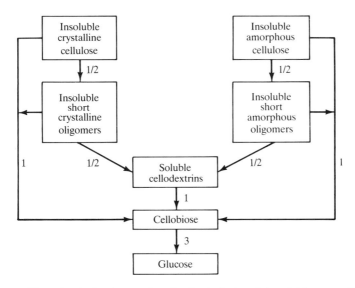

Fig. 8.11 Hypothetical scheme for the hydrolysis of insoluble crystalline and amorphous cellulose. The synergistic action of (1) cellobiohydrolases and (2) endoglucanases yields cellobiose, subsequently hydrolysed to glucose by (3) β-glucosidases (from Harder and Hens 1989).

the disaccharide cellobiose, but are ineffective in degrading the chemi-cally-modified carboxymethyl celluloses.

3. β-glucosidases (EC 3.2.1.21) which convert cellobiose and short-chain oligosaccharides into glucose.

However, there are several important objections to this model. Firstly, in cellulase enzyme preparations obtained from *Trichoderma*, multiple protein species encoding each enzyme activity are found in the crude extracellular protein of cultures. Some of these result from differences in glycosylations introduced on to the surface of the extracellular fungal proteins during their passage from the endoplasmic reticulum where the polypeptides are synthesized to the Golgi apparatus. Here they are encapsulated in mem-brane packets for secretion at the hyphal growth tip. An alternative method of generation of isoenzymes is by protease action on the *N*-termini of secreted proteins which is in itself part of the normal maturation process. However, the study of molecular genetics has shown that in *Trichoderma* there are two genes coding for cellobiohydrolases, *cbh1* and *cbh2*, and probably a minimum of four genes encoding endoglucanases, *egl1*, *egl2*, *egl3*, *egl4*. The gene products of these differ in structure and in catalytic activity with the net effect being that different enzymes act synergistically when presented with native cellulose substrates. In particular, the two cellobiohydrolases (1 and 2) act synergistically with each other and also with different subsets of the endoglucanases on crystalline substrates. The β-glucosidases, which are not involved in the depolymerization do not show the synergism but enhance the activity of the depolymerizing enzymes by removing oligosaccharides which exert a strong product repression on the glucohydrolases. Furthermore, more than 70% of the protein in crude cellulase is the cellobiohydrolase 1 isoenzyme, which would not be predicted by this model.

Further study, using carefully defined substrates has revealed that dividing the depolymerizing enzymes into *exo*- and *endo*acting species was simplistic since the purified major cellobiohydrolase 1 isoenzyme can degrade highly crystalline, ordered cellulose without addition of endoglu-canase and also lacks specificity for the penultimate glycosidic bond of oligocellodextrins (van Tilbeurgh *et al.*, 1982). The co-operation of the two cellobiohydrolases is also difficult to interpret in terms of chain-end attack or *exo–exo* co-operation (Fagerstam and Petterson, 1980). Consequently, an alternative theory has been proposed that relates the ability of cellulases to bind to celluloses and their ability to hydrolyse crystalline domains. Enzymes which do not bind tightly to native celluloses appear to restrict their action to the less crystalline zones in the native polymer.

Chanzy and Henrissat (1983) have shown that the highly crystalline cellulose from the algae *Valonia macrophysa* is a particularly useful substrate for analysis of the enzyme attack. Enzymic hydrolysis reduces

20 nm-broad cellulose microcrystals into subelements with a lateral dimension of only a few nanometers. Colloidal gold labelling of cbh1 with 4–6 nm particles has allowed electron microscopic visualization of the attack of cellobiohydrolase 1 on the cellulose surface. Results have shown that the cbh1-Au complex has a preferential adsorption on the 110 face of the microcrystals (Henrissat and Chanzy, 1986). X-ray and electron diffraction studies of digested crystals, which become fibrillated during enzyme hydrolysis with a dramatic reduction in fibre thickness but little reduction in length, suggest that the enzyme may catalyse selective hydrolysis of certain edges of planes in the native crystal which would then be disrupted by release of internal forces. Further it has been proposed that the alternate cellobiohydrolase *iso*enzyme, cellobiohydrolase II, acts as a true *exo*enzyme, shortening fibres in a manner that would lead to apparent synergism in product formation.

8.4.4   Molecular studies of *Trichoderma* cellulases

A significant advance in our understanding of the structure and function relationship in *Trichoderma* cellulases has been obtained through the work of Knowles and his colleagues at the Technical Research Centre of Finland (reviewed in Knowles *et al.*, 1987). In these studies, remarkable structural information that is of general interest in research on enzyme structure and activity has been obtained. This group has isolated the genes encoding the two cellobiohydrolases (*cbh1, cbh2*) and two of the endoglucanases (*egl1, egl3*). Furthermore, in collaborative experiments they have examined aspects of the glycoprotein structure that relate to the enzyme action. What has proved particularly exciting is that all of the cellulases share a region of high primary sequence homology despite having quite dissimilar overall amino acid sequences (Fig. 8.12). In two of the genes this region of 70% homology is found at the *C*-termini of the polypeptide (cbh1, egl1) whereas in the other two (cbh2, egl2) this region is at the *N*-termini. Very similar arrangements are found in cellulase genes of *Cellulomonas fimi* where there is a 50% homologous region. Although these regions of homology differ between the bacterial and the fungal cellulase enzymes both are joined to the rest of the protein by sequences that are rich in hydroxyl amino acids and proline. Consequently it is apparent that the cellulases must share some common functional domain structure with strong conservation protecting this aspect of the protein tertiary structure.

   The *Trichoderma* cellulase homologous domains have been removed from active enzyme by limited proteolysis which has resulted in impairment of hydrolysis of microcrystalline cellulose with little effect on breakdown of small soluble substrates. This has suggested that the hydrolytic active site must be localized in a core protein whereas the terminal domain must have a

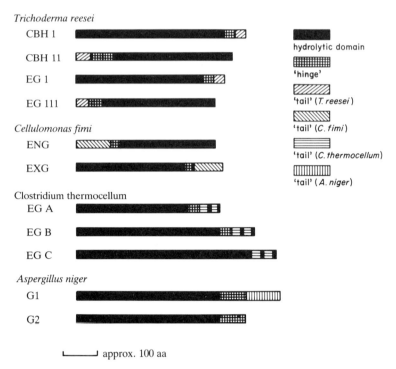

Fig. 8.12 Schematic structures of different cellulase and glucoamylase genes. Each enzyme consists of a catalytic domain probably linked *via* a flexible 'hinge' region to a tail domain that is most likely responsible for disrupting the structure of the semicrystalline substrate (the drawing is only roughly to scale) (taken from Knowles *et al.*, 1987).

role in substrate binding or attack on the crystalline face. Similar experiments have shown that in the glucoamylase of *A. niger* a C-terminal domain appears to be responsible for binding of the enzyme to raw starch and this is also linked to the hydrolytic domain by a region rich in the hydroxyl amino acids, serine and threonine (Svensson *et al.*, 1983; Boel *et al.*, 1984).

Secondary structure prediction of the cellulases suggests a predomination of β-sheet structures and small angle X-ray diffraction analysis suggests that the cellobiohydrolase 1 molecule is tadpole-shaped with a flexible terminal tail linked by a hinge structure in which serine and threonine residues are heavily glycosylated (*O*-glycosylation) (Fig. 8.13) (Knowles *et al.*, 1987). The role of glycosylations has itself been the subject of some conjecture since non-glycosylated core enzyme is as active as native enzyme in hydrolysing oligocellodextrins. However, heavily overglycosylated cellulases have been produced by cloning *Trichoderma* genes into the yeast *S*.

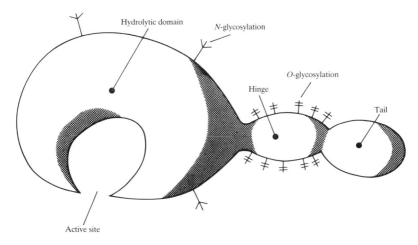

Fig. 8.13 A hypothetical model of a cellulase. This model is based on the analysis of amino acid sequence data and on low angle X-ray diffraction studies of *Trichoderma* cellobiohydrolase 1 (from Knowles *et al.*, 1987).

cerevisiae and cellobiohydrolase II, produced in the heterologous host with excessive glycosylation, has a lower catalytic activity against non-crystalline β-glucan and reduced affinity to crystalline cellulose, suggesting a change in substrate binding (Penttila *et al.*, 1987).

Cleavage of the cellulose β-glycosidic bond is thought to occur by an acid catalysis mechanism. In lysozyme, which also cleaves β-1,4 glycoside bonds a conserved active site Glu residue donates a proton from its sidechain carboxyl group to the bond to be cleaved to form a carbonium ion intermediate stabilized by a conserved Asp residue. However, analysis of other carbohydrate depolymerizing enzymes, such as the amylase from *A. niger* fails to reveal any homology around Asp and Glu acid residues which suggests that the active site geometry may differ greatly between enzymes, requiring the profound changes in tertiary structure which can be assumed from the lack of primary sequence homology.

To summarize, experimental evidence suggests that the *Trichoderma* cellulases are good models for the enzymes of polysaccharide depolymerization. The study of cellobiohydrolases (with a tertiary structure that is currently being revealed by X-ray crystallographic study) suggests that the enzyme has two functional domains, one carrying the catalytic activity and the second giving an affinity for a particular face of the cellulose crystal (Fig. 8.13). Removal of the latter eliminates hydrolysis of the crystalline part of the cellulose microfibril, thought to occur by stripping of polymer chains from the crystal surface, but maintains activity against small soluble oligocellodextrins. The two domains appear to be linked by a flexible region.

A further complexity is that it is likely that the proteolytic activity present in the proteins secreted, mentioned earlier as producing protein hetero-geneity, may also result in changes in the protein structure during enzyme action, changing the affinity of the enzyme towards an increase in hydrolysis of soluble cellulose breakdown products with reaction time.

8.5 INDUSTRIAL APPLICATIONS OF PLANT CELL-WALL DEGRADING ENZYMES

In contrast to the highly purified enzymes that must be used in elegant studies of substrate affinity and catalytic action, industrial enzymes are crude mixtures of total extracellular protein produced by growth of fungi on low-cost substrates such as whey, molasses and corn steep liquor. Typical are three industrial enzyme preparations studied by Harder and Hens (1989) shown in Table 8.2, each of which contains more than ten differing activities. Yet this shotgun approach in application of enzymes is frequently appropri-ate in a situation where enzymes are replacing thermomechanical processing of food raw materials. It is therefore useful to consider in some depth a current process in which incorporation of enzyme processing has made possible a technological advance.

8.5.1   The ALKO Process for whole-grain processing of barley

ALKO is a company, wholly owned by the Finnish state, with a monopoly for the production and retailing of distilled spirits in Finland. The largest proportion of the market in this Scandinavian country is for vodka, which is essentially ethanol from which other flavour congeners have been removed by distillation and filtration through charcoal. Currently vodka production in Finland is wholly as a byproduct of starch manufacture from barley, this polymer being used in the dominant Finnish pulp and paper industry for modifying paper properties.

Barley is widely grown in Finland as in the cooler parts of North Europe because it gives a good yield and can be planted either in the autumn (winter barley) or in the spring. The best prices are paid for grain that can be used in the malting and brewing industries which set close specifications for the cereal that they will purchase. Barley unsuitable for these purposes has traditionally been used for animal feed, especially for monogastric animals such as pigs, because it has a high starch content (55–62%) and can contain 10–15% of storage proteins. However, the exact chemical composition of the grain can vary dramatically between crops, between varieties, and with climatic differences. Moreover the cell wall has a very rigid structure which causes the grain to have a high indigestible fibre content and to be difficult and energy-consuming to grind. Furthermore, the starch is not present as a

Table 8.2 Specific enzyme activities of cell wall degrading enzymes in three industrial enzyme preparations (units/g) (from Harder and Hens, 1989)

Enzyme-activity	Substrate	A	B	C
Polygalacturonase	Polygalacturonic acid	1892	3314	1878
Pectin lyase	Pectin DE 90	43	53	74
Pectinesterase	Pectin DE 65	548	448	66
Combined pectolytic activity	Pectin DE 75	198	290	274
Cellulase	CMC	998	180	1228
Cellulase	Avicel	99	1	22
Arabanase; linear	1-5-α-L-arabinan	9	10	16
Arabanase: branched	1,3;1,2;1,5-α-L-arabinan	14	14	16
α-L-Arabinofuranosidase	PNP-arabinofuranoside	35	37	333
Galactomannanase	Galactomannan	3	4	9
Mannanase	1,4-β-D-mannan	7	0	4
Galactanase	1,4-β-D-galactan	11	58	91
Xylanase	1,4-β-D-xylan	4	0	2

homogeneous polymer but as two granule forms – large type A granules and smaller type B granules (Fig. 8.14). The type B granules by virtue of their high surface-to-volume ratio contain significantly higher quantities of granule-surface proteins which limit the applications for use of this starch fraction in paper treatment.

In the ALKO process for whole-grain utilization of barley (Fig. 8.15) (Stelwagen *et al.*, 1989) enzymes are used in two stages – breakdown of the grain cell walls and liquefaction/saccharification of starch. The products are
1. high-quality A granule starch, low in protein content
2. potable ethanol for vodka production
3. barley fibre, molasses and storage proteins which are used in animal feed production, although the protein is suitable for a range of food applications.

Initially, the grain is dry-milled and screened to reduce the fibre content of the resulting flour which is then soaked at 60°C with addition of cellulases and hemicellulases to degrade cell wall material further, releasing starch and proteins into the liquid phase. The resulting slurry is transferred to a starch separator which generates two streams – an overflow that yields the protein fraction and a cyclone for separation of the larger A granules from the smaller B granules which appear in the cyclone overflow. The B-granules are then liquefied by treatment with α-amylase and saccharified by gluco-amylase to form low molecular weight sugars. Finally, the hydrolysed starch is fermented with distilling yeasts, and the resulting ethanol distilled to yield the neutral spirit required for vodka production.

The resulting A starch is very consistent in particle size (Fig. 8.16) and is suitable for the following applications: paper coating and sizing; production

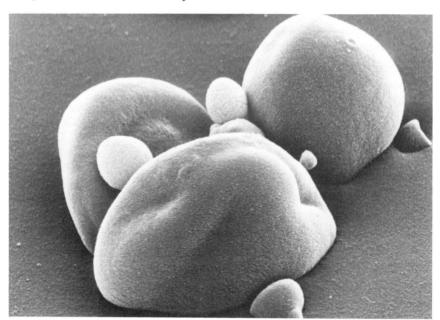

Fig. 8.14 Electron micrograph of barley starch showing large A granules and smaller B granules (from Tester 1988).

of high fructose and maltose syrups; and in brewing. Good water binding properties and low gelatinization temperature also make this polymer attractive to the meat industry which requires such polymers in sausage manufacture. The B-starch stream is also relatively pure, containing a very high concentration of fermentable solids with nitrogen essential for yeast nutrition, and a low content of cell-wall material which causes fouling of still surfaces. The barley protein is highly digestible by monogastrics (pigs and poultry) and ruminants, although supplementation with lysine is required to achieve a good amino acid balance in the diet.

8.5.2 Applications of enzymes in wet-milling of maize

In wet-milling of maize, grain is pretreated by steeping to break down the kernels. In this 'steeping', cell walls are broken down producing a feedstock from which germ, fibre, protein (gluten) and starch and also the solubilized cellwall components, referred to as cornsteep liquor, are extracted. Important parameters in wet-milling of maize are
1. yield of the valuable corn starch
2. quality of the cornsteep liquor

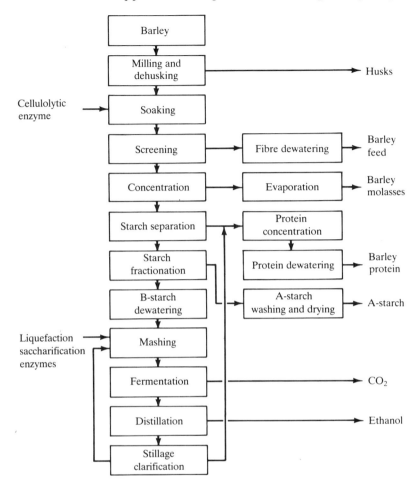

Fig. 8.15 Basic flowsheet for ALKO process for starch-ethanol-feed production from whole barley (from Stelwagen *et al.*, 1989).

3. time required for the steeping, which in the traditional process takes between 36 and 60 h.

Maize also contains significant amounts of phytic acid which initially dissolves in the cornsteep liquor but subsequently complexes with proteins and metals to form a sludge or reacts with metal surfaces causing fouling. Phytic acid is also important because pigs, in whose feed cornsteep liquor is widely utilized, are unable to metabolise this organophosphorus compound which passes through the gut elevating the phosphate level of pig slurry, for which in many countries the farmer must pay a penalty imposed for effluent treatment.

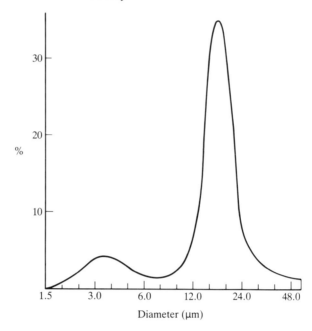

Fig. 8.16   Particle size distribution of starch obtained from the ALKO whole-barley process (from Stelwagen *et al.*, 1989).

The yield of corn starch is commonly reduced by retention of the polymer to fibre or protein. ALKO (Stelwagen *et al.*, 1989) is consequently now marketing an enzyme preparation that will enhance starch recovery whilst reducing both phytic acid content of steep liquor and steeping time. In laboratory tests the company has shown that enzyme treatment can result in a reduction of steeping time to 12 h with no significant reduction in starch yield. Starch yield was also increased from 94.4 to 96.5% with a corresponding increase in protein content of gluten and reduction in starch content of fibre using enzyme addition in the conventional steeping process (Table 8.3). Moreover after 10 h of incubation with phytase, phytic acid in cornsteep liquor had been reduced to less than 35% of the original value.

8.5.3   Application of enzymes in monogastric animal feeds

The concept that enzymes could be used to break down cell-wall material in monogastric animal feeds has been proposed since the 1920s. Both pigs and poultry are largely unable to hydrolyse non-starch polysaccharides using either their own enzymes or those of symbiotic micro-organisms. Consequently addition of enzymes to feed can increase the digestion capacity of

Table 8.3 Effect of the cell-wall degrading enzyme preparation ECONASE EP 434 on conventional steeping in wet-milling of maize (From Stelwagen *et al.*, 1989)

	Yield in % of dry weight	
	Control[a]	ECONASE EP 434[b]
Dry substance in CSL	5.28	5.61
Germ	7.34	7.12
Fibre	9.70	9.21
bound starch in fibre[c]	19.01	17.16
Starch	64.09	65.49
protein in starch[c]	0.37	0.37
Gluten	7.31	6.24
protein in gluten[c]	46.57	51.43
Dry substance in supernatant	2.21	2.25
Starch recovery	94.4	96.5
Total dry substance recovery	95.91	95.92

[a] conventional lab steeping for 48 h at 50°C and with 0.2% SO_2.
[b] conventional lab steeping with ECONASE EP 434 (70 PU/g corn).
[c] expressed as a percentage of the total fraction.

animals and improve digestibility and utilization of feed components, resulting in an increased rate of weight gain. Furthermore enzymes are thought to decrease the effects of antinutritional substances such as trypsin inhibitors, tannins and glucosinolates and to increase the release of starch, protein and fat from feeds (Krogdahl, 1986).

The use of enzyme supplements in feed manufacture can be straight-forward, with addition of dry enzyme, produced in large-scale submerged fermentations directly to feeds. In the range of chemical, thermal and mechanical treatments utilized in feed compounding, enzymes have tended to be denatured, reducing activity. The possible advantages of enzyme addition to the agronomist can be summarized as improved availability of the nutrients in feed, increasing weight gain with increased release of nutrients in the foregut and small intestine thereby reducing the amount of fermentable material arriving at the large intestine, which tends to cause postweaning diarrhoea. Furthermore poorly-digestible, low-cost plant materials can be utilized in feed manufacture and these can also extend storage life of feeds.

Using enzymes modified to assist in stabilization towards heat treatment, in the pelleting, drying and cooling process, recoveries of enzyme activity of over 80% are obtained and these enzymes are subsequently partly protected by the feed particles from denaturation in the low pH environment of the

Table 8.4 The unused energy in different animal feed-stuffs (from Junnila *et al.*, 1989)

Feed stuff	The share of unused energy (%)
Soybean oil meal	21.6
Rape seed oil meal	40.5
Sunflower meal	51.4
Barley	22.3
Oats	32.6
Wheat	15.9
Maize	13.5
Rye	17.2
Pea	20.9
Barley hull	52.9
Oat hull	66.7
Wheat bran	37.6
Rye bran	36.9
Meat and bone meal	41.8
Grass meal	53.9
Green meal	59.4

foregut. The potential for increasing energy yield from feeds is high as is shown in Table 8.4.

The Finnish Sugar Company is currently marketing enzyme mixes of hemicellulases, cellulases, glucanases and proteases for addition to poultry feeds (Junnilla *et al.*, 1989). A typical application is the specific enzyme premix, *Avizyme* which contains the cell-wall degrading enzymes with some protease and amylase activity. Addition of such enzyme cocktails at 0.1% to diets principally of barley, wheat and soybean meal results in a 4–5% increase in the rate of weight gain (Table 8.5) with a reduction in the problem of sticky droppings through barley feeding, caused by high glucan and hemicellulose, which cause management problems and reduce nutrient adsorption.

8.6 EXOGENOUS ENZYMES IN CHEESEMAKING

8.6.1 The process

One of the major traditional applications of enzymes in the food industry is in the clotting of milk in cheese manufacture and in the maturation of the treated curd to yield the final product. Milk contains two broad classes of protein, caseins, which are maintained in micelles as a colloidal suspension

Table 8.5 The effect of the enzyme preparation '*Avizyme*' on the performance of chicken on barley-canola seed diets (from Junnila *et al.*, 1989)

	Control	Avizyme	Difference (%)
Diet 1 (50% barley)			
Wg/chick/d (g)	43.18	45.10	+ 4.5
FCE	1.86	1.78	+ 4.5
Diet 2 (50% barley + 10% full-fat canola seed)			
Wg/chick/d (g)	43.30	46.21	+ 6.7
FCE	1.87	1.81	+ 3.3
Diet 3 (45% barley + 10% full-fat canola seed)			
Wg/chick/d (g)	43.95	45.84	+ 4.3
FCE	1.84	1.80	+ 2.2
Diet 4 (40% barley + 10% full-fat canola seed)			
Wg/chick/d (g)	42.82	44.84	+ 4.7
FCE	1.85	1.83	+ 1.1

Broilers given 2 × 4 diets (prot. 22.8%, ME 12.7 MJ/kg), 4 replicates, 25 chickens/replicate, duraction 42 days.
Composition of the diets: barley 40–50%, wheat 5–12%, soybean meal 19–23%, canola seed 0–10%, herring meal 10%, lard 4–7%.

in a solution of the second-class whey proteins, and lactose. In raw milk lipid is also present, predominantly as triglycerides sequestered from the aqueous environment by inclusion in milk-fat globule membranes.

8.6.2 Milk caseins

Casein micelles are formed by aggregation of small submicelles which are formed from three classes of individual casein polypeptide (Schimdt, 1982). The composition of the casein fraction (approximately 80% of total protein) is: α-caseins (2350) 50%; β-caseins (2400) 30%; κ-caseins (1900) 15%. These molecules aggregate because they have a tertiary structure similar to a rugby ball with a high percentage of hydrophobic amino acids at one end and hydrophilic modifications to surface hydroxylated amino acids, serine and threonine at the opposing 'end' (Fig. 8.17(a)). Two of the casein classes, α and β, have phosphate groups esterified to serines (*8.1*), whereas the third, the κ-caseins, carry the trisaccharide unit, *N*-acetyl neuraminyl-galactosyl-(1,6)-*N*-acetylgalactosamine (*8.2*) linked to threonine residues. It is currently thought, largely as a result of the model proposed by Slattery and Evard (Slattery, 1979), that submicelles cross-link by divalent calcium cations and citrate ions forming bridges between phosphoserines on the α- and β-caseins (Fig. 8.17(b)). However, the casein trisaccharide units are unable to participate in this interlinking of submicelles and with agglomeration the influence of these hydrophilic non-bonding components of the submicelles increases as the surface area of the micelle gets larger and thus the opportunity for binding of further submicelles decreases (Fig. 8.17(c)). The result of such polymerization are casein micelles that vary in diameter between 50 and 300 nm primarily in relation to the calcium ion concentration of the whey.

In the initial stages in cheese manufacture, microbial starter cultures of *Lactococcus* and *Lactobacillus* ferment lactose to lactic acid, reducing the pH of the solution. At a pH below 6.0 proteolytic enzymes are added, traditionally chymosin as rennet from the milkfed calf stomach. This enzyme specifically cleaves the bond between residues 105 and 106 of κ-casein, releasing the glycosylated portion of the protein into solution as glyco-macropeptide, and retaining the insoluble paracasein in the submicelle. Thus there is no further barrier to polymerization of α- and β-caseins which link (Fig. 8.18), at a frequency dependent upon the calcium ion concentration of the solution, to form a curd which after further mechanical, thermal and maturation processes becomes cheese.

The original objective of cheese manufacture was to develop a method of storing milk nutrients through the winter when feedstuffs were of limited availability. The lipids and proteins are preserved by a combination of low pH, lowered water activity and salting. During the maturation process, enzymes and some micro-organisms act on the proteins and lipids of the curd to develop the flavour and texture regarded as typical for that cheese type (Fox, 1989; Scott, 1986).

8.6.3 Enzymes in cheese manufacture: Lipases

Lipases and esterases are enzymes that hydrolyse fatty acid esters of glycerol (EC 3.1.1.3). They are differentiated on the basis of whether they act on soluble substrates, esterases, or insoluble substrates, lipases. They are important in cheese making, notably in the production of cheese flavour products from milk lipids. The enzymes can release one, two or three fatty acids yielding diglycerides, monoglycerides and glycerol as well as free fatty acids. Different lipases have differing specificities, releasing individual fatty acids at varying rates. Calf pregastric esterase for example, a normal component of rennet pastes, is specific for C4:0 to C10:0 fatty acids. Microbial lipases derived from *Mucor* are now in widespread use in the cheese industry.

However, only limited attack on lipids occurs in the production of most cheeses. Exceptions are in blue cheeses such as Blue Stilton and Danish Blue

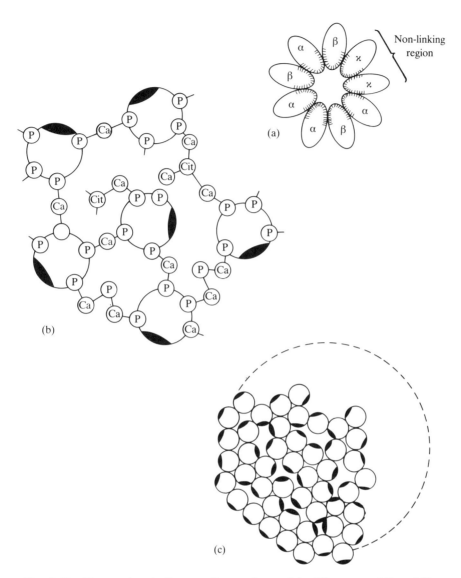

Fig. 8.17 The casein micelle according to the model of Slattery and Evard (from Coultate, 1984). (a) Cross-section of a typical submicelle showing the distribution of the three types of casein molecules. Predominantly hydrophilic areas are shaded. (b) Cross-link formation between submicelles. The non-linking regions are shown in black; P = phosphate, Ca = calcium, Ci = citrate. (c) Formation of a full-sized micelle. As the curvature of the exposed surface decreases there is less opportunity for the binding of further submicelles.

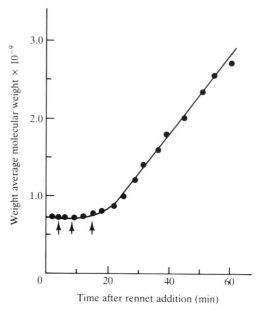

Fig. 8.18 Growth of molecular weight of a micellar fraction of original molecular weight 7×10^8 when treated with rennet. The three arrows represent the times for 50%, 75% and 95% splitting of the κ-casein. The solid line represents the calculated course of the reaction (from Dalgleish 1982).

and in the manufacture of hard Italian cheeses such as Romano and Parmesan, where porcine pregastric esterase is added at the curd stage, giving these cheeses special flavours and properties.

8.6.4 Enzymes in cheese manufacture: Proteases

Proteolysis by enzymes secreted by fungi such as *Penicillium candidum* which is encouraged to envelop soft cheeses, is of central importance in major textural changes such as the ripening and softening of Camembert resulting in a pH increase as the cheese matures. The products of proteolysis – peptides and amino acids – themselves contribute flavours and provide substrates that are broken down to yield sensorially-active degradation products such as amines, thiols, and thioesters. Since ripening of cheeses, particularly of low moisture, slow ripening varieties such as Cheddar, is expensive it is perceived as being desirable to enhance the rate of the process by adding exogenous enzymes, provided that the complex maturation processes do not become unbalanced, and considerable research is in progress to make this possible.

Consequently, one of the primary strategies is the addition of exogenous proteinases to cheese curds to enhance the rate of flavour and texture development. The primary proteinase in any cheese is the coagulant, which can be added at greatly varying enzyme:substrate ratios. Normally, in the case of calf rennet *c.* 6% of added protease is retained in the curd after the subsequent mechanical dewatering stages. Retention of the protease increases with reduction in whey pH and cheeses made with excessive acidity undergo more rapid proteolysis than normal, thought to be in part related to solubilization of colloidal calcium phosphate. Chymosin, however is no longer the only coagulant in use, since demand for enzyme is much greater than supply of calf stomach. Consequently a significant part of the market was taken by microbial rennets, notably from fungi of the *Mucor* family, although chymosin produced by genetic manipulation with expression of the bovine gene in micro-organisms is now available. *Mucor* rennet behaves differently from chymosin following clotting because it is retained by the curd at somewhat lower concentrations, at 4–6%, independent of the pH at drainage.

Many proteolytic enzymes will induce clotting of milk, production of micellar instability being related to the ability to cleave casein at approximately the region of the critical 105–106 bond (Dalgleish, 1982). Consequently the action of trypsin which hydrolyses κ-casein at 14 different sites, including the bonds at residues 97, 111, or 112, yields polypeptides with sufficiently similar behaviour to lead to clotting. However the multiplicity of alternate cleavage sites leads to the disruption of the curd clot almost as soon as it is formed. Chymosin is an acid protease that has evolved a very limited specificity, since multiplicity of hydrolysis sites would reduce the value of the natural clotting. In industrial terms this selectivity maximizes the yield of insoluble curd, limiting release of soluble polypeptides into whey, and minimizing production of small peptides with terminal amino acids which may confer bitterness. However it is necessary that chymosin, or substituted renneting enzymes, should have an action upon α- and β-caseins, since this is a central part of cheese maturation.

Chymosin is known to attack both α- and β-caseins at specific sites producing a distinct pattern of peptide formation that is the maturation process. The primary site of cleavage of β-caseins is known to be in the region 189–193 with reports of the bond 189–190 or 192–193 being the specific target. α-Casein also appears to have primary sites of hydrolysis at Phe23–Phe24 and Phe24–Phe25, subsequent degradation of the major peptides depending upon the pH and ionic strength. What is critical for all coagulants is that the rate of attack upon the central bond in κ-casein is more rapid than the attacks on other casein bonds: in the case of chymosin this bond is attacked two orders of magnitude faster than the alternate bonds.

Pepsin is also present in the stomach of the milkfed calf, estimated to contain 88–94% chymosin and 6–12% pepsin. Chymosin is traditionally obtained by washing the fourth stomach (abomasum) of the newly slaughtered calf, then slicing the tissue into strips which are extracted in 12–20% sodium chloride solution. The resulting extract is filtered, and enzyme purified by 'salting down'. The polypeptide is present as a proenzyme, which on acidification is cleaved at the N-terminus to become the active enzyme. Porcine pepsin is used commercially together with chymosin (50/50), since chymosin has a pH optimum of 6.2–6.4 whereas pepsin has a more acidic pH optimum (1.7–2.3), and the enzymes are to an extent complementary. Consequently in production of curds and cheeses of differing texture both precise enzyme mixture, enzyme:substrate ratio and milk pH at point of enzyme addition are critical factors.

The major microbial rennets used are extracellular enzyme preparations obtained by culture of *Mucor miehei* and *M. pusillus*. The *M. miehei* enzyme attacks substrate polypeptides at Phe-Val,Leu-Tyr, Phe-Phe, and Phe-Tyr bonds. This enzyme degrades caseins at pH 5.5–7.5 attacking both α- and β-caseins, but with low incidence of bitter peptide formation. The enzyme is frequently used as mixtures with porcine pepsin to produce a curd of optimum properties. A further strategy used in modulation of activity is manipulation of milk ion content since elevated sodium ion contents increase coagulation time whereas addition of calcium ions, as calcium chloride, reduces clotting time. *Mucor pusillus* protease is more proteolytic than either chymosin or the *M. miehei* enzyme altogether it generates large product peptides, with hard curds which tend to lose fat into the whey, reducing yield. Consequently blending of the calf and microbial enzymes is frequently carried out in dairy practice.

8.6.5 Enzymes in accelerated cheese ripening

The work of Kosikowski and his coworkers (Kosikowski and Iwasaki, 1975) in the early 1970s appears to be the earliest report of addition of enzymes that assist in the ripening of Cheddar cheese. In these studies combinations of 41 available enzymes – acid and neutral proteinases, lipases, bacterial decarboxylases and lactases – are reported. Most of the enzymes gave little enhancement of cheese flavour, whereas others produced bitter flavours, resulting from peptides with unfavourable terminal amino acids. However over a period this group showed that combinations of selected proteinases and peptidases together with *Mucor* coagulant could produce increased cheese flavour after 1 month maturation at 20°C although a higher temperature of 32°C resulted in an over-ripe, burnt flavour with secretion of fluids from the cheese. However, enzyme ripened cheeses were shown to have considerable potential for use in production of processed cheeses where blending of highly flavoured cheese with low-grade natural cheese is carried out prior to the heat-treatment used to produce the flavour-stable final product; this forms a large part of the market in the United States.

Law and his colleagues at the Food Research Institute, in Reading, United Kingdom, have carried out a programme of research over the past decade which has shown some of the complexities of using enzymes to develop cheese flavour (Law, 1980; Law and Kolstad, 1983; Law and Wigmore 1982, 1983). Cheeses which were treated with proteinase–peptidase enzyme preparations from bacterial cultures could exhibit the flavour of a 16-week Cheddar cheese after only an 8-week incubation. However when the incubation was prolonged perceived differences in flavour decreased, until at 22 weeks there was no significant differences between enzyme-treated and control cheeses. Furthermore it was found that excessive addition of enzyme resulted in production of strong off-flavours, including bitterness. Further studies on the activity of acid, neutral and alkaline proteinases in developing flavour and texture in Cheddar revealed that bitterness was a problem observed with addition of many microbial enzymes. Neutral proteinases were shown to have potential advantages by virtue of their instability, which limits the extent of their activity. Incorporation of these maturation proteases into curds increased formation of small peptides and amino acids of up to 400% but increases of over 150% resulted in production of bitter off-flavours. One widely held theory for this is that *endo*- and *exo*peptidases may be in imbalance in extracellular enzyme preparations. Thus the use of additional preparations of intracellular proteinases from cheese starter bacteria has been explored to address this problem. The addition of cell-free extracts from lactococci has the effect of improving the texture of cheese since use of neutral protease alone, although resulting in a 50% reduction of ripening time, yields cheese with a softer body that is more brittle than controls. Increasing the level of lactococcal extracts, appears to increase proteolysis but at a rate independent of increase in cheese flavour intensity, suggesting that further transformations of amino acids to sensorially important compounds is the rate-limiting step in many enzyme enhanced maturations. This combined enzyme–neutral protease/cell-free extract preparation '*Accelase*' is produced commercially by Imperial Biotechnology in London and has been used in large-scale commercial trials (Fox, 1989).

Proteinase/peptidase preparations have also been described for accelerating the ripening of Dutch, Tilsit and similar cheeses (Fox, 1989). In this case proteinases from *Penicillium candidum*, related to those used in soft cheese maturation, were used in combination with peptidases applied as crude extracts of cheese starter bacteria.

8.6.6   The importance of enzymes in cheese-flavour development

Lipolysis is very important in the development of strongly-flavoured cheeses such as Parmesan, blue cheeses and Feta. In Italian cheese production crude preparations of porcine pregastric esterase, lipases from *Penicillium roqueforti, P. candidum*, and *M. miehei* have all been employed. In the ripening of blue cheeses *P. roqueforti* lipases, secreted from the growing organism, injected into the processed curd are central in developments of the typical peppery flavour, produced by partial β-oxidation of fatty acids in lipolysis yielding high concentrations of methyl ketones. Exogenous lipases, such as preparations from an *Aspergillus*, can be utilized to assist the fungus in developing flavour in less than 50% of the normal maturation time. This appears to take place partly as a result of proteinases in the *Aspergillus* enzyme but also through an apparent stimulation of fungal growth through the enhanced concentration of fatty acids in the cheese.

Enzymes are also employed to produce synthetic cheeses for food formulation applications. 'Blue cheese' is a popular ingredient particularly in salad dressings and similar convenience foods in the United States. Cheese is not actually required for the product and a cheaper source of the flavour is produced by mixing free- or spray-dried whey with cream that has been incubated at 20°C for 48 h with an *A. oryzae* lipase, spores of *P. roqueforti* and bacterial starter cultures. An alternative has been the isolated lipases of *P. roqueforti*.

Christian Hansen A/S in Denmark have marketed (under the trade name '*Flavorage*') a preparation containing a particularly useful lipase from *A. oryzae* which has an exceptionally high specificity for C6–C8 fatty acids and a proteinase (Fox, 1989). This lipase releases longer-chain fatty acids than certain of the other lipases in cheese maturation enzymes, accelerating the production of the flavour of an 'aged' Cheddar. A further important factor that has made attractive the use of this lipase is that it has the valuable property of forming 200 nm micelles in aqueous media so that 94% of the enzyme added to milk is trapped in the curd, whereas 90% of the proteinase is lost into the whey.

The problem of enzyme loss into the whey, rather than the desirable retention in the curd, has attracted much attention. A typical strategy in modifying flavour development enzymes so that they are added to the milk but retained in the cheese is that developed by Olsen and his coworkers at the University of Wisconsin in Madison (Braun and Olsen, 1986a, b). Enzymes, cell-free extracts and substrates for production of particular aroma components such as diacetyl and acetoin, are encapsulated in milkfat globules which are added to cheese milks prior to renneting, continuing their action through the maturation period. A problem in such enzymic actions is

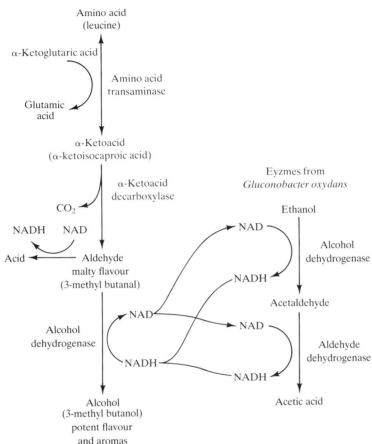

ENZYMES FROM
LACTOCOCCUS LACTIS VAR. *MALTIGENES*

Fig. 8.19   Lipid packaged combined cell-free extract system from *Lactococcus lactis* var. *maltigenes* and *Gluconobacter oxydans* for generating flavour compounds (3-methylbutanal, 3-methylbutanol and acetic acid) and recycling nicotinamide adenine dinucleotide/reduced nicotinamide adenine dinucleotide (NAD/NADH) (from Braun and Olson 1986b).

the need to regenerate enzyme cofactors such as NAD. This has been approached by addition of lipid-packaged cellfree extracts of two micro-organisms, *Lactococcus lactis* var. *maltigenes*, a component of cheese starter cultures which generates flavour and aroma compounds, and *Gluconobacter oxydans* which converts ethanol to acetic acid, regenerating NADH (Fig. 8.19). The net result is that typical cheese volatile aroma compounds such as acetic acid, 3-methylbutanal and 3-methylbutanol and amino acid breakdown products such as alcohols and aldehydes are produced, in ratios that can be modified by changing the environmental conditions utilized in the cheese maturation.

8.7 CONCLUSIONS

Since most food products are biological materials presented either as whole cells or tissues, or as purer preparations of biological molecules, it is not surprising that enzymes have proven to be a valuable tool in processing food products. The complexity of many of these food components, such as proteins and starches, makes it inevitable that it is only when pure preparations of well-defined enzymes are available that their detailed structure can be analysed and they can be processed in a precise manner. The development of a range of products from starch has illustrated how the use of enzymes can solve technical problems in the food industry and make available new ingredients for food processing. It would be surprising if this trend did not continue not only in starch processing, but also in the processing of proteins, lipids, structural polysaccharides such as cellulose and pectin, and in the production of new flavours and pigments for the food industry.

The use of more sophisticated processing techniques will increasingly demand not only enzymes with a given catalytic activity but also enzymes whose properties are appropriate for the process conditions. These may be obtained from different organisms such as thermotolerant bacteria, or in the future possibly by a combination of molecular genetics and protein engineering. Such a development will however, be dependent upon an improved knowledge of the molecular basis of enzyme action and properties.

A further development is the use of enzymes in biosynthesis. For example, lipases can be used in the synthesis of commercially important esters, many of which are of interest in the food industry as flavour compounds and emulsifiers. Some work has been carried out on the use of lipases to produce flavour esters such as citronellyl butyrate using a lipase from *A. niger*. Such reactions normally employ low-water non-aqueous systems, as in the production of citronellyl butyrate, where optimal conversion is obtained with 30% water in the reaction mixture. Lipases can also be used to increase the value of cheap fats by transesterification. The conversion of the midfraction of corn oil to a product which can substitute for cocoa butter by transesterification with stearic acid has been carried out commercially using lipases from *Rhizopus japonicus* and *Mucor hiemei* (West, 1988). However, at present, enzymes such as lipases are expensive, costing between £50 and £1000 per kg, limiting their use to the production of products valued in excess of £5 per kg.

REFERENCES

van Arsdell, J. N., Kwok, S., Schweickart, V. L., Ladner, L. B., Gelfand, D. H. and Innis, M. A. (1987) *Biotechnology*, **4**, 60.

Boel, E., Hjort, I., Svensson, B., Norris, F., Norris, K. E. and Fiil, N. P. (1984) *EMBO J.*, **3**, 1097.

Boel, E., Hansen, M. T., Hjort, I., Hoegh, I. and Fiil, N. P. (1982) *EMBO J.*, **3**, 1581.

Braun, S. D. and Olson, N. F. (1986a) *J. Dairy Sci.*, **69**, 1202.

Braun, S. D. and Olson, N. F. (1986b) *J. Dairy Sci.*, **69**, 1209.

Bucke, C. (1983) in *Microbial Enzymes and Biotechnology*, (ed. W. M. Fogarty), Applied Science Publishers, London, p. 93.

Carlson, M., Tanssing, R., Kustin, S. and Botstein, D. (1983) *Mol. Cell. Biol.*, **3**, 439.

Chandler, P. M., Zwar, J. A., Jacobsen, J. V., Higgins, T. J. V. and Inglis, A. S. (1984) *Plant Mol. Biol.*, **3**, 407.

Chanzy, H. and Henrissat, B. (1983) *Carbohydr. Polym.*, **3**, 161.

Clarke, A. J. and Svenssen, B. (1984) *Carlsberg Res. Commun.*, **49**, 111.

Cole, G. E., McCabe, P. C., Inlow, D., Gelfand, D. H., Ben-Bassat, A. and Innis, M. (1988) *Biotechnology*, **6**, 417.

Coultate, T. P. (1984) *Food: the Chemistry of its Components*, Royal Society of Chemistry, London.

Dalgleish, D. G. (1982) in *Developments in Dairy Chemistry – 1*, (ed. P. F. Fox), Elsevier Applied Science, London, p. 157.

Erratt, J. A. and Nasim, A. (1987) in CRC *Crit. Revs. Biotechnol.*, **5**, 95.

Fagerstam, L. G. and Petterson, L. G. (1980) *FEBS Lett.*, **119**, 97.

Fengel, D. and Wegener, G. (1984) *Wood, Chemistry, Ultrastructure, Reactions*, Walter de Gruyter, New York.

Fogarty, W. M. (1983) in *Microbial Enzymes and Biotechnology* (ed. W. M. Fogarty), p. 1.

Fogarty, W. M. and Kelly, C. T. (1979) in *Progress in Industrial Microbiology* (ed. M. J. Bull), **15**, p. 87.

Fogarty, W. M. and Kelly, C. T. (1983) in *Microbial Enzymes and Biotechnology* (ed. W. M. Fogarty), Applied Science Publishers, London, p. 131.

Fox, P. F. (1989) *Food Biotechnol.*, in press.

Gillard, B. K., White, R. C., Zingaro, R. A. and Nelson, T. E. (1980) *J. Biol. Chem.*, **255**, 8451.

Godfrey, T. (1983) in *Industrial Enzymology* (ed. T. Godfrey and J. Reichelt) MacMillan, Nature Press, New York, p. 466.

Godfrey, T. and Reichelt, J. (1983) *Industrial Enzymology, The Application of Enzymes in Industry*, MacMillan, Nature Press, New York.

Goring, D. A. I. (1977) in *Cellulose Chemistry and Technology* (ed. J. C. Arthur Jnr.), A. C. S. Symp. Series 48, American Chemical Society, Washington.

Griffin, P. J. and Fogarty, W. M. (1973) *J. Appl. Chem. Biotechnol.*, **23**, 301.

Harder, A. and Hens, H. J. H. (1989) *Food Biotechnol.*, in press.

Hayashi, S. and Nakamura, S. (1981) *Biochim. Biophys. Acta*, **657**, 40.

Hemmingsen, S. H. (1979) in *Applied Biochemistry and Bioengineering, 2, Enzyme Technology*, (ed. L. B. Wingarde, E. Katchalski-katzir E. and Goldstein) Academic Press, New York, p. 157.

Hideka, H. and Adachi, T. (1980) in *Mechanisms of Saccharide Polymerisation and Depolymerisation* (ed. J. J. Marshall), Academic Press, New York, p. 101.

Innis, M. A., Holland, M. J., McCabe, P. C., Cole, G. E., Wittman, V. P., Tal, R., Watt, K. W. K., Gelfand, D. H., Holland, J. P. and Meade, J. H. (1985) *Science*, **228**, 21.

Junnilla, M., Helander, E. and Inborr, J. (1989) *Food Biotechnol.*, in press.

Kainuma, K., Wako, K., Kobayashi, S., Nogami, A. and Suzuki, S. (1975) *Biochim. Biophys. Acta*, **410**, 333.

Kennedy, J. F., Cabalda, V. M. and White, C. A. (1988) *Trends in Biotechnol.*, **6**, 184.

Kerr, A. J. and Goring, D. A. I. (1975) *Cell Chem. Technol.*, **9**, 563.

Kjolberg, D. and Manners, D. J. (1963) *Biochem. J.*, **86**, 258.

Knowles, J., Lehtovaara, P. and Teeri, T. (1987) *Trends in Biotechnol.*, **5**, 255.

Kondo, H., Nakatani, H., Matsuno, R. and Hiromi, K. (1980) *J. Biochem.*, **87**, 1053.

Kosikowski, F. V. and Iwasaki, T. (1975) *J. Dairy Sci.*, **58**, 963.

Krogdahl, A. (1986) in *Proc. 7th Eur. Poultry Conf.*, Paris (ed. L. M. Larbier), pp. 239–248.

Law, B. A. (1980) *Dairy Ind. Int.*, **45**, 5, 15, 17, 19, 20, 22.

Law, B. A. and Kolstad, J. (1983) *J. Dairy Res.*, **50**, 519.

Law, B. A. and Wigmore, A. S. (1982) *J. Dairy Res.*, **49**, 137.

Law, B. A. and Wigmore, A. S. (1983) *J. Dairy Res.*, **50**, 519.

Lee, E. Y. C. and Whelan, W. J. (1971) in *The Enzymes* (ed. P. D. Boyer), **5**, 191.

McGee, H. (1984) *On Food and Cooking*, Allen & Unwin, London.

MacGregor, E. A. (1988) *J. Protein Chem.*, **7**, 399.

McKay, G. A. and Tavlamides, L. L. (1979) *J. Mol. Catal.*, **6**, 57.

Meade, J. H., White, T. J., Shoemaker, S. P., Gelfand, D. H., Chang, S. and Innis, M. A. (1987) in *Food Biotechnology* (ed. D. Knorr), p. 393.

Montenecourt, B. S. (1983) *Trends in Biotechnol.*, **5**, 156.

Nijpels, H. H. (1981) in *Enzymes and Food Processing*, (eds G. G. Birch, N. Blakeborough and K. J. Parker), Applied Science Publishers, London, p. 89.

Norman, B. E. (1981) in *Enzymes and Food Processing*, (eds G. G. Birch, N. Blakeborough and K. J. Parker), Applied Science Publishers, London, p. 15.

Paterson, A. (1988) in *Physiology of Industrial Fungi*, (ed. D. R. Berry), Blackwell Scientific, Oxford, p. 101.

Peppler, H. J. and Reed, G. (1987) in *Biochemistry: A Comprehensive Treatise, 7A* (eds. H. J. Rehm and G. Reed) VCH, Weinheim, BRD, p. 547.

Penttila, M. E., Andre, L., Saloheimo, M., Lehtovaara, P. and Knowles, J. K. C. (1987) *Yeast*, **3**, 175.

Pilnick, W. and Rambouts, F. M. (1981) in *Enzymes and Food Processing*, (ed. G. G. Birch, N. Blakeborough and K. J. Parker), Applied Science Publishers, London, p. 105.

Pressey, R. and Avants, J. K. (1973) *Biochim. Biophys. Acta*, **309**, 363.

Preston, R. D. (1986) in *Cellulose: Structure, Modification and Hydrolysis*, (eds R. A. Young and R. M. Rowell), Wiley Interscience, New York, p. 3.

Roels, J. A. and van Tilberg, R. (1979) in *Immobilised Microbial Cells*, (ed. K. Venkatsubramanian), American Chemical Society, Washington, p. 147.

Rohyt, J. F. and Ackerman, R. J. (1971) *Arch. Biochem. Biophys.*, **145**, 101.

Rogers, J. C. and Milliman, C. (1983) *J. Biol. Chem.*, **258**, 8169.

Rose, I. A. (1975) *Adv. in Enzymol.*, **43**, 491.

Saha, B. D. and Zeikus, J. G. (1987) *Process Biochem.*, (**June**), 78.

Schmidt, D. G. (1982) in *Developments in Dairy Chemistry – 1*. (ed. P. F. Fox), Elsevier Applied Science, London, p. 61.

Scott, R. (1986) *Cheesemaking Practice* (2nd edn.), Elsevier Applied Science, London.

Shukla, T. P. (1975) *Crit. Rev. Food Technol.*, **5**, 325.

Siraishi, F., Kawakami, K., Yuasa, A., Kojima, T. and Kusunoki, K. (1987) *Biotechnol. Bioeng.*, **30**, 374.

Slattery, C. W. (1979) *J. Dairy Res.*, **46**, 253.

Stelwagen, P., Lehmussaari, A. and Vaara, T. (1989) *Food Biotechnol.*, in press.

Svensson, B., Larsen, K., Svendson, I. and Boel, E. (1983) *Carlsberg, Res. Commun.*, **48**, 529.

Takasaki, Y. and Yamanobe, T. (1981) in *Enzymes and Food Processing* (eds G. G. Birch, N. Blakeborough and K. J. Parker), Applied Science Publishers, London, p. 73.

Takrama, J. and Madsen, N. B. (1988) *Biochemistry*, **27**, 3308.

Tester, R. (1988) Ph.D. Thesis, University of Strathclyde.

Thoma, J. A. (1976) *Carbohydr. Res.*, **48**, 85.

Tubb, R. S. (1986) *Trends in Biotech.*, **4**, 98–104.

van Tilbeurgh, H., Claeyssens, M. and de Bruyne, C. K. (1982) *FEBS Lett.*, **149**, 152.

West, S. I. (1988) *Chemistry in Britain* (**Dec.**), 1220.

Wovcha, M.G. and Brooks, K. E. (1980) *Br. Pat.*, 2,031,905.

Yamashita, I., Suzuki, K. and Fukui, S. (1986) *Agric. Biol. Chem.*, **50**, 475.

Yamasaki, Y., Suzuki, Y. and Ozawa, J. (1977) *Agric. Biol. Chem.*, **41**, 2149.

Yamashita, I., Hatano, T. and Fukui, S. (1984) *Agric. Biol. Chem.*, **48**(6), 1611.

Yuen, S. (1974) *U.S. Pat.*, 3,793,461.

9 | Enzymology and protein chemistry in the wider area of biology

Keith E. Suckling

9.1 INTRODUCTION

Biochemistry, with its major component fields which include enzymology, occupies a central position in the spectrum of the physical and biological sciences, providing a link between them as well as a common ground on which the two major divisions of natural science may be practised together. The greater part of this book has considered how the ideas and concepts that have arisen out of just one area of biochemical study, that of enzymology, have been of value in stimulating work in other, mainly chemical fields. We have seen that looking over the shoulder of the biologist has been an inspiration to many chemists, who hope to make use of the elusive and exquisite selectivity and catalytic power of enzymes.

To the biochemist, and more especially to the biologist, however, the study of enzymes is just a part of a much larger subject which comprises every kind of biological process. In one sense biochemistry could be regarded as a sort of integrated enzymology since the great majority of biochemical processes depend upon enzymes for catalysis and upon other proteins with many enzyme-like properties. In biology it is important to study not only the catalytic mechanisms of enzymes, but also to understand how the rates of the reactions they catalyse are regulated. Mechanisms for regulation can include modification of the activity of an existing enzyme as well as modifying its rate of synthesis and degradation. Enzymology is thus expanded into the context of the whole cell and organism.

To discuss the impact of enzymology on biochemistry itself and on neighbouring disciplines is a bit like trying to analyse a simply stated but all-embracing topic such as the influence of writing on human culture; the impact is so all-pervasive that such an attempt runs the risk of appearing trivial. To the well-read biochemist, little of what follows will be new. Those readers whose background is more exclusively chemical or molecular in nature may find the broadening of horizons at the end of this book to be both a reference point for biochemical techniques used earlier and also a useful open-ended close.

9.1.1 Areas of biochemical study

In a general sense enzymology is part of a wider field of biochemical study, that of proteins. Since enzymes catalyse chemical reactions their study has been predominant in the minds of many chemists, but there is much important chemistry to be found in non-enzyme proteins. Some of the wide range of proteins with important biological functions discussed in this chapter are illustrated in Table 9.1. Enzyme-catalysed reactions themselves combine not only in metabolic pathways, by which small molecules are synthesized and degraded, but also in significant physiological processes such as blood clotting, muscle contraction and the regulation of many cellular processes, for example those associated with cell division, gene expression and the response to extracellular stimuli (see section 9.8).

Enzymes are more and more used as tools in many biological fields. They can be coupled to other techniques such as microscopy and immunology in a very powerful way. Perhaps the greatest impact of using enzymes as experimental tools has been in the areas of genetic engineering which depends upon the specific properties of many purified enzymes to synthesize and to degrade sequences of DNA and RNA.

In many of these areas the main object of study is the behaviour of the larger biological system: resolution to the molecular level is not of first interest. However, there are now many areas of biology where the time is

Table 9.1 Non-enzymic protein systems dicussed in Chapter 9

System	Function
Antibody	Specific non-covalent binding to antigen (section 9.6)
Plasma lipoproteins	Binding of insoluble lipid molecules for transport in the blood (section 9.7)
Acetylcholine receptor	Binds acetylcholine released from a neuromuscular junction to initiate muscle contraction (section 9.8)
Repressor protein	Interacts with a specific region of DNA to regulate transcription (section 9.5)

now ripe for a detailed molecular understanding to be achieved. When we consider the interactions between macromolecules such as DNA and proteins the concepts of what we might call 'small substrate molecule' enzymology may prove to be limiting. A number of proteins whose substrates are macromolecules such as nucleic acids have now been studied in great detail (section 9.5). These and studies of other proteins have allowed the natural economy of design of proteins to become more and more apparent. One finds proteins with widely different functions using similar structural motifs (section 9.4). This has allowed a protein structure based approach to taxonomy and evolution to develop.

As soon as we consider the differences between species the roles of genes come to mind. We are now witnessing a continuous exchange of information between the structure of genes and the proteins whose structures they determine. This is being accelerated by the application of the techniques of genetic engineering (sections 9.2, 9.7 and 9.8).

We have seen that chemists have used all kinds of non-protein systems in attempts to identify the source of the catalytic activity of enzymes or for other purposes such as in synthesis (section 4.8). Proteins are, however, not the only biological catalysts. In the last few years it has become clear that certain types of RNA are able to carry out true catalysis of very significant biological processes (section 9.6). As with some of the other systems that we shall discuss in this chapter, this discovery requires fundamental changes in the dogma of biology.

In the following sections we shall discuss some selected aspects of these recent developments in areas that are part of, or border on, the study of enzymes. A number of themes recur. Particularly strong is the influence of molecular biology and the growing body of knowledge about the main types of protein structures that occur in nature. Most of this work could only be (and was) predicted when the first edition of this book was written. There is an increasing number of biological systems accessible to a detailed molecular understanding.

9.2 STUDIES OF ENZYMES BY TECHNIQUES OF MOLECULAR BIOLOGY

The power of the newer techniques of molecular biology to enable carefully designed studies on protein structure and function has already been exemplified in this book (sections 4.8 and 7.2). The principles of the techniques of modifying proteins by genetic means, known as site-directed mutagenesis, are relatively simple. If one knows the amino acid sequence of a protein, the sequence of bases in the DNA that codes for that protein can be deduced from the genetic code. It is then possible to synthesize a modified DNA making small but specific changes in the DNA using a combination of chemical and biochemical methods. The DNA so produced can be incorporated into a cell, often a bacterium or yeast using a suitable carrier (known as a vector) such as a bacteriophage (a viral DNA source that infects bacteria) or a plasmid (a small circular loop of DNA that can be transferred between bacteria and incorporated into the bacterial genome). The foreign DNA is treated as endogenous material by the host cell and is transcribed to RNA and a modified protein synthesized as a consequence. In favourable cases up to half of the total protein in the host cell can be the modified molecule. The effects of the modification induced in the DNA on the properties of the protein can be studied once it has been purified from the rest of the cell culture system. Often the key step is obtaining sufficient information on the sequence and the structure of the protein of interest.

The impact of the techniques of site-directed mutagenesis on studies of enzymes has been substantial (reviewed by Knowles, 1987) for several reasons. First, as we have seen earlier in this book, there is a strong theoretical background of physical organic chemistry that can describe the nature of enzymic catalysis in some detail. Secondly, many enzymes have been studied in great depth from the point of view of their catalytic mechanism (kinetics and chemical modification studies) and structure (high resolution X-ray diffraction studies). Enzymes which are already understood with this degree of familiarity have been prime targets for site-directed modification. Some of the best studied include tyrosyl-tRNA synthetase (Fersht, 1987) and dihydrofolate reductase (Benkovic *et al.*, 1988).

In addition to a good basic knowledge of the structure of an enzyme, a penetrating study by site-directed mutagenesis requires the gene for the enzyme and a suitable system for expressing the gene, that is a cell culture system that can synthesize the modified enzyme in quantity using the synthetic gene as the blueprint. The modifications that can be made are, of course, limited to the naturally occurring amino acids, the 19 amino acids other than the one to be altered. The most revealing changes that could be made would be those that alter the characteristics of a side chain of an amino acid without causing a deformation of the overall folding of the polypeptide chain (as chemical modification studies, which insert larger groups, may frequently do). Thus changes have been made in the polarity of groups (e.g. aspartate to asparagine, a very non-perturbing modification) or in functional groups, without too much change in bulk (e.g. cysteine to serine or histidine to asparagine). Within these limitations it is remarkable what has been achieved in a relatively short period of time both in terms of understanding enzymes and in raising new challenges.

Even the simplest of modifications seems to raise theoretical problems to answer. For example, if we try to alter the specificity of trypsin so that it hydrolyses peptide bonds adjacent to amino acid residues bearing a negative charge, rather than the positively charged lysine or arginine that is the natural preference, the result is not exactly what is expected. Changing the aspartate (negative charge) that is located at the bottom of the pocket in the enzyme structure which binds the amino acid side chain to a lysine (positive charge) results in a protein that certainly does not hydrolyse adjacent to positively charged sites, or indeed negatively charged ones. The mutant enzyme tends to prefer neutral peptides (Graf *et al.*, 1987). Clearly the new lysine residue did not occupy the position in the protein that was expected.

In another case a major theoretical basis for the regulation of enzyme activity was brought into question (Lau and Fersht, 1987). Phosphofructo-kinase from *Escherichia coli* is inhibited by phosphoenolpyruvate (the inhibitor) and activated by ATP and GTP (the effectors). These modifi-cations in activity are thought to occur by binding of the effector at specific sites in the enzyme, causing confrontational changes to states of different activity. If it is assumed that the enzyme can exist in two states, one more active than the other, it is possible to predict the kinetic behaviour of the enzyme in the presence of varying concentrations of effector and inhibitor. Such theoretical treatments have been widely used for many years. When one residue at the effector binding site was altered (glutamate to alanine) a complete change in the profile of activation was observed. Phosphoenol-pyruvate became an effector rather than an inhibitor. Such a change cannot be accounted for by the classical theoretical treatments of enzyme catalysis and control. Such theories do not address the intimate molecular changes occurring remote from the active site.

9.3   THEORETICAL TREATMENTS OF ENZYME CATALYSIS

The modification of proteins by site-directed mutagenesis has given the opportunity for a very critical analysis of many of the fundamental concepts of enzyme catalysis and mechanisms that have been developed over the last 30 years. The power of computational techniques has increased substan-tially, as has the experience with molecules of increasingly large size. Many more structures of proteins are known in detail from X-ray diffraction studies than was the case 10 years ago. These three factors have contributed to many penetrating studies of enzymes, proteins and their interactions with substrates and, frequently neglected, their environment. It continues to be conceptually convenient to examine the importance of the different classes of interaction such as hydrophobic and polar interactions known to occur in enzymes separately from each other. Thus many of the experiments on tyrosyl-tRNA synthetase were directed towards an analysis of the energetics associated with hydrogen bond formation during the first stage of the catalytic mechanism (Fersht, 1987). Another challenge to a theoretical understanding is the role played by interactions between charged residues in protein structure and function. Successful attempts have been made to calculate the energies associated with such interactions and also the energies of interaction between charged groups and the environment of the protein (reviewed by Warshel, 1987) using detailed three-dimensional structures as a basis.

Some interesting conclusions have emerged from these studies. One is that proteins have sufficient polar and polarizable groups within them so that they appear to be able to 'solvate' charged groups without the need for counterions or solvent in the immediate vicinity. In these theoretical treatments, long-range polar interactions could be quantified very effectively, but shorter-range interactions, such as those that are probably very significant in the function of enzymes, were found to require a more precise microscopic model. The models used in the studies referred to here tended to be simple symmetrical models that could not predict the surprising result obtained when a simple ion pair that stabilizes part of the tertiary conformation of an enzymes was reversed. When a strong ion pair in aspartate aminotransferase (Asp^- . . . Arg^+) was inverted by genetic engineering to Arg^+ . . Asp^-, the interaction of the former system was found to be 2.7 kcal mol^{-1} greater than the latter, although the apparent chemical nature of the interactions in both forms are identical. With surprises such as this we are reminded that for quantitative analysis we are dealing with extremely complex systems that force us to concentrate on very limited aspects of the many factors that may contribute to their function (Hwang and Warshel, 1988).

Some researchers have presented arguments that many of the mechanistic models that have been developed over the last 30 years must now be regarded as misleading. In one study it had been concluded that the most significant energetic factor in enzyme catalysis is the binding energy between the substrate and enzyme (Dewar and Storch, 1985). This promotes the extrusion of solvent and allows functional groups to get within bonding distances. Other mechanisms (strain, transition state stabilization etc) are described as 'intriguing but unnecessary' (Menger, 1985). Others would argue that studies of mutated enzymes clearly show the relevance of the concept of tight binding of the transition state (see also catalytic antibodies, section 9.6). In another approach to a theoretical study of enzyme-catalysed reactions it was argued that since enzymes tend to exclude solvent from their reactive sites, the nature of the chemistry that takes place during bonding changes is closer to that in gas-phase reactions than to reactions in solution. A consequence of this argument is that studies of enzyme-catalysed reaction mechanisms by methods using poor substrates or enzyme-inhibitor complexes may well be misleading. The basic energies required would be those of gas-phase reactions, which are not widely available. The current profusion of experimental data of modified enzymes and protein structures ensures that this will continue to be an area for a lively debate.

9.4 PROTEIN STRUCTURE, HOMOLOGY AND GENETIC RELATIONSHIPS

We have already mentioned that the increased amount of detailed information from X-ray diffraction studies of proteins has allowed the development of more generalized concepts of protein structure. A number of common structural features can be discerned in the assembly of ordered areas of proteins (areas of secondary structure) into the functional tertiary structure (for example Figs 9.1, 9.2 and 9.6). Membrane-spanning regions (section 9.8) are increasingly inferred from primary sequence data, without the backup of three-dimensional information. It has also become clear that there are a number of commonly recurring themes in overall protein structure. For example the β-barrel arrangement in Fig. 9.1 is shared by triose phosphate isomerase, glucose isomerase and enolase (from yeast, the fourteenth example of this type (Chothia 1988)). Other examples of related proteins to be discussed in this chapter are some DNA-binding proteins (Fig. 9.2) and cell-surface receptor molecules of different types (Fig. 9.7).

The question immediately comes to mind as to how diversity of function has evolved within a relatively small number of common designs. In fact the study of evolution and taxonomy has benefited enormously from the molecular insights offered by the comparison of the structure of related

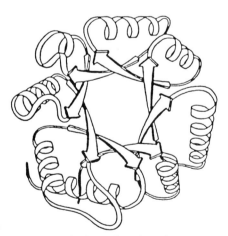

Fig. 9.1   The barrel structure of triose phosphate isomerase, a structure shared by many proteins. The coiled ribbons represent α-helices joined by ribbons of less ordered structure to β-strands (reproduced with permission from Dr J. Richardson).

proteins both at the three-dimensional level and the level of the structure of the gene. Successful folding patterns for proteins appear to be very adaptable to detailed changes in function, sometimes in a very bizarre way (Sawyer, 1987). For example, β-lactoglobulin (a protein from cow's milk), human plasma retinol binding protein and bilin-binding protein from insects, proteins with widely different functions, all show a common cross-hatching structural pattern of 8 strands of antiparallel β-sheet. This pattern has now been found in at least 10 other proteins including apoliprotein D, which is found in high density lipoprotein in plasma (*cf.* section 9.7). DNA sequences are known for four of these proteins, and, even more remarkably, the structure of the genes seem to be similar. In eukaryotic cells, the DNA containing the linear sequence of bases coding for the amino acid sequence of the protein is found in separated lengths, or coding regions known as exons. The regions of DNA that separate the exons do not contribute to the protein sequence and are known as introns or intervening sequences. All of these sequences, introns and exons are transcribed into RNA in the nucleus of the cell and the introns are then

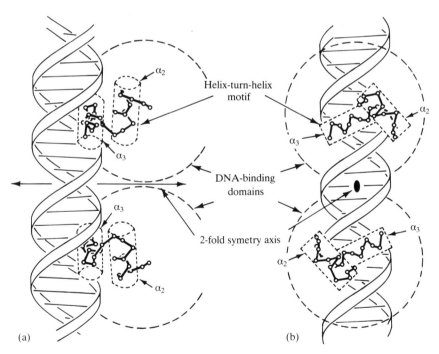

Fig. 9.2   The helix-turn-helix motif found in DNA-binding proteins discussed in the text. The helix-turn-helix motif (α₂ and α₃) lies in the major groove of the DNA helix. The two views are at right angles to each other (Matthews, 1988. Reprinted by permission from Nature, **355**, 294. Copyright © 1988 Macmillan Magazines Ltd).

removed from the resulting RNA by a specific editing process (section 9.6). The positions and, presumably, functions of the coding sequences (exons) and the intervening sequences (introns) of the four proteins just mentioned are related. All this suggests a common evolutionary origin and a diversification of application of this effective structural motif. These structures all have the function of binding sparingly soluble or labile molecules.

Other studies are relating the functions of specific domains in proteins to exons in the corresponding genes (reviewed by Pain, 1986). For example, one subdomain and one exon appear to be sufficient for the oxygen-binding activity of myoglobin. Thus the broad generalization might be that an exon in a gene corresponds to a domain or a functional unit in a protein and that this is equivalent to one ancestral unit in the evolution of a gene. Homologies are not the rule, however (Ogata *et al.*, 1987). Proteins with apparently identical function can evolve with quite different structures. Monellin is an intensely sweet protein found in certain African berries. Thaumatin is another protein with a similar sweetness. Both proteins must be equivalent as far as recognition by the taste receptors in the mouth, and both, with other sweet compounds, are recognized by antibodies raised against thaumatin. The striking observation is that there is no homology in the amino acid sequence in these two proteins and no similarities between their backbone structures.

The diversity of function of proteins of very similar structure can be even more striking. The crystallins, the proteins of the lens of the eye, are very closely related to enzymes of known metabolic function. For example, ε-crystallin in birds and reptiles is very closely related to lactate dehydrogenase and τ-crystallin in birds, reptiles and fish is related to enolase (Doolittle, 1988).

9.5 INTERACTIONS BETWEEN PROTEINS AND DNA

The growing intimacy between studies of protein structure and related structures of nucleic acids has been apparent from many of the examples we have discussed so far. The most direct illustrations of this are to be found in the interactions of specific proteins with DNA. These interactions are necessary for the chemical manipulation of DNA by the cell, the synthesis of RNA, and for the expression of the signals that determine if genes are active or not.

One of the best examples of protein–DNA interactions is the repressor–operator complex of bacteriophage 434 (Aggarwal *et al.*, 1988, and earlier references therein). The repressor protein binds to a specific control sequence in the DNA, the operator, and prevents the binding of RNA polymerase, the enzyme necessary for the synthesis of RNA from the DNA template. The purified repressor was crystallized with a synthetic piece of DNA corresponding to the operator region and the crystal structure determined with a resolution to between 3.2 and 4.5 Å. The characteristic structure for interaction with DNA had been determined by earlier studies at lower resolution. This is described as a helix-turn-helix motif that fits neatly into the grooves of the DNA structure (Matthews (1988), Fig. 9.2). It was found that hydrogen bonds between the peptide nitrogens of the protein fit closely with the phosphate groups of the DNA. Three glutamine residues on one of the helices form van der Waals and hydrogen-bond contacts with the five outer base pairs of the operator. The DNA at the centre of the interaction is wound more tightly than in the unbound structure. The repressor recognizes the DNA in a specific conformation and a specific sequence of bases. In the more recent studies the details of the twisting of the DNA molecule induced by the binding of the repressor were revealed. The α_3 subunit (Fig. 9.2) was shown to determine the position of the repressor in the major groove of the DNA. The specificity of the binding was accounted for by unusual three-centre hydrogen bonds.

A similar detailed study at 2.5 Å resolution has been reported on the repressor protein from bacteriophage lambda. The precise orientation of the repressor is assured by an extensive hydrogen bonding network between the protein and the sugar backbone. Several charged amino acid side chains form bonds with sites in the major groove of the DNA. There are also some hydrophobic interactions. Certain amino acid side chains appear to co-operate to recognize a single base in the interior of the DNA molecule. The contacts of the repressor with the backbone enhance the specificity of the binding by positioning the residues that interact with the nucleotide bases (Jordan and Pabo, 1988).

In another study the crystal structure of a repressor protein known as *trp* bound to DNA has been determined at 1.8 Å resolution in the presence of the effector tryptophan and in its absence (Zhang *et al.*, 1987; Otwinowski *et al.*, 1988). This has allowed the changes that occur when the tryptophan binds to this repressor, which result in tighter binding of the repressor to the protein, to be determined. The indole ring of the tryptophan bound to the repressor is an integral part of a structure that stabilizes an otherwise unstable conformation. In the two previous examples, groups from the protein were shown to interact directly with bases in the DNA. The *trp* repressor was shown to be different. No evidence was found for direct hydrogen bonds or for non-polar contacts between the protein and the bases. The sequence of bases in the DNA structure may be sensed by the repressor indirectly through hydrogen-bonded water bridges.

The interactions between proteins and DNA must be characterized by both short-range interactions between functional groups and by larger-scale interactions that depend upon the overall folding of the protein and the coiling of the DNA (Ptashne, 1989). Even longer-range interactions are important in the expression of many genes, where it is known that the regulatory elements in the DNA are situated many base pairs away from the genes whose expression they modify. Other DNA-protein interactions concern the packing of the DNA molecule within the nucleus. This is achieved by the association of the DNA molecule with basic proteins known as histones. DNA-histone complexes are large enough to be observed in the electron microscope. Thus studies of interactions between proteins and DNA take us from the most intimate hydrogen-bonded interactions to the overall shape of some of the largest molecular aggregates known (reviewed by Travers and Klug, 1987).

9.6 NOVEL CATALYSTS

We have become used to thinking of enzymes as the sole biological catalysts. However, two novel types of catalyst, one naturally occurring and one synthetic, have caused us to broaden our views of biological catalysts. In the synthetic example two areas of protein chemistry have been brought together to great effect. Antibodies are a class of proteins that are synthesized by specialized blood cells to bind specifically to a foreign compound or antigen. One of the breakthroughs in this field of immunology in the last decade has been the ability to use cell biological techniques to produce single antibody molecules directed against a specific region of an antigen molecule. These monoclonal antibodies differ from those produced in the intact animal as part of its immune response (polyclonal antibodies) in that the serum from an immunized animal contains many different antibody molecules directed against different parts of the same antigen. Monoclonal antibodies, in contrast, are unique protein molecules, and bind to very specific regions of small antigen molecules. In principle these binding sites could become part of an active site if a catalytic function could also be incorporated. Such a combination of antibody and enzyme activity has now been achieved (Lerner and Tramontano, 1988). The key concept has been discussed a number of times in this book, namely that enzymes bind tightly to the transition state for the catalysed reaction. Thus for the hydrolysis of an ester the transition state would be the tetrahedral intermediate. Clearly one cannot raise an antibody to a transition state, but a suitable model, such as a phosphonate ester could be used (Fig. 9.3). Antibodies produced against the phosphonate have a rate acceleration of up to 16 500-fold over the uncatalysed hydrolysis.

The combination of enzyme activity and the specificity of binding of antibodies offers a feast to the molecular imagination of new catalysts for reactions not found in nature using proteins that can readily be produced in

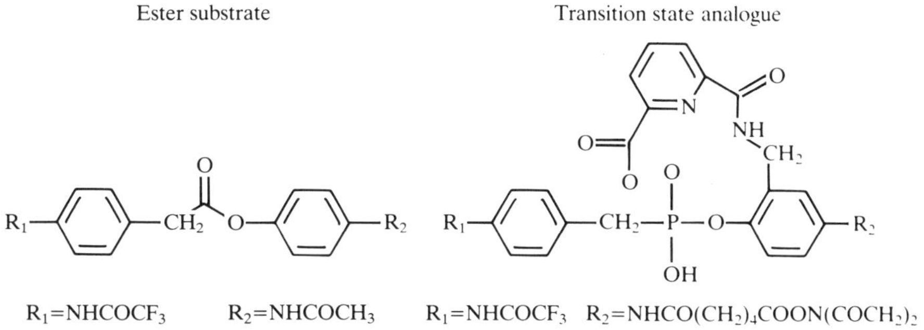

Ester substrate

Transition state analogue

R$_1$=NHCOCF$_3$ R$_2$=NHCOCH$_3$ R$_1$=NHCOCF$_3$ R$_2$=NHCO(CH$_2$)$_4$COON(COCH$_2$)$_2$

Fig. 9.3   The ester substrate and phosphonate ester transition state analogue used in producing a catalytic antibody (Massey, 1988. Reprinted by permission from Nature, **328**, 457. Copyright © Macmillan Magazines Ltd).

cell culture. Since in principle cultures can be carried out in bulk, this is seen as a major opportunity for the biotechnology industry.

Proteins as catalysts have been the theme of this book. It is now clear that in living systems catalysis is not exclusive to proteins. As so often happens, new biology is leading to new chemistry and a further understanding of the fundamentals of catalysis. The chemistry in question here is the versatile reactivity of phosphate esters. Exchanges between phosphate esters are energetically neutral and the scope of such reactions is great because much crucial biology depends upon specific phosphodiesters. The primary RNA transcript of a gene in eukaryotic cells consists of a sequence of coding regions, exons, and intervening sequences, introns. In order for the RNA to be converted into its active form to serve as a messenger for protein synthesis or as a component of the ribosome, the introns must be removed specifically. Several laboratories have shown over the last few years that this process is catalysed by the RNA itself. In other words, the splicing reactions are autocatalytic. Some RNA molecules can also cleave sections out from their chains and others have been shown to act as true catalysts on other molecules. There is already a substantial literature on 'ribozymes'. Such systems have further stimulated interest in the central role of RNA in cell biology and its possible key role in evolution, linking the rather restricted chemistry of DNA to the functional versatility of proteins (Steitz, 1988). Some examples of the catalytic activity of RNA are shown in Fig. 9.4. They include self-splicing (Cech, 1987) and catalytic activity on another molecule (Uhlenbeck, 1987).

9.7   APOLIPOPROTEIN B

The major part of this book has concerned enzymes. However many important proteins with key biological functions are not enzymes. In such

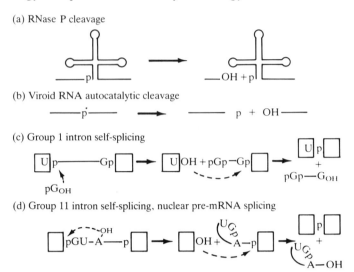

(a) RNase P cleavage

(b) Viroid RNA autocatalytic cleavage

(c) Group 1 intron self-splicing

(d) Group 11 intron self-splicing, nuclear pre-mRNA splicing

Fig. 9.4 Examples of the catalytic activity of RNA. (a) The cleavage of the 5'- leader sequence of a pre-tRNA by ribonuclease P. (b) The autocatalytic cleavage of plant viroids, virusoids and linear satellite RNAs. (c) Two-step self-splicing reactions of introns in nuclear rRNA genes, mitochondrial mRNA and tRNA genes and chloroplast tRNA genes. (d) Two-step splicing reaction of introns in structural genes of fungal and plant mitochondrial DNA and in structural genes of tRNA in chloroplasts, and of nuclear pre-mRNAs. Boxes correspond to exons, lines to introns (Green, 1988. Reprinted by permission from Nature, **336**, 716. Copyright © Macmillan Magazines Ltd).

systems we are therefore not concerned with understanding how a protein catalyses a chemical reaction, but rather with other physiological processes which relate to binding to other macromolecules or molecular assemblies, a process that often initiates specific cellular responses. The concepts of protein structure and function and the methods that have been devised to study enzymes are equally applicable in these systems. For the first example of this type we will start with the largest known monomeric protein, apolipoprotein B (apoB). This protein has a molecular weight of 512 000. Its function is to enable insoluble lipids such as cholesterol and cholesteryl esters to be transported in the blood in the form of a particle known as a lipoprotein. A typical structure is shown in Fig. 9.5. There are several classes of lipoprotein. ApoB is particularly found associated with a class known as low-density lipoprotein (LDL). LDL has acquired a sinister reputation since it has become very clear that elevated levels of LDL in the blood are strongly connected with atherosclerosis, the disease which leads to heart attacks, strokes and about one-third of the mortality in western societies.

In LDL apoB has several functions (Suckling and Groot, 1988). One is to

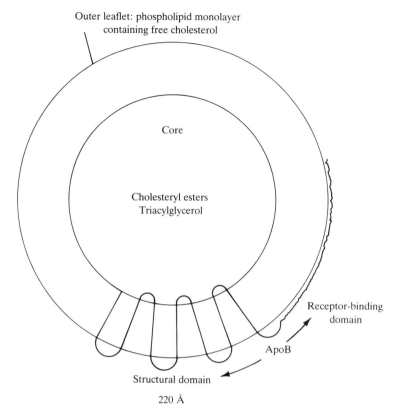

Fig. 9.5 Structure of low density lipoprotein. LDL contains one apolipoprotein, apoB. The domains in the apoB structure deduced from studies outlined in the text are shown (Suckling and Groot, 1988, reproduced with permission).

stabilize the structure of the lipoprotein particle by providing a surface that can interact with the aqueous medium of the blood plasma and another that can interact with the apolar lipids that form the outer leaflet of the particle. Another most significant function is recognition by a specific receptor on the surface of cells. This leads to the uptake of the particle into a cell by the process of endocytosis (Table 9.2). In this way LDL is removed from the blood, particularly by the liver. In people who do not have sufficient receptors for LDL this process is less effective and LDL accumulates in the blood leading to premature atherosclerosis. There are common defects found in the LDL receptor, occurring in one person in every 500, very frequent for a genetic disorder.

With this background we can look into the structure of the protein in more detail (Knott *et al.*, 1988; Scott *et al.*, 1988). Such a large protein with amphipathic properties proved very difficult to purify. It tends to aggregate

Table 9.2 Events in endocytosis of lipoprotein

1. Lipoprotein binds to receptor
2. Receptors aggregate in coated pit on cell surface
3. Coated pit is taken into cell forming an endosome
4. Some of the components of the coated pit are released and travel back to the cell surface
5. The remainder of the endosome fuses with a lysosome
6. Hydrolysis of the protein and lipid components of the lipoprotein takes place with in the lysosome and the products of hydrolysis are released to the cytoplasm

into intractable clusters when it is removed from its lipid environment. Because of its size and the difficulty in handling, the classical methods of protein structure determination could not be used. Instead two groups used approaches that combined classical methods of sequencing of proteins with more recent molecular biological techniques. The classical methods involved cutting the apoB into defined fragments using proteases and then determining the sequence of amino acids in the fragments by sequential chemical degradation. This could only give part of the overall sequence. The complementary molecular biological approach involved isolating the gene coding for apoB and determining the sequence of nucleotide bases in the DNA. From this sequence the primary structure, the order of the amino acids in the protein, could be deduced. The complete polypeptide was shown to consist of 4536 amino acids.

The primary structure of the apoB tells us nothing directly about its three dimensional structure, information which is now required to enable us to understand how this protein carries out its several functions. However, we have seen in earlier sections that there are recurring themes in protein structures. Many proteins are associated with biological membranes and are localized in the essentially hydrophobic membrane by a helical region of hydrophobic amino acids of a defined length. Such membrane-spanning regions can often be discerned in the sequence of proteins for which no more structural information is available. Many of these correlations have been incorporated into computer programs that allow one to predict certain structural features of a folded, three-dimensional polypeptide chain. Some of these methods were applied to apoB. Regions of the structure were found which probably are responsible for binding lipid and hence stabilizing the structure of the particle. Other areas were defined that are probably responsible for binding to the LDL receptor. In contrast to the binding to the lipid, this binding is a polar interaction between charged amino acid side chains.

When this extremely large structure had been determined, a further very significant physiological question could be addressed that led to the discovery of a completely new biological process (Scott *et al.*, 1988). ApoB as we have described it is synthesized in the liver and secreted into the blood from the liver in a precursor lipoprotein to LDL known as VLDL. Another species of apoB is synthesized and secreted in similar particles by the intestine. This form of apoB is about 48% of the size of the liver form; accordingly the liver form is known as apoB-100 and the intestinal form apoB-48. It is the normal rule that a different gene codes for each different protein, and although apoB-100 and apoB-48 have many similarities, this is what one would have expected to find. However, a very thorough search could find no evidence for two different apoB genes. It appeared that the same gene was responsible for the synthesis of apoB-100 and apoB-48. A very detailed comparison of the sequence of the RNA coding for apoB-48 with that for apoB-100 showed very surprising results. A single nucleotide had been changed: a uracil had been changed to a cytosine. In the context of the surrounding sequence this modification changed the sequence from specifying the amino acid glutamine to one specifying the end of the sequence, stop. This could not have occurred by the normal splicing process of removing introns (sections 9.4 and 9.6); a completely novel mechanism must be operating in the nucleus of the intestinal cell to produce this modification and this mechanism at the time of writing is unknown.

The story of apoB is striking for the technical feat of determining the structure of an exceptionally large protein and explaining much of its function, and for the unexpected way that knowledge about the structure of the two forms of the protein led to the discovery of a completely new biological process.

9.8   RECEPTORS

The apoB receptor is just one example of a very wide range of cell-surface receptors that allow cells to sense their environment and to adjust their response to it. Receptors are integral membrane proteins and the structures of several of them have been determined in the last few years. In a similar way to apoB we can consider how these structures relate to each other and to the function of the cell-surface receptor. The ligands that bind to cell-surface receptors are of a varied chemical nature. They include small molecules such as acetylcholine, glycine and γ-aminoglutamic acid (GABA). These ligands are involved in processes in the nervous system, regulating the movement of ions across the plasma membrane of a cell through receptor-activated channels. Other receptors respond to polypeptide hormones. Of these polypeptides perhaps the best known is insulin but there is a large class of peptide growth factors, for example the epidermal growth factor (EGF) that have an important and varied physiological role. Table 9.3 summarizes the characteristics of a number of cell-surface receptors.

Table 9.3 Examples of cell-surface receptors

Ligand	Role and mechanism of action of receptor
Receptors mediating ion channels	
Acetylcholine	Opens sodium and potassium channel in response to neurotransmitter (Popot and Changeux, 1984)
Glycine	Glycine is an inhibitory neurotransmitter in spinal cord and brain stem. Interaction with receptor opens chloride channels (Grenningloh, 1987)
GABA	γ-aminobutyric acid mediates neuronal inhibition and opens chloride channel in rest of brain (Schofield *et al.*, 1987)
Receptors acting through G proteins	
β-adrenergic (adrenaline)	Binding of the hormone to the receptor activates adenyl cyclase by interaction with the G protein (Gilman, 1987)
Peptide receptors	
Epidermal growth factor	Stimulates cell proliferation. Activates tyrosine kinase activity (Carpenter, 1987)
Low density lipoprotein	Binds LDL and leads to receptor-mediated endocytosis of the particle *via* coated pits in the cell surface (Suckling and Groot, 1988)
Insulin	Numerous physiological roles, e.g. regulates glucose uptake into non-hepatic cells. Initial response is tyrosine kinase activation (Andersen, 1989)

The main biological question that we want to answer when studying these receptors is how the binding of the ligand elicits the specific intracellular response. Many possible mechanisms are known, including association with other plasma membrane proteins to activate intracellular enzymes, and covalent modification of specific proteins (usually by phosphorylation). Inhibitory mechanisms are also known. So in the structure of the receptor we are looking for a number of characteristics, a membrane-spanning region (probably hydrophobic with about 24 amino acids), a ligand binding site (outside the cell) and some function on the intracellular domain of the receptor that responds to the binding of the ligand and initiates the intracellular response. We now have information, at least in outline, on these characteristics for a number of receptors.

To determine the structure of the receptors a strategy has generally been used that is similar to that applied to apoB. In outline, a number of peptides from the purified receptor are sequenced and the combined information on amino-acid sequences is used to define oligonucleotides with a sequence of bases corresponding to the amino acid sequence of the peptide. These oligonucleotides are synthesized, usually using an automated chemical method, and used as probes for corresponding sequences in the whole DNA of a suitable cell type in culture. The oligonucleotides will only bind tightly, or hybridize, to sequences in the DNA that are exactly complementary to them. In this way fragments of the gene for the receptor can be identified. Given sufficient fragments and overlaps between them, the whole sequence of the gene can be deduced. The sequence is translated into an amino acid sequence which can now be used to deduce structural features of the protein, for example by using the computer correlation routines mentioned above, or simply by comparing the sequences with other proteins of similar function.

It has become clear that there exist several families of receptors. These include peptide growth factors such as epidermal growth factor and also the LDL receptor (Fig 9.6(a)). The insulin receptor is a more complex, but related structure (Fig. 9.6(b)). There appears to be a family of receptor structures for small molecules regulating ion channels. Thus the GABA glycine and acetylcholine receptors all appear to be related (Fig. 9.6(c)). Two classes of receptors with different mechanisms of action are known in communication between cells of the central nervous system, those described here and those that couple to so-called G proteins. Such receptors include those that respond to adrenaline, the polypeptide hormone glucagon and other peptide hormones such as the adrenocorticotropic hormone (ACTH). Binding of the hormone to the receptor causes an interaction with the G protein. These proteins are located in the plasma membrane of the cell and are so called because they bind GTP. They then initiate a number of cellular responses, depending upon the nature of the G protein (Gilman, 1987), for example, motivation of adenylate cyclase.

As with other classes of protein, there appears to be an economy of design in the structure of the ion-channel activating receptors (Unwin, 1986). Interestingly, the three-dimensional structure of the acetylcholine receptor in membranes has been determined at low resolution, something that has rarely been done for proteins of this type (Brisson and Unwin, 1985). This has allowed the arrangement of the polypeptide chains spanning the membrane that form the ion-conducting channel to be observed, and results agree with the data from the amino acid sequence of the protein.

Having determined many of the structures we can see if they shed any light on the mechanism of action of the receptor–ligand complex. Many of the receptors that bind the polypeptide hormones transmit this information into the interior of the cell by the activation of an enzymatic function associated with their cytoplasmic domain. Thus the primary mechanism for the generation of the intracellular message (the second message) following the binding of EGF to its receptor is stimulation of a tyrosine kinase activity (Fig. 9.6(a)), that is the ability to esterify with a phosphate group the phenolic hydroxyl group of a tyrosine residue on a protein. Receptors of this kind commonly phosphorylate themselves (autophosphorylation) and this may be a necessary first step before other proteins can be phosphorylated.

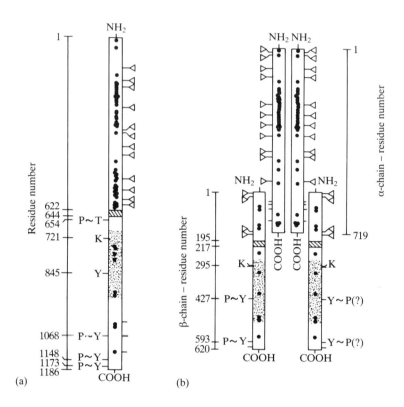

(a)

(b)

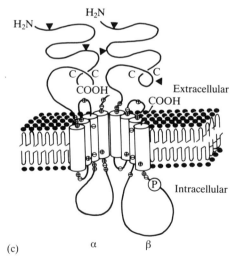

(c)

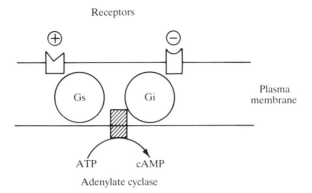

Fig. 9.7 Action of G proteins. On binding of a hormone or pharmacological agonist to its receptor on the outside face of the plasma membrane (stimulatory +, inhibitory −) the G protein (Gs, stimulatory, Gi, inhibitory) couples the receptor and the adenylate cyclase, which is situated on the inside surface of the plasma membrane of the cell. This stimulates or inhibits the formation of cyclic AMP within the cell. The concentration of cyclic AMP in turn regulates the activity of many intracellular processes through protein phosphorylation.

These proteins, for example the lipcortins, probably provide the link between the receptor–ligand complex and the regulation of enzyme activity and gene expession deeper within the cell. Details of these mechanisms are being studied by modification of the receptor structure by site-directed mutagenesis. The activation of cellular functions that results produces an overall response that can last for seconds or hours. Interestingly a number of protein products of oncogenes, that is proteins that are thought to be expressed in malignant cells and participate in the promotion of their

Fig. 9.6 Structures of receptors. (a) Epidermal growth factor. (b) Insulin. (c) GABA receptor. In (a) and (b) − = sequences for N-linked glycosylations, ◁ = probable sites of N-linked oligosaccharide chains, ● = cysteine residues, Y = tyrosine residues, P∼Y = phosphotyrosine residues, T = threonine residues, K = lysine residues, crosshatched area = membrane-spanning region, stippled area = sequences similar to the oncogene *src* kinase. The disulphide bridges between the subunits of the insulin receptor have not been shown. (c) The GABA receptor is shown schematically associated with the cell membrane. The membrane-spanning helices in each subunit are shown as cylinders. The looping of the polypeptide chain in the extracellular region is hypothetical but shows a disulphide bond predicted by the structural analysis from the protein sequence. Triangles show potential N-glycosylation sites. Charged residues at the ends of domains are shown. The letter P indicates the site for a cyclic AMP-dependent phosphorylation on a serine residue (Carpenter, 1987 (a) and (b). Reproduced with permission from *Annu. Rev. Biochem.*, **56**, © Annual Reviews Inc. Schofield *et al.* 1988 (c). Reprinted with permission from *Nature*, **288**, 226. Copyright © 1988 Macmillan Magazines Ltd).

proliferation, are also tyrosine kinases and show structural homology with this class of receptor protein. One way in which the tyrosine kinase activity may be expressed is that binding of EGF to a monomer of the receptor initiates the aggregation of receptors into clusters in which the phosphorylation process can take place (Carpenter, 1987).

In the case of the ion-channel receptors the transmembrane segments contain an arrangement of amino acids that appears to promote conduction of the movement of ions through the pore that they form. An ion-selective filter consisting of charges opposite to that of the ion to be transported is present at the mouth of the channel. It is thought that the ligand binds to one of the extracellular domains of the receptor and that this causes a change in the orientation of the chains such that the mouth of the ion channel becomes fully open. Such changes could be modified by drugs. As with a number of these systems, a more detailed molecular analysis can now be made by site-directed mutagenesis. Such a study is lengthy in a system such as the GABA receptor because the modified receptor must be synthesized in a cell and incorporated into the membrane of a cell in order to test its function. In the case of enzymes, where the rate of a chemical reaction can be measured directly, testing the modified protein is much easier.

9.9 CONCLUSIONS

The boundaries between areas of chemistry and biology become more and more obscure. A development such as genetic engineering, originating in molecular biology, is avidly taken up and becomes a key tool in a deeper study of the relationship between the structure and function of proteins as catalysts, as receptors or as molecules with other functions. Incisive experiments can be designed to combine the power of the new techniques with the detailed structures of some systems determined by X-ray crystallography. These interactions not only lead to a deeper fundamental understanding of biology and chemistry but also to practical advances. Catalytic antibodies offer a great opportunity for the biotechnology industry. Major diseases will be understood better because of the application of molecular techniques to key processes. Often we are not dealing with enzyme-catalysed reactions alone but also with specific binding of messenger molecules with receptors. The penetration and breadth of the concepts of enzyme chemistry resonate throughout biology.

REFERENCES

Aggarwal, A. K., Rodgers, D. W., Drottar, M., Ptashne, M. and Harrison, S. C. (1988) *Science*, **242**, 899.
Andersen, A. S. (1989) *Nature*, **337**, 12.

Benkovic, S. J., Fierke, C. A. and Naylor, A. M. (1988) *Science*, **239**, 1105.

Brisson, A. and Unwin, P. N. T. (1985) Nature **315**, 473.

Carpenter, G. (1987) *Ann. Rev. Biochem.*, **56**, 881.

Cech, T. R. (1987) *Science*, **236**, 1532.

Chothia, C. (1988) *Nature*, **333**, 598.

Dewar, M. J. S. and Storch, D. M. (1985) *Proc. Natl. Acad. Sci. USA*, **82**, 2225.

Doolittle, R. F. (1988) *Nature*, **336**, 18.

Fersht, A. R. (1987) *Biochemistry*, **26**, 8031.

Gilman, A. G. (1987) *Ann. Rev. Biochem.*, **56**, 615.

Graf, L. Craik, S., Patthy, A., Roczniak, S., Fletterick, R. J. and Rutter, W. J. (1987) *Biochemistry*, **26**, 2616.

Grenningloh, G., Rientz, A., Schmitt, B., Methfessel, C., Zensen, M., Beyreuther, K., Gundelfinger, E. D. and Betz, H. (1987) *Nature*, **328**, 215.

Hwang, J.-K. and Warshel, A. (1988) *Nature*, **334**, 269.

Jordan, S. R. and Pabo, C. O. (1988) *Science*, **242**, 893.

Knott, T. J., Pease, R. J., Powell, L. M., Wallis, S. C., Rall, S. C., Innerarity, T. L., Blackhart, B., Taylor, W. H., Marcel, Y., Milne, R., Johnson, D., Fuller, M., Lusis, A. J., McCarthy, B. J., Mahley, R. W., Levy-Wilson, B. and Scott, J. (1986) *Nature*, **323**, 734.

Knowles, J. R. (1987) *Science*, **236**, 1252.

Lau, F. T. and Fersht, A. R. (1987) *Nature*, **326**, 811.

Lerner, R. A. and Tramontano, A. (1988) *Scientific American*, **258**, 42.

Matthews, B. W. (1988) *Nature*, **355**, 294.

Menger, F. M. (1985) *Acc. Chem. Res.*, **18**, 128.

Ogata, C., Hatada, M., Tomlinson, G., Shin, W.-C. and Kim, S.-H. (1987) *Nature*, **328**, 739.

Otwinowski, Z., Schevitz, R. W., Zhang, R.-G., Lawson, C. L., Joachimiak, A., Mamorstein, R. Q., Luisi, B. F. and Sigler, P. B. (1988) *Nature*, **335**, 321.

Pain, R. (1986) *Nature*, **320**, 216.

Popot, J. L. and Changeux, J. P. (1984) *Physiol. Rev.*, **64**, 1162.

Ptashne, M. (1989) *Scientific American*, **260**, 24.

Sawyer, L. (1987) *Nature*, **327**, 659.

Schofield, P. R., Darlison, M. G., Fujita, N., Burt, D. R., Stephenson, F. A., Rodriguez, H., Rhee, L. M., Ramachandran, J., Reale, V., Glencorse, T. A., Seeburg, P. H. and Barnard, E. A. (1987) *Nature*, **328**, 221.

Scott, J., Wallis, S. C., Pease, R. J., Knott, T. J. and Powell, L. (1988) in *Hyperlipidemia and Atherosclerosis* (eds K. E. Suckling and P. H. E. Groot), Academic Press, 47.

Steitz, J. A. (1988) *Scientific American* , **258**, 36.

Suckling, K. E. and Groot, P. H. E. (1988) *Chemistry in Britain*, 436.

Travers, A. and Klug, A. (1987) *Nature*, **327**, 280.

Uhlenbeck, O. C. (1987) *Nature*, **328**, 596.

Unwin, P. N. T. (1986) *Nature*, **323**, 12.

Warshel, A. (1987) *Nature*, **330**, 15.

Zhang, R.-G., Joachimiak, A., Lawson, C. L., Schevitz, R. W., Otwinowski, Z. and Sigler, P. B. (1987) *Nature*, **327**, 591.

Index

Abzymes 162
Acetoacetate decarboxylase 16
  stereochemistry 19
Acetoin 77
Acetyl choline receptor 369
Acetyl cholinesterase 252
Acid catalysis in glycoside hydrolysis
  331
Acid proteases 193
  mechanism of action 194
Acids, hard and soft 230
Acylases in industrial production of
  L-aminoacids 117
Acylation, selective with crown ether
  complexes 143
Addition reactions, micellar control of
  146
Adenosine triphosphate (ATP)
  recycling 102, 133
AIDS 249
Alcohol dehydrogenase 100
  horse liver 108
  in organic solvents 155
  models for 52 et seq.
  reactivity extension possibilities 152
  yeast, genetic engineering of 157
Aldol reactions 125

Aldolases 126
Alko process 332
Allylic pyrophosphates 291
Aluminium 252
Alzheimer's disease 207, 252
Aminoacid oxidase 68
Aminoacids, unnatural, synthesis 119
Amylases 309 et seq.
α-amylase, structure 314
α-amylases 310
α-amylases, genes for 313
α-amylases substrate interaction with
  313
β-amylases, genes for 313
Anaemia 233, 248
Analogy in synthetic uses of enzymes
  106
Angiotensin converting enzyme 5, 189
Angiotensin converting enzyme,
  inhibitor synthesis 118, 127
Antibodies, catalytic 45, 162, 362
Antibodies 353, 360
Anticancer drugs 176, 228
Antimalarial agents 173
Antitumour activity 249
Apolipoprotein B 363
Arthritis, rheumatoid 229, 253

Ascorbic acid 108
Aspartase in industrial synthesis of
    aspartic acid 114
Aspartate aminotransferase, site
    directed mutagenesis of 357
Aspergillus oryazae acylase in amino
    acid manufacture 120
Asymmetric Michael addition, crown
    ether complexes 143
Asymmetric oxidation by HLADH 109
Asymmetric oxidations 70
Asymmetric reduction
    by HLADH 108
    by nicotinamide coenzyme models 60
Asymmetric synthesis 3
Atherosclerosis 200, 363
Atomic absorption spectrometry 261
Augmentin 185
Auranofin 253

β-barrel 358
Bacteriophage 361, 360
Barley 332
Benzene cis-glycol 158
Benzene dioxygenase 158
Benzoin 77
Beta galactosidase 319
Beta methylaspartase, stereospecific
    addition catalysed by 115
Bilaphos, biosynthesis of 275
Bile acid sequestrants 200
Binaphthyl, chiral 117, 129, 143
Binding energy in enzymic catalysis 363
Binding sites 24
    multiple 40
Bioavailability of drugs 200
Biochemical differences between
    species 173
Biological technologies 98
Biomimetic chemistry 42, 133
Biosynthesis
    natural products 265 *et seq.*
    catecholamines 172
Biosynthetic enzymes 267 *et seq.*
Biosynthetic pathways 98
    genetic manipulation of 108
Biotin 34
Bi-product inhibitors 191
Blocked mutants in biosynthetic studies
    283
β-blockers 97

β-blockers, synthesis 124
Blood group determinants 126
Blood pressure, control of 189
Boranes, chiral 104, 129
Borohydride, complex reagents 129
Brewing 332
Brønsted relationship 28

Cadmium 239, 251
Caeruloplasmin 230, 233, 238, 239, 241
Calcium 250
    in protein structure 339
Calixarenes 136
Calmodulin 250
Calsequestrin 250
Cancer 234, 230
    chemotherapy 133
Candida cyclindraceae lipase (CCL) 117
Captopril, synthesis 127
Carbapenems 186
Carbohydrates, acylation using lipases
    155
Carbon acids, proton transfer from 27
Carbon–carbon bond formation by
    enzymes 125
Carboxypeptidase 231, 248
Carboxypeptidase A 14, 190
Carboxypeptidases, penicillin binding
    182
Catalysis
    acid 54
    acid–base 26
    and association
    by enzymes
        binding energy in 357
        mechanism 8 *et seq.*
        theoretical treatments of 22 *et seq.*,
            356
    by metal ions 52
    concepts 9
    intramolecular 55
    non-covalent interactions 56
    synthetic 42
Catalytic antibodies 45, 162
Catalytic efficiency of enzymes 31
Catecholamines 172
Cefoxitin 185
Cell surface receptors 367
Cell-free system
    for generating flavour compounds 347
    in biosynthetic studies 270, 275

Cellulases
  fungal 327
  molecular studies of 329
Cellulose 321
  hydrolysis 326
Celluloses, structure 320, 323
Cephalosporin 187
  biosynthesis of 269
Cephalosporin A, as an elastase
    inhibitor 215
Cephalosporium acremonium,
    isopenicillin N synthetase 160
Cheese flavour development 346
Cheese making 338 *et seq.*
Cheese ripening, accelerated 344
Chemically modified enzymes 151
Chemiluminescence 74
Chemo-enzymatic synthesis 164
Chirality 20
Chlorination
  anisole 141
  phenol, biomimetic in micelles 146
  steroids 138
Cholesterol biosynthesis, inhibitors of
    200 *et seq.*
Cholestyramine 200
Chorismate mutase 163
Choroperoxidase 150
Chromium 244
Chymosin 340
Chymotrypsin 108
  in organic solvents 153
Circular dichroism 258
Claisen rearrangement, catalytic
    antibodies 163
Clavulanic acid 185, 275
Clofibrate 200
Cobalt 246
Coenzyme A 84
Coenzyme models 50 *et seq.*
Coenzyme recycling 100 *et seq.*, 347
Computer graphics 198, 221
Computer programs 366
Condensation reactions
  thiazolium ion mediated 75
  thiol esters in 86
Conformation, changes in 15
Copper containing proteins 240
Copper in biological systems 238 *et seq.*
Corrin 247
Corynebacterium spp. oxidase 108

Coupled enzymes in cofactor recycling
    101
Covalent bonds, in selectivity of
    biomimetic systems 138
Crown ethers 136, 142 *et seq.*
  in nicotinamide coenzyme models 59,
    61
Crystallin 360
Curtin–Hammett principle 15
Cyclodextrins 42, 135, 141 *et seq.*, 320
  in nicotinamide coenzyme models 58
  pyridoxal models 81
Cycloguanil 174
Cyclophanes
  in flavin models 71
  in nicotinamide coenzyme models 60
  in pyridoxal models 81
Cytochrome P-450 149, 236

Daunosamine 107
Deazaflavins 63
Debranching of polysaccharides 317
Decarboxylation, by thiazolium salts 77
Dehydrogenases, in chiral intermediate
    preparation 116
Diaminopyrimidines 173
Diamond lattice model 109
Diels–Alder reaction 142
Diet, human 318
Diffusion and reaction rates 32
Dihydrofolate reductase
  in synthesis 131
  inhibitors of 173 *et seq.*
Dihdyrofolate reductases, inhibitor
    resistant 180
Dipodascus uninucleatus reductase 128
Distilled spirits 332
DNA 354, 358, 362
DNA photolyase 70
DNA probes 368
DNA, recombinant, in biosynthetic
    studies 299
Dopamine β-hydroxylase 172
Drug development 172
Drug toxicity 201
Drug transport 180

Effective molarity 25, 36
Elastase
  human leucocyte, inhibitors of 207
    et seq.

Elastase—*cont.*
  porcine pancreatic, X-ray analysis
    212
Elastin 208
Electrochemical recycling of NADH
    103
Electron spin resonance, EPR 261
Electron transfer reactions 57
Elimination reaction, pectin lyase 324
Emphysema, pulmonary 207
Emulsin 2
Enalapril 192
Enantiomers 21
  and biological activity 97
Enantioselective reduction 60, 108, 143
Endocytosis 366
Energy diagram 10, 272
Enolase 356
Entropy, and enzymic catalysis 22
Enzyme catalysis, specificity 15
Enzyme models 50 *et seq.*
  rate enhancements in 42
Enzyme stereoselectivity 114
Enzymes and conventional reagents in
    competition 113 *et seq.*
Enzymes
  immobilized 104, 154
  in animal feeds 356
  in cheesemaking 338 *et seq.*
  in organic solvents 106, 117, 153
    et seq.
  in synthesis 4, 95 *et seq.*
  in industrial synthesis 105
  in synthesis 4
  stability of 104
Enzymic catalysis
  and association 24
  and entropy 22
Enzymic efficiency 31
Epoxidation by cytochrome P-450
    model 149
Epoxides, chiral 123
Erwinia herbicola reductase 108
Erythromycin 278
Esterases in synthesis 117
Esters, synthesis with enzymes in
    organic solvents 153
Evolution 31, 358, 363
  divergent 187
EXAFS 249, 257
Exons 359

Farnesyl pyrophosphate synthetase 284
Fatty acids, biosynthesis of 277
Ferredoxin 237
Ferritin 233
Fischer, Emil 2
Flavin coenzymes
  models 64 *et seq.*
  models, as photosensitizers 69
  models, asymmetric oxidation by 70
  models, for oxygenases 71
  models, oxidation by 66
Flavin mononucleotide (FMN) in NAD
    recycling 101
Flavins
  5-deaza 63
  in chemically modified enzymes 151
Flavour components of cheeses 347
Fluoroketones, in enzyme inhibitors
    196, 210
Folic acid 171
Food industry 315 *et seq*
Formate dehydrogenase 103
Functional group modification in
    synthesis by enzymes 128 *et seq.*

GABA receptor 367, 369
Gene cloning
  in biosynthetic studies 281, 283, 291
  in biosynthesis 268
Genes for
  α-amylases 313
  amylases 316
  cellobiohydrolases 328
  cellulases 329
  glucose oxidase 318
  lipoproteins 367
Genes, structure 359
Genetic disorder 365
Genetic engineering 6, 98, 108, 156
Genetic relationships 359
Glucoamylases 315
  mechanism of 316
Glucose 6-phosphate dehydrogenase
    101
Glucose isomerase 318, 358
Glucose oxidase 319
Glucose syrups, production of 315
Glutathione 84, 244
Glyceraldehyde 3-phosphate
    dehydrogenase 54
Glycine receptor 369

Glyoxalase I 87
Gold 253
Grandisol, synthesis 117

Haemocyanin 241
Haemoerythrin 237
Haemoglobin 234
    chemically modified 151
    EPR spectrum 261
    Raman spectrum 260
Haemoproteins 67
Haemosiderin 234
Hammond postulate 10
Hydroxysteroid dehydrogenase 113
Helix-turn-helix motif in protein
        structure 361
Hemicelluloses 321
Hexokinase 102
Histones 362
Homology 357, 358
Host–guest interactions 58
Human diet 318
Hybridization of oligonucleotides 369
Hydrogen bonding
    in catalysis by enzymes 366
    in protein structure 361
Hydrogenase 102
Hydrolases 116
Hydroxymethylglutarylcoenzyme A
        reductase 203
Hydroxymethylbilane 267
Hypercholesterolaemia 203

Immobilization, of NAD 103
Immobilized cells 114
Immobilized enzymes 103, 104, 154
    in amino acid manufacture 120
    in food industry 318, 321
Industrial enzymes 332
Industrial oxidation, enzymes in 151
Industrial production of L-aminoacids
        117
Industrial synthesis of aspartic acid 114
Industrial use of enzymes in synthesis
        105
Inhibitor binding, to dihydrofolate
        reductase 177
    molecular basis for selectivity 178
Inhibitors, bi-product 191
Insect attractants, synthesis 118, 119
Insulin 244

Insulin receptor 369
Integral membrane proteins 367
Intermediates in reactions 11
Intermediates
    chiral, preparation of 116 et seq.
    radical 138, 160
    reactive 12
    tetrahedral 163
Intramolecular catalysis 34
Intramolecular effects 55
Intramolecular reactions 25
Introns 359
Invertase 321
Invertin 2
Iron in biological systems 232 et seq.
Iron sulphur proteins 67
Isopenicillin N synthetase 160, 269
    mechanism of 274
Isoprene rule 284
Isosteres for peptides 195
Isotope effects, in biosynthetic studies
        271
Isotopic labels, chiral 114
Isotopically labelled compounds,
        stereoselective synthesis 103

Kinetics in investigating mechanism 12

β-lactam antibiotics 180 et seq., 215
    biosynthesis 269
    stereochemistry of biosynthesis of 114
    synthesis using enzymes 159
β-lactamases 181, 185 et seq.
β-lactams, synthesis 119
Lactate dehydrogenase 100, 360
    model for 54
LDL receptor 369
Lead 252
Leucotrienes, synthesis of 113, 118
Leucovorin 133
Linear free energy relationships 29
Lipases
    in cheese manufacture 340
    in organic solvents 153
    non-natural substrates 154
Lipcortin 371
Lipoic acid 84
Lipoprotein 353
Lipoproteins 363 et seq.
Lipoproteins, plasma 203
Lithium 256

Lock and key hypothesis 3, 219
Lovastatin (mevinolin) 205
Low density lipoprotein 364
Low density lipoprotein receptor, gene 203
Luciferase, models for 73
Lysozyme 331

Macrolides, biosynthesis of 281
Magnesium 250
Maize 334
Malting 332
Maltose syrups 314
Manganese 246
Mechanism
    description of 9
    of enzymic catalysis 8 *et seq.*
    and stereochemistry 20
Membrane proteins 366
Membrane reactor 103
Memory effects 83
Metal complexes, stability of 231
Metal ions as drugs 253 *et seq.*
    biological function 228
    catalysis by 52
    classification of 230
    coordination geometries 231
Metallothinein 238, 251
Methotrexate 174
Methoxatin 68
Methyl transferases 281
Methysalicylate synthase 278
Mevastatin 203
Mexiprostil 132
Micelles 57, 67, 76, 78, 82, 136, 145 *et seq.*
    casein 339
    functionalized 147
Michael addition, asymmetric 143
Microorganisms in synthesis 128
    reactivity patterns 106
Microscopic reversibility 13
Milk products 319
Milk proteins 339
Model reactions 34
Molecular biology 221, 354
    advances in 156 *et seq.*
Molecular dynamics 220
Molecular models 198, 221
Molybdenum 67, 239, 248

Molybdenum complexes, in synthesis 140
Monellin 360
Monensin 377
Monobactam 181
Monoclonal antibodies 362
More O'Ferral-Jencks diagram 10
Mucor rammanianus reductase 128
Multiple binding 40
Mutations, conservative 355
Myocrisin 253
Myoglobin 360

Na-K ATPase 243
Natural products, screening of 203
New enzymes 158
Nicotinamide coenzyme binding, in dihydrofolate reductase 176
Nicotinamide coenzyme recycling 133
Nicotinamide coenzymes 21, 40
Nicotinamide coenzymes, models 51
    as electron donors 57
    asymmetric reduction by 60
    host-guest 58
    metal ion catalysis in 52
    non-covalent interactions in 56
    oxidation by 61
    proton catalysis in 54
Nicotinamide coenzymes, recycling 100
Nicotinamide nucleoside analogues, synthesis 119
Nitration, of phenol, biomimetic 148
Nitrogenase 248
NMR 238, 257, 286
    in studies of enzyme inhibition 212, 220
    in biosynthetic studies 265 *et seq.*, 296, 299
Non-covalent interactions in catalysis 56
Non-enzymic proteins 353
Novel catalysts 362

Operator 366
Organometallic complexes in stereoselective synthesis 126
Osteoarthritis 207
Oxidation
    biomimetic 149
    by nicotinamide coenzyme models 61

Oxygenases, flavin-dependent, models for 71

Papain, chemically modified 151
Partition function 23
Pasteur 1
Pectate lyase 325
Pectin 322
Pectin esterase 326
Pectin lyase 324
Pectinases 323
Penicillin binding proteins 182
Penicillin G 181
Penicillin V 185
Penicillin, biosynthesis of 269
Penicillins, synthesis 159
Pentalenene synthetase 298
Pentalenolactone, biosynthesis of 293
Pepsin 34
  in cheese manufacture 343
Pepstatin 195
Peptidases
  in synthesis 117
  stereoselectivity in organic solvents 153
  stereoselectivity in organic solvents 154
Peptide isosteres 195
Peptides
  cyclic 199
  synthesis with enzymes in organic solvents 153
Phenoxypropanoate herbicides 96
Pheromones, synthesis 118, 119
Phosphofructokinase, site directed mutagenesis of 366
Phosphonates
  as enzyme inhibitors 192
  as tetrahedral intermediate analogues 162
Phosphonothricin 275
Photochemical chlorination, of steroids 138
Photochemical reactions 57, 69, 85
Photolyase, DNA 70
Photosensitizers 69
Physical methods in bioinorganic chemistry 257
Physical organic chemistry 8, 22
Pig liver esterase (PLE) 117, 131

Pig pancreatic lipase (PPL) 117
  in chiral epoxide synthesis 125
Pinitol, synthesis 159
Plant cell wall breakdown by enzymes 321 et seq.
Plasmids 182, 361
Plastocyanin 241
Polyether antibiotics 277
Polyethers, biosynthesis of 279
Polygalacturonase 326
Polyketides, biosynthesis of 277 et seq.
Polymer supported cyclodextrin 141
Polymer supported reactions 76
Polymeric reagents 83
Polyphenylene 158
Porphyrin ligand 228
Porphyrins 5
  biosynthesis of 267
Potassium 250
Prephenic acid 163
Probes, DNA 368
Prochirality 21
Proguanil 174
Proline racemase 38
Prostaglandin E analogue, synthesis 130
Prostaglandins, synthesis of 113, 119, 128
Protease inhibitors, endogenous 207 et seq.
Protease N 155
Proteases
  acid 193
  in cheese manufacture 342
  metallo 190
  serine 187, 207
Protecting groups 125
Protein chemistry, advances in 156 et seq.
Protein folding 357
Protein structure 358 et seq.
Proteins, interactions with DNA 360
Proton transfer 26, 29, 54
  in inhibitor binding 177
Pseudomonas putida, benzene dioxygenase 158
Pullulanase 317
Pyridoxal 242
Pyridoxal model, biomimetic synthesis 142
Pyridoxal phosphate 80 et seq.

Pyrimethamine 174
Pyrimidine based drugs 173
Pyrroloquinoline quinone (PQQ) 68
Pyruvate decarboxylase 20, 39, 41, 77
    inactivation 78
Pyruvate dehydrogenase 38

Quantum mechanical calculations 222

Raman spectroscopy 237, 258
Rate determining step 13, 31
Rational drug design 176
Reaction intermediate analogue 40
Reaction profile, for penicillin
    biosynthesis 272
Receptors 353, 367 et seq.
Receptors
    as targets for drug design 171
    cell surface 367
    models for 145
    structure determination of 368
Red cells
    EPR of 261
    NMR of 258
Regulation of DNA expression 362
Renin 189, 193
Renin–angiotensin system, inhibitors of
    189 et seq.
Rennet 340
Rennet, microbial 343
Reporter substrates 36
Repressor 360
Reverse transcriptase 249
Rhodium complexes, enantioselective
    reductions with 128
Ripening, of fruit and vegetables 33
RNA 248
    as a catalyst 363
    editing 360
Robinson, Sir Robert 5
Ruthenium porphyrin complex,
    biomimetic oxidation by 149

Second messenger 250, 369
Selectivity
    covalent control of 138
    in synthesis 95 et seq.
Sequence homology in cellulases 329
Serine proteases 187, 208
Sesquiterpene antibiotics 293
    biosynthesis of 286

Sharpless epoxidation 123, 131
Sialic acids 126
Site directed mutagenesis 30, 32, 221,
    268, 274, 354, 371
Snake venoms 190
Sodium 250
Sodium pump 243
Solubility of substrates 131
Species specificity 176
Starch
    hydrolysis 319
    liquefaction of 333
    processing of 308
    saccharification of 333
    structure of 309
Statistical mechanics 22
Steady state 12
Stellacyanin 241
Stereochemical studies, in biosynthesis
    272, 285, 287, 294
Stereochemistry
    and enzymic specificity 15
    and reaction mechanism 20
Stereoelectronic effect 15
Stereogenic elements in molecules 21
Stereoselective synthesis
    of isotopically labelled compounds
        103
    organometallic complexes in 126
Stereoselectivity 2, 96
Steroids, selective functionalization of
    138
Storage, of iron 233
Structure–activity relationship 214
Substitution reactions 11, 57
Substrate binding, in organic solvents
    153
Subtilisin, chemically modified 151
Sulphanilamide 171
Sulphonamides 214
Superoxide dismutase 240, 246, 248
Supramolecular catalysis 44, 136
Synthesis, asymmetric 3
Synthesis, enzymes in 95 et seq.
Synthesis, selectivity in 95 et seq.

Target enzyme, choice of 216
Tentacle molecules 146
Terpenes, cyclic, biosynthesis of 284
Terpenoid cyclases 284

Tetrahedral adducts, in serine protease inhibition 210
Tetrahydrofolic acid 131
Thalidomide 96
Thaumatin 360
Thermoanaerobium brockii alcohol dehydrogenase 112
Thermodynamics 22
Thermophilic organisms 105, 112, 327
Thiamine diphosphate 20, 41
Thiamine pyrophosphate (diphosphate, TPP), models for 75
Thiazolium ions 68
  in condensation reactions 75
Thiazolium salts 75 *et seq.*
Thiazolium salts and flavins 79
Thiazolium salts
  in asymmetric synthesis 77
  in decarboxylation 77
Thienamycin 181, 186
Thiol coenzymes 84
Thiol esters 86
Thiol ligands 239, 253, 260
Thymidylate synthetase 173
Tocopherol 107
Toxic metals 227, 251
Toxicity 201
  and metabolism 234
Trace element deficiency 228
Transamination, pyridoxal mediated 81
Transferrin 233
Transglucosidases 316
Transition state
  binding 357
  stabilization 37
Transition state analogue
  and catalytic antibodies 162
  w.r.t. penicillin action 181
Transition state analogues 36
  for acid proteases 195
Transition state and monoclonal antibodies 362
Transketolase, inactivation 78
Transpeptidases
  bacterial 181
  mechanism of action 183
Transport
  in biosynthesis 267
  of folates 180
Trichodiene synthetase 291

Trimethoprim 174
Triose phosphate isomerase 32, 358
Triparanol 201
Troponin C 250
Trp repressor 361
Trypsin
  in cheese manufacture 343
  site directed mutagenesis of 355
Tylactone synthase 281
Tylosin 281
Tyrosine hydroxylase 172
Tyrosine kinase 369
Tyrosyl t-RNA synthetase 30, 356

Uptake
  of copper 239
  of iron 232

Vanadium 243
Vanadium complexes, in synthesis 140
Vector 355
Viscosity 33
Vitamin B_1 75
Vitamin B_6 80
Vitamin B_{12} 247
Vitamin C 108
Vitamin K synthesis, biomimetic 142

X-ray crystallography 219, 330, 357
  of acid proteases 194
  of bacterial carboxypeptidase-transpeptidase 187
  of dihydrofolate reductase 176
  of porcine pancreatic elastase 212
Xanthine oxidase 67, 249
Xylose isomerase 324

Yeast 2
  alcohol dehydrogenase, genetic engineering of 157
  condensation reactions mediated by 106
  dehydrogenases 117
  genes for amylases 316
  reductases 106

Zeolites 152
Zinc 248
Zinc proteases 190